Modeling Prevalent Cancers

This volume offers a comprehensive exploration of prevalent cancers and the innovative modeling techniques driving advancements in treatment strategies. In today's landscape of cancer research and therapeutic development, understanding the complexities of cancer biology and translating these insights into effective treatments is paramount. This book fills a crucial gap in the literature by providing a detailed examination of the diverse modeling approaches utilized in the study of prevalent cancers, including breast, lung, colorectal, prostate, ovarian, and melanoma.

Yash Sharma is Assistant Professor in the Department of Biotechnology, IILM University, Greater Noida, Uttar Pradesh, India. He has enriched academic and research experience of 7 years. After M.Tech., he worked as JRF and SRF in a DST SERB-funded project "Formulation of Tobacco Stem Bioactive Compound Loaded Nanoparticle for Wound Healing" at Amity University, Noida. During his Ph.D., he was awarded the ICMR-SRF fellowship on a project entitled "Anticancer Activity of Seed Cake Extract of *O. Sanctum* against OSCC Cell Line and It's Tumor Suppressor Gene Analysis." He is an experienced researcher focused on drug discovery and pre-clinical studies on cell lines and in vivo models.

Kumud Bala is currently working as Professor at Amity Institute of Biotechnology and Deputy Dean of Student Welfare at Amity University Uttar Pradesh, Noida. She attained her Ph.D. degree from Patna University, and her area of expertise is Cancer Immunology, Pharmacology, Plant Phytochemistry, Clinical Microbiology, Neuroendocrinology, Therapeutics, and Molecular Diagnostics. She has more than 27 years of teaching and research experience. She has published more than 50 publications including Research Papers, Review Articles, Books, and Book Chapters in Scopus-indexed and peer-reviewed journals of high impact. Currently, she is running more than four government-funded projects and has completed two projects funded by ICMR and DST.

Sudhir Kumar Pal is Assistant Professor in the School of Engineering at IILM University, Greater Noida. He is Ph.D. from VIT, Vellore and completed his master's as a Bioinformatician from DY Patil University, Navi Mumbai and bachelor's with a specialization in Biotechnology from AAIDU, Prayagraj. During all his academic career, he has gained specialization in the field of Structural Biology and Bioinformatics. He also has 2 years of research experience from foreign universities like the University of South Bohemia and Masaryk University.

Monu Monu is Assistant Professor in the Department of Biotechnology at IILM University, Greater Noida, and holds a Ph.D. in Biological Sciences from the CSIR-Institute of Genomics and Integrative Biology, Delhi, India. His research focuses on identifying disease-associated specific proteins to understand rheumatoid arthritis (RA) pathogenesis and develop diagnosis methods. After completing his M.Tech. in Biotechnology, he has significantly contributed to multiple CSIR-, ICMR-, and DST-funded projects covering diverse topics such as SARS-CoV-2 ocular infections, arthritis, cardiovascular disease, and *Mycobacterium tuberculosis*.

Modeling Prevalent Cancers

Understanding and Innovating Therapeutic Approaches

Edited by Yash Sharma, Kumud Bala, Sudhir Kumar Pal, and Monu Monu

CRC Press
Taylor & Francis Group
Boca Raton London New York

CRC Press is an imprint of the
Taylor & Francis Group, an **informa** business

First edition published 2026
by CRC Press
2385 NW Executive Center Drive, Suite 320, Boca Raton FL 33431

and by CRC Press
4 Park Square, Milton Park, Abingdon, Oxon, OX14 4RN

CRC Press is an imprint of Taylor & Francis Group, LLC

ISBN: 9781032951386 (hbk)
ISBN: 9781032951393 (pbk)
ISBN: 9781003583349 (ebk)

DOI: 10.1201/9781003583349

Typeset in Times
by codeMantra

Contents

List of Contributors

Astha Agarwal
Department of Biotechnology
IILM University
Greater Noida, India

Soumya Aggarwal
School of Basic & Applied Sciences
IILM University
Greater Noida, India

Tanya Aggarwal
Department of Biotechnology
IILM University
Greater Noida, India

Ramasamy Amsaveni
Department of Biotechnology
Karpagam Academy of Higher Education
Coimbatore, India

Kumud Bala
Amity Institute of Biotechnology
Amity University
Noida, India

Juni Banerjee
Department of Biotechnology and Bioengineering
Institute of Advanced Research
Koba Institutional Area
Gandhinagar, India

Shuvomoy Banerjee
Department of Biotechnology and Bioengineering
Institute of Advanced Research
Koba Institutional Area
Gandhinagar, India

Rohit Bhardwaj
Department of Pharmaceutical Sciences
Faculty of Pharmaceutical Sciences
Gurukula Kangri (Deemed to be University)
Haridwar, India

Veluswamy Bhuvaneshwari
Department of Biotechnology
PSGR Krishnammal College for Women
Coimbatore, India

Roma Chandra
Department of Biotechnology
IILM University
Greater Noida, India

Sorabh Chaudhary
Department of Biology
SRM Institute of Science and Technology
Delhi-NCR Campus
Ghaziabad, India

Muskan Chauhan
Department of Biological Sciences & Engineering (BSE)
Netaji Subhas University of Technology (NSUT)
New Delhi, India

Garima Gupta
Department of Biotechnology
IILM University
Greater Noida, India

Pranjay Gupta
Department of Biotechnology
IILM University
Greater Noida, India

Astha Jadav
Department of Biotechnology and Bioengineering
Institute of Advanced Research
Koba Institutional Area
Gandhinagar, India

Santosh Kumar Mishra
Department of Life Sciences
Sharda School of Biosciences and Technology
Sharda University
Greater Noida, India

Yatender Kumar
Department of Biological Sciences & Engineering (BSE)
Netaji Subhas University of Technology (NSUT)
New Delhi, India

Khushboo Kumari
School of Life Sciences
Jawaharlal Nehru University
New Delhi, India

Khushi Kumari
Department of Biotechnology
IILM University
Greater Noida, India

Anamika Kushwaha
Department of Chemical Engineering
Chungbuk National University
Cheongju, Chungbuk, Republic of Korea

Monu Monu
Department of Biotechnology
IILM University
Greater Noida, India

Roma Pandey
Department of Biotechnology
IILM University
Greater Noida, Uttar Pradesh, India

Reena A. Rajput
Department of Biotechnology and Bioengineering
Institute of Advanced Research
Koba Institutional Area
Gandhinagar, India

Satyendra Kumar Rajput
Department of Pharmaceutical Sciences
Faculty of Pharmaceutical Sciences
Gurukula Kangri (Deemed to be University)
Haridwar, India

Madhav Sharan
Department of Biotechnology
IILM University
Greater Noida, India

Nidhi Sharma
Amity Institute of Pharmacy
Amity University
Noida, India

Yash Sharma
Department of Biotechnology
IILM University
Greater Noida, India

Sourav Shukla
Department of Biotechnology and Bioengineering
Institute of Advanced Research
Gandhinagar, India

Anjali Singh
Institute of Applied Medicine and Research
Ghaziabad, India

Avinash Singh
Department of Biotechnology
MIET
Meerut, India

Samriddhi Singh
Department of Biotechnology
IILM University
Greater Noida, India

Vandana
Department of Biotechnology
IILM University
Greater Noida, India

Pankaj Verma
Department of Biotechnology
IILM University
Greater Noida, India

Introduction

THE MAGNITUDE OF A MODERN EPIDEMIC

Cancer stands as one of humanity's most formidable health challenges, casting a shadow over millions of lives worldwide and fundamentally reshaping how we approach medical research, treatment, and care. This complex group of diseases, characterized by the uncontrolled growth and spread of abnormal cells, represents not just a medical condition but a global crisis that demands our urgent attention, innovative thinking, and collaborative efforts across disciplines (Nwosu, 2024).

The statistics paint a sobering picture: cancer continues to be a leading cause of morbidity and mortality across all continents, affecting people regardless of age, gender, or socioeconomic status. From the densely populated cities of Asia to the rural communities of Africa, from the developed nations of North America to the emerging economies of South America, cancer's reach is truly global. What makes this challenge even more daunting is the projected increase in cancer incidence over the coming decades, driven by aging populations, lifestyle changes, environmental factors, and improved diagnostic capabilities that reveal previously undetected cases (The global challenge of cancer, 2020).

Yet within this challenging landscape lies hope. The past several decades have witnessed unprecedented advances in our understanding of cancer biology, leading to revolutionary treatments and significantly improved outcomes for many patients. The journey from a universally fatal diagnosis to a condition that, in many cases, can be successfully treated or managed as a chronic disease represents one of modern medicine's greatest triumphs. This transformation has been possible through the relentless pursuit of knowledge, the development of sophisticated research methodologies, and the unwavering dedication of researchers, clinicians, and patients who have contributed to our collective understanding of these diseases (Chehelgerdi et al., 2023; Kashyap et al., 2023; Liu et al., 2025).

THE DIVERSE LANDSCAPE OF CANCER TYPES

Cancer is not a single disease but rather a collection of related conditions, each with its unique characteristics, challenges, and treatment approaches. This diversity is perhaps one of the most striking aspects of cancer research and treatment, as it requires specialized knowledge and tailored approaches for each type (Benz, 2017; de Visser & Joyce, 2023).

Breast cancer exemplifies this complexity, manifesting in multiple subtypes that behave differently and respond variably to treatment. The estrogen receptor-positive (ER^{+}) subtype, which represents most of the breast cancer cases, responds well to hormonal therapies that block estrogen's effects on cancer cells (Miziak et al., 2023). In contrast, human epidermal growth factor receptor 2-positive (HER^{2+}) breast

DOI: 10.1201/9781003583349-1

cancers require targeted therapies that specifically address the overexpression of this protein (Yersal & Barutca, 2014). Perhaps the most challenging is triple-negative breast cancer (TNBC), which lacks the three most common receptors targeted by current therapies, making treatment more complex and often requiring more aggressive approaches (Krzyszczyk et al., 2018; Orrantia-Borunda et al., 2022).

Lung cancer remains one of the deadliest forms of cancer globally, largely due to its tendency to be diagnosed at advanced stages when treatment options are more limited. The disease's strong association with smoking has made it a focal point for prevention efforts, yet the increasing incidence among non-smokers has highlighted the role of other factors, including environmental pollutants, genetic predisposition, and occupational exposures. The development of targeted therapies based on specific genetic mutations has revolutionized treatment for certain lung cancer patients, offering hope where traditional chemotherapy had limited success (Caini et al., 2022; Smith, 2024).

Colorectal cancer presents another fascinating case study in cancer diversity. This disease demonstrates how genetic mutations accumulate over time, with key alterations in genes such as APC, KRAS, and BRAF driving the transformation from normal tissue to malignant tumor (Li et al., 2021; Ottaiano et al., 2021). Understanding these molecular pathways has not only improved our comprehension of how colorectal cancer develops but has also led to more personalized treatment approaches based on the specific genetic profile of each patient's tumor (Armaghany et al., 2012; Li et al., 2024; Parmar & Easwaran, 2022; Testa et al., 2018; Yan et al., 2024).

Prostate cancer affects a significant portion of the male population, particularly as men age. What makes this cancer unique is its often-indolent nature – many men die with prostate cancer rather than from it. This characteristic has led to important discussions about the benefits and risks of screening, the concept of active surveillance for low-risk tumors, and the development of more precise methods to distinguish aggressive from slow-growing cancers (Bleyer et al., 2020; de Vos et al., 2024; Rebello et al., 2021; Salinas et al., 2014).

Ovarian cancer represents one of the most challenging cancers to treat, largely due to its tendency to remain undetected until it has spread throughout the abdominal cavity. The difficulty in early detection, combined with the cancer's aggressive metastatic potential, has made it a focus of intense research efforts aimed at developing better screening methods and understanding the mechanisms of metastasis (Arora et al., 2025; Jasen, 2009).

The field also encompasses less common but equally important cancers, such as **retinoblastoma**, a rare eye cancer that primarily affects infants and young children. Despite its rarity, retinoblastoma has been instrumental in advancing our understanding of cancer genetics, as it was one of the first cancers linked to a specific genetic mutation. The study of **oral squamous cell carcinoma** has highlighted the role of lifestyle factors, particularly tobacco and alcohol use, in cancer development, while also demonstrating the importance of early detection in improving outcomes (Hameed et al., 2024; Kaur & Patel, 2025; Lohmann & Gallie, 1993).

Liver cancer, particularly hepatocellular carcinoma, represents a global health challenge with significant regional variations in incidence and risk factors. The strong association with hepatitis B and C infections has made this cancer a target for prevention through vaccination and antiviral treatments, demonstrating how understanding cancer causes can lead to effective prevention strategies (Abdelhamed & El-Kassas, 2024; Ivancovsky Wajcman et al., 2025; Llovet et al., 2021; Shen et al., 2023; Yang et al., 2019).

THE EVOLUTION OF CANCER RESEARCH METHODOLOGIES

The fight against cancer has been marked by continuous evolution in research methodologies, each advancement building upon previous discoveries to create an increasingly sophisticated understanding of these diseases (Laplane & Maley, 2024). This progression from simple observational studies to complex, multidimensional research approaches reflects not only technological advances but also our growing appreciation for cancer's complexity (Swanton et al., 2024).

Traditional approaches to cancer research began with careful clinical observations and basic laboratory studies. Early researchers relied primarily on examining cancer tissues under microscopes, conducting population studies to identify risk factors, and testing treatments through clinical trials. While these methods provided fundamental insights into cancer behavior and treatment, they were limited in their ability to capture the full complexity of cancer biology (Swanton et al., 2024).

***In vitro* models** revolutionized cancer research by allowing scientists to study cancer cells in controlled laboratory environments. The development of established cell lines provided researchers with consistent, reproducible systems for studying cancer behavior and testing potential treatments (Anu Varshini et al., 2025). These models enabled high-throughput screening of thousands of compounds, dramatically accelerating the pace of drug discovery. The evolution from simple two-dimensional cell cultures to more sophisticated three-dimensional systems, including spheroids and organoids, has brought laboratory models closer to the complexity of actual tumors, providing more relevant insights into cancer behavior and treatment responses (Mirabelli et al., 2019; Momoli et al., 2025).

***In vivo* models** using laboratory animals have been indispensable for understanding how cancer behaves in living organisms. Xenograft models, where human cancer cells are implanted into immunocompromised mice, have allowed researchers to study human cancer biology in a living system (Holen et al., 2017). Genetically engineered mouse models have provided insights into how specific genetic changes lead to cancer development, while patient-derived xenografts have enabled the study of individual patient tumors, paving the way for personalized medicine approaches (Onaciu et al., 2020; Zhang et al., 2024).

***In silico* modeling** represents the newest frontier in cancer research, leveraging the power of computers to simulate cancer behavior and predict treatment responses (Edelman et al., 2010). These computational approaches can integrate vast amounts of data from genomic studies, clinical trials, and laboratory experiments to identify patterns and relationships that would be impossible to detect through traditional methods alone (Trisilowati & Mallet, 2012). Machine learning and artificial intelligence techniques are increasingly being applied to predict which patients will respond to specific treatments, identify new drug targets, and optimize treatment combinations (You et al., 2022).

THE PROMISE OF INTEGRATED RESEARCH APPROACHES

The future of cancer research lies not in any single methodology but in the intelligent integration of multiple approaches. Modern cancer research increasingly relies on combining insights from cell culture studies, animal models, computational predictions, and clinical observations to create a comprehensive understanding of cancer biology and treatment.

This integrated approach recognizes that cancer is fundamentally a systems-level disease that cannot be fully understood by examining individual components in isolation. The tumor microenvironment, immune system interactions, genetic heterogeneity, and treatment resistance mechanisms all contribute to cancer's complexity and must be considered together to develop effective treatments.

Personalized medicine has emerged as a natural consequence of this integrated approach, recognizing that each patient's cancer is unique and requires individualized treatment strategies (Masucci et al., 2024). By combining genomic profiling of tumors with computational models that predict treatment responses and clinical data that track outcomes, researchers and clinicians can increasingly tailor treatments to individual patients, maximizing effectiveness while minimizing side effects (Dlamini et al., 2022).

Prevention and early detection remain critical components of the cancer control strategy. Understanding the molecular mechanisms of cancer development has led to improved screening methods and the identification of individuals at high risk who might benefit from enhanced surveillance or preventive interventions (Golemis et al., 2018; Pashayan et al., 2020). The integration of genetic testing, imaging technologies, and biomarker analysis is creating new opportunities for detecting cancers at their earliest, most treatable stages (Jin et al., 2025).

LOOKING FORWARD: CHALLENGES AND OPPORTUNITIES

As we stand at the threshold of a new era in cancer research and treatment, we face both unprecedented opportunities and significant challenges. The explosion of genomic data, the development of immunotherapies that harness the body's own immune system to fight cancer, and the emergence of precision medicine approaches offer tremendous hope for improving outcomes for cancer patients.

However, significant challenges remain. Cancer's ability to evolve and develop resistance to treatments continues to pose problems for long-term management. The high cost of many new treatments raises questions about accessibility and healthcare equity. The complexity of cancer biology means that despite decades of research, we still have much to learn about how these diseases develop, progress, and can be most effectively treated.

Perhaps most importantly, the global nature of the cancer challenge requires international collaboration and the sharing of knowledge, resources, and expertise. Cancer affects people in all parts of the world, but research resources and treatment capabilities are not equally distributed. Addressing this disparity will require coordinated efforts to build research capacity, share data and findings, and ensure that advances in cancer treatment benefit all populations.

The pages that follow in this book will explore these themes in greater detail, examining specific cancer types, research methodologies, and treatment approaches. Each chapter builds upon the foundation laid here, providing deeper insights into the complexities of cancer while highlighting the remarkable progress that has been made and the promising directions for future research and treatment.

The story of cancer research is ultimately a story of human ingenuity, perseverance, and hope. It represents our species' remarkable ability to confront seemingly insurmountable challenges through careful observation, creative thinking, and collaborative effort. As we continue to unravel the mysteries of cancer, we move closer to a future where these diseases no longer cast the shadow they do today over human health and well-being.

REFERENCES

Abdelhamed, W., & El-Kassas, M. (2024). Hepatitis B virus as a risk factor for hepatocellular carcinoma: There is still much work to do. *Liver Research (Beijing, China)*, 8(2), 83–90. https://doi.org/10.1016/j.livres.2024.05.004

Armaghany, T., Wilson, J. D., Chu, Q., & Mills, G. (2012). Genetic alterations in colorectal cancer. *Gastrointestinal Cancer Research*, 5(1), 19–27.

Arora, T., Mullangi, S., Vadakekut, E. S., & Lekkala, M. R. (2025). *Epithelial Ovarian Cancer*. StatPearls Publishing, Treasure Island, FL.

Anu Varshini, A. M., Usha Nandini, S., Ramesh Kumar, V., Thangam, T., & Parthasarathy, K. (2025). New frontiers in anti-cancer drug testing: The need for a relevant In vitro testing model. *NAM Journal*, 1, 100003. https://doi.org/10.1016/j.namjnl.2024.100003

Benz, E. J. (2017). The Jeremiah Metzger lecture cancer in the twenty-first century: An inside view from an outsider. *Transactions of the American Clinical and Climatological Association*, 128, 275–297.

Bleyer, A., Spreafico, F., & Barr, R. (2020). Prostate cancer in young men: An emerging young adult and older adolescent challenge. *Cancer*, 126(1), 46–57. https://doi.org/10.1002/cncr.32498

Caini, S., Del Riccio, M., Vettori, V., Scotti, V., Martinoli, C., Raimondi, S., Cammarata, G., Palli, D., Banini, M., Masala, G., & Gandini, S. (2022). Quitting smoking at or around diagnosis improves the overall survival of lung cancer patients: A systematic review and meta-analysis. *Journal of Thoracic Oncology*, 17(5), 623–636. https://doi.org/10.1016/j.jtho.2021.12.005

Chehelgerdi, M., Chehelgerdi, M., Allela, O. Q. B., Pecho, R. D. C., Jayasankar, N., Rao, D. P., Thamaraikani, T., Vasanthan, M., Viktor, P., Lakshmaiya, N., Saadh, M. J., Amajd, A., Abo-Zaid, M. A., Castillo-Acobo, R. Y., Ismail, A. H., Amin, A. H., & Akhavan-Sigari, R. (2023). Progressing nanotechnology to improve targeted cancer treatment: Overcoming hurdles in its clinical implementation. *Molecular Cancer*, 22(1), 169. https://doi.org/10.1186/s12943-023-01865-0

de Visser, K. E., & Joyce, J. A. (2023). The evolving tumor microenvironment: From cancer initiation to metastatic outgrowth. *Cancer Cell*, 41(3), 374–403. https://doi.org/10.1016/j.ccell.2023.02.016

de Vos, I. I., Remmers, S., Hogenhout, R., Roobol, M. J., & ERSPC Rotterdam Study Group. (2024). Prostate cancer mortality among elderly men after discontinuing organised screening: Long-term results from the European randomized study of screening for prostate cancer rotterdam. *European Urology*, 85(1), 74–81. https://doi.org/10.1016/j.eururo.2023.10.011

Dlamini, Z., Skepu, A., Kim, N., Mkhabele, M., Khanyile, R., Molefi, T., Mbatha, S., Setlai, B., Mulaudzi, T., Mabongo, M., Bida, M., Kgoebane-Maseko, M., Mathabe, K., Lockhat, Z., Kgokolo, M., Chauke-Malinga, N., Ramagaga, S., & Hull, R. (2022). AI and precision oncology in clinical cancer genomics: From prevention to targeted cancer therapies-an outcomes based patient care. *Informatics in Medicine Unlocked*, 31, 100965. https://doi.org/10.1016/j.imu.2022.100965

Edelman, L. B., Eddy, J. A., & Price, N. D. (2010). In silico models of cancer. *Wiley Interdisciplinary Reviews. Systems Biology and Medicine*, 2(4), 438–459. https://doi.org/10.1002/wsbm.75

Golemis, E. A., Scheet, P., Beck, T. N., Scolnick, E. M., Hunter, D. J., Hawk, E., & Hopkins, N. (2018). Molecular mechanisms of the preventable causes of cancer in the United States. *Genes & Development*, 32(13–14), 868–902. https://doi.org/10.1101/gad.314849.118

Hameed, S., Yu, A. C., Almadani, B., Abualkhair, S., Ahmad, K., & Zauli, G. (2024). Genetic risk factors and clinical outcomes in childhood eye cancers: A review. *Genes*, 15(3), 276. https://doi.org/10.3390/genes15030276

Holen, I., Speirs, V., Morrissey, B., & Blyth, K. (2017). In vivo models in breast cancer research: Progress, challenges and future directions. *Disease Models & Mechanisms*, 10(4), 359–371. https://doi.org/10.1242/dmm.028274

Ivancovsky Wajcman, D., Nicolàs, A., Picchio, C. A., van Selm, L., Dusheiko, G., Younossi, Z. M., Dillon, J. F., Alqhatani, S. A., Razavi, H., Colombo, M. G., Kautz, A., Dore, G. J., & Lazarus, J. V. (2025). Prioritising viral hepatitis elimination to prevent hepatocellular carcinoma: A public health approach for effective preventive hepatology. *JHEP Reports*, 7, 101436. https://doi.org/10.1016/j.jhepr.2025.101436

Jasen, P. (2009). From the "silent killer" to the "whispering disease": Ovarian cancer and the uses of metaphor. *Medical History*, 53(4), 489–512.

Jin, D., Khan, N. U., Gu, W., Lei, H., Goel, A., & Chen, T. (2025). Informatics strategies for early detection and risk mitigation in pancreatic cancer patients. *Neoplasia*, 60, 101129. https://doi.org/10.1016/j.neo.2025.101129

Kashyap, B. K., Singh, V. V., Solanki, M. K., Kumar, A., Ruokolainen, J., & Kesari, K. K. (2023). Smart nanomaterials in cancer theranostics: Challenges and opportunities. *ACS Omega*, 8(16), 14290–14320. https://doi.org/10.1021/acsomega.2c07840

Kaur, K., & Patel, B. C. (2025). *Retinoblastoma*. StatPearls Publishing, Treasure Island, FL.

Krzyszczyk, P., Acevedo, A., Davidoff, E. J., Timmins, L. M., Marrero-Berrios, I., Patel, M., White, C., Lowe, C., Sherba, J. J., Hartmanshenn, C., O'Neill, K. M., Balter, M. L., Fritz, Z. R., Androulakis, I. P., Schloss, R. S., & Yarmush, M. L. (2018). The growing role of precision and personalized medicine for cancer treatment. *Technology*, 6(3–4), 79–100. https://doi.org/10.1142/S2339547818300020

Laplane, L., & Maley, C. C. (2024). The evolutionary theory of cancer: Challenges and potential solutions. *Nature Reviews. Cancer*, 24(10), 718–733. https://doi.org/10.1038/s41568-024-00734-2

Liu, Y., Zhang, Y., Li, H., & Hu, T. Y. (2025). Recent advances in the bench-to-bedside translation of cancer nanomedicines. *Acta Pharmaceutica Sinica B*, 15(1), 97–122. https://doi.org/10.1016/j.apsb.2024.12.007

Li, J., Ma, X., Chakravarti, D., Shalapour, S., & DePinho, R. A. (2021). Genetic and biological hallmarks of colorectal cancer. *Genes & Development*, 35(11–12), 787–820. https://doi.org/10.1101/gad.348226.120

Li, Q., Geng, S., Luo, H., Wang, W., Mo, Y.-Q., Luo, Q., Wang, L., Song, G.-B., Sheng, J.-P., & Xu, B. (2024). Signaling pathways involved in colorectal cancer: Pathogenesis and targeted therapy. *Signal Transduction and Targeted Therapy*, 9(1), 266. https://doi.org/10.1038/s41392-024-01953-7

Llovet, J. M., Kelley, R. K., Villanueva, A., Singal, A. G., Pikarsky, E., Roayaie, S., Lencioni, R., Koike, K., Zucman-Rossi, J., & Finn, R. S. (2021). Hepatocellular carcinoma. *Nature Reviews. Disease Primers*, 7(1), 6. https://doi.org/10.1038/s41572-020-00240-3

Lohmann, D. R., & Gallie, B. L. (1993). Retinoblastoma. In Pagon, R. A., Adam, M. P., Ardinger, H. H., Wallace, S. E., Amemiya, A., Bean, L. J., Bird, T. D., Ledbetter, N., Mefford, H. C., Smith, R. J., & Stephens, K. (Eds.), *GeneReviews(®)*. University of Washington, Seattle.

Masucci, M., Karlsson, C., Blomqvist, L., & Ernberg, I. (2024). Bridging the divide: A review on the implementation of personalized cancer medicine. *Journal of Personalized Medicine*, 14(6), 561. https://doi.org/10.3390/jpm14060561

Mirabelli, P., Coppola, L., & Salvatore, M. (2019). Cancer cell lines are useful model systems for medical research. *Cancers*, 11(8), 1098. https://doi.org/10.3390/cancers11081098

Miziak, P., Baran, M., Błaszczak, E., Przybyszewska-Podstawka, A., Kałafut, J., Smok-Kalwat, J., Dmoszyńska-Graniczka, M., Kiełbus, M., & Stepulak, A. (2023). Estrogen receptor signaling in breast cancer. *Cancers*, 15(19), 4689. https://doi.org/10.3390/cancers15194689

Momoli, C., Costa, B., Lenti, L., Tubertini, M., Parenti, M. D., Martella, E., Varchi, G., & Ferroni, C. (2025). The evolution of anticancer 3D in vitro models: The potential role of machine learning and AI in the next generation of animal-free experiments. *Cancers*, 17(4), 700. https://doi.org/10.3390/cancers17040700

No authors listed. (2020). The global challenge of cancer. *Nature Cancer*, 1(1), 1–2. https://doi.org/10.1038/s43018-019-0023-9

Nwosu, N. (2024). Cancer: A disease of modern times? *Cureus*, 16(11), e74666. https://doi.org/10.7759/cureus.74666

Onaciu, A., Munteanu, R., Munteanu, V. C., Gulei, D., Raduly, L., Feder, R.-I., Pirlog, R., Atanasov, A. G., Korban, S. S., Irimie, A., & Berindan-Neagoe, I. (2020). Spontaneous and induced animal models for cancer research. *Diagnostics*, 10(9), 660. https://doi.org/10.3390/diagnostics10090660

Orrantia-Borunda, E., Anchondo-Nuñez, P., Acuña-Aguilar, L. E., Gómez-Valles, F. O., & Ramírez-Valdespino, C. A. (2022). Subtypes of breast cancer. In Mayrovitz, H. N. (Ed.), *Breast Cancer*. Exon Publications, Australia. https://doi.org/10.36255/exon-publications-breast-cancer-subtypes

Ottaiano, A., Santorsola, M., Caraglia, M., Circelli, L., Gigantino, V., Botti, G., & Nasti, G. (2021). Genetic regressive trajectories in colorectal cancer: A new hallmark of oligo-metastatic disease? *Translational Oncology*, 14(8), 101131. https://doi.org/10.1016/j.tranon.2021.101131

Parmar, S., & Easwaran, H. (2022). Genetic and epigenetic dependencies in colorectal cancer development. *Gastroenterology Report*, 10, goac035. https://doi.org/10.1093/gastro/goac035

Pashayan, N., Antoniou, A. C., Ivanus, U., Esserman, L. J., Easton, D. F., French, D., Sroczynski, G., Hall, P., Cuzick, J., Evans, D. G., Simard, J., Garcia-Closas, M., Schmutzler, R., Wegwarth, O., Pharoah, P., Moorthie, S., De Montgolfier, S., Baron, C., Herceg, Z., … & Widschwendter, M. (2020). Personalized early detection and prevention of breast cancer: ENVISION consensus statement. *Nature Reviews. Clinical Oncology*, 17(11), 687–705. https://doi.org/10.1038/s41571-020-0388-9

Rebello, R. J., Oing, C., Knudsen, K. E., Loeb, S., Johnson, D. C., Reiter, R. E., Gillessen, S., Van der Kwast, T., & Bristow, R. G. (2021). Prostate cancer. *Nature Reviews. Disease Primers*, 7(1), 9. https://doi.org/10.1038/s41572-020-00243-0

Salinas, C. A., Tsodikov, A., Ishak-Howard, M., & Cooney, K. A. (2014). Prostate cancer in young men: An important clinical entity. *Nature Reviews. Urology*, 11(6), 317–323. https://doi.org/10.1038/nrurol.2014.91

Shen, C., Jiang, X., Li, M., & Luo, Y. (2023). Hepatitis virus and hepatocellular carcinoma: Recent advances. *Cancers*, 15(2), 533. https://doi.org/10.3390/cancers15020533

Smith, E. (2024). Lung cancer: A deadly form of cancer worldwide. *African Journal of Respiratory Medicine*, 19(4), 855.

Swanton, C., Bernard, E., Abbosh, C., André, F., Auwerx, J., Balmain, A., Bar-Sagi, D., Bernards, R., Bullman, S., DeGregori, J., Elliott, C., Erez, A., Evan, G., Febbraio, M. A., Hidalgo, A., Jamal-Hanjani, M., Joyce, J. A., Kaiser, M., Lamia, K., ..., & Hanahan, D. (2024). Embracing cancer complexity: Hallmarks of systemic disease. *Cell*, 187(7), 1589–1616. https://doi.org/10.1016/j.cell.2024.02.009

Testa, U., Pelosi, E., & Castelli, G. (2018). Colorectal cancer: Genetic abnormalities, tumor progression, tumor heterogeneity, clonal evolution and tumor-initiating cells. *Medical Sciences : Open Access Journal*, 6(2), 31. https://doi.org/10.3390/medsci6020031

Trisilowati, D., & Mallet, D. G. (2012). In silico experimental modeling of cancer treatment. *ISRN Oncology*, 2012, 828701. https://doi.org/10.5402/2012/828701

Yang, J. D., Hainaut, P., Gores, G. J., Amadou, A., Plymoth, A., & Roberts, L. R. (2019). A global view of hepatocellular carcinoma: Trends, risk, prevention and management. *Nature Reviews. Gastroenterology & Hepatology*, 16(10), 589–604. https://doi.org/10.1038/s41575-019-0186-y

Yan, L., Shi, J., & Zhu, J. (2024). Cellular and molecular events in colorectal cancer: Biological mechanisms, cell death pathways, drug resistance and signalling network interactions. *Discover. Oncology*, 15(1), 294. https://doi.org/10.1007/s12672-024-01163-1

Yersal, O., & Barutca, S. (2014). Biological subtypes of breast cancer: Prognostic and therapeutic implications. *World Journal of Clinical Oncology*, 5(3), 412–424. https://doi.org/10.5306/wjco.v5.i3.412

You, Y., Lai, X., Pan, Y., Zheng, H., Vera, J., Liu, S., Deng, S., & Zhang, L. (2022). Artificial intelligence in cancer target identification and drug discovery. *Signal Transduction and Targeted Therapy*, 7(1), 156. https://doi.org/10.1038/s41392-022-00994-0

Zhang, C., Sui, Y., Liu, S., & Yang, M. (2024). In vitro and in vivo experimental models for cancer immunotherapy study. *Current Research in Biotechnology*, 7, 100210. https://doi.org/10.1016/j.crbiot.2024.100210

1 Breast Cancer Models

Ramasamy Amsaveni and
Veluswamy Bhuvaneshwari

1.1 INTRODUCTION

Breast cancer represents a heterogeneous group of diseases, with each subtype demonstrating unique genetic profiles, biological behaviors, and responses to treatment (Yersal and Barutca, 2014). It is the most commonly diagnosed cancer worldwide among women and one of the leading causes of cancer-related deaths. In 2022, breast cancer affected approximately 2.3 million women globally, resulting in an estimated 670,000 deaths. The disease occurs worldwide and can affect women of any age, though it is more commonly diagnosed in older women (Torre et al., 2017). The prevalence of breast cancer varies significantly across different regions, often influenced by socioeconomic factors. In countries with a high Human Development Index (HDI), approximately 1 in 12 women will be diagnosed with breast cancer at some point in their lives, and about 1 in 71 women is expected to die from the disease (Tao et al., 2015). In contrast, countries with lower HDI show different patterns, where fewer women are diagnosed, but those who do are more likely to face higher mortality rates due to limited access to timely treatment.

The strongest risk factor for breast cancer is being female, with nearly 99% of cases occurring in women (Momenimovahed and Salehiniya, 2019). Breast cancer in men, although rare, also requires treatment based on principles similar to those used for women. Certain risk factors increase the likelihood of developing breast cancer, including advancing age, obesity, excessive alcohol consumption, family history, and exposure to certain types of radiation. Other factors include reproductive history, such as the age when menstruation began or the age at first pregnancy, as well as lifestyle factors like smoking and the use of hormone therapy after menopause (Irigaray et al., 2007). Although a family history of breast cancer raises the risk, most breast cancer cases occur in individuals with no known family history. Inherited genetic mutations, particularly in the BRCA1, BRCA2, and PALB2 genes, are also known to increase breast cancer risk significantly. Women who test positive for these genetic mutations might consider preventive measures, such as preventive surgery or chemoprevention (Huber-Keener, 2022). The World Health Organization (WHO) has initiated a Global Breast Cancer Initiative (GBCI) aimed at reducing breast cancer mortality by 2.5% annually. This initiative aims to prevent up to 2.5 million deaths by 2040, focusing on three key objectives: promoting health for early detection, ensuring timely diagnosis, and providing comprehensive breast cancer management. Research efforts have driven the creation of various models that allow scientists to explore the intricate biology of breast cancer, discover new therapeutic targets, and develop personalized treatment approaches. Breast cancer research has made significant progress in recent years,

DOI: 10.1201/9781003583349-2

largely due to the development of more sophisticated and biologically relevant *in vitro* models that simulate the tumor microenvironment (TME). These models are critical for understanding the pathophysiology of breast cancer, studying metastasis, and screening for novel therapies. From engineered three-dimensional (3D) tumor models to organoid-based systems, recent innovations are helping to bridge the gap between preclinical research and clinical application. Below, we summarize key advancements in breast cancer modeling, citing recent studies that highlight the potential of new technologies and their implications for cancer therapy (Thakuri et al., 2018).

The diversity in breast cancer subtypes, such as estrogen receptor-positive (ER^+), human epidermal growth factor receptor 2-positive (HER^{2+}), and triple-negative breast cancer (TNBC), necessitates models that can accurately mimic the intricate biological and molecular characteristics of each subtype. To address these needs, researchers employ *in vitro* (cell-based), *in vivo* (animal-based), and *in silico* (computational) models. Each type of model provides unique advantages, contributing essential insights into cancer biology and treatment (Pistollato et al., 2020).

1.2 BREAST CANCER SUBTYPES

Breast cancer's epidemiological profile varies significantly across different populations, influenced by genetic, environmental, and lifestyle factors. The disease manifests in several major subtypes, primarily defined by the expression or absence of certain receptors. These include ER^+, HER^{2+}, and TNBC. Each subtype of breast cancer has unique molecular features that affect its progression and response to treatment. These subtypes are classified according to their molecular profiles, with the main categories being ER^+, HER^{2+}, and TNBC (Parise and Caggiano, 2019).

1.2.1 Estrogen Receptor-Positive (ER^+)

ER^+ breast cancer is defined by the presence of estrogen receptors on the surface of cancer cells, which enables the tumor to use estrogen as a growth signal. This subtype accounts for approximately 70%–80% of all breast cancers, making it the most common subtype. The presence of estrogen receptors on these cells allows for targeted hormone therapies to effectively reduce tumor growth by disrupting the interaction between estrogen and its receptor (Almeida et al., 2020).

Key hormonal therapies for ER^+ breast cancer include tamoxifen and aromatase inhibitors (AIs):

- Tamoxifen is a selective estrogen receptor modulator (SERM) that binds to estrogen receptors on cancer cells, blocking estrogen from attaching and activating the receptors. It is commonly prescribed for both premenopausal and postmenopausal women and is taken over long durations, often 5–10 years, to reduce recurrence risk (O'Regan and Jordan, 2002).
- AIs work by inhibiting the enzyme aromatase, which converts androgens into estrogen in postmenopausal women. Common AIs include anastrozole, letrozole, and exemestane, which lower estrogen levels, effectively "starving" ER^+ cancer cells (Lui, 2017).

ER^+ breast cancer typically has a better prognosis and slower progression than other subtypes, as it grows at a slower rate and responds well to hormonal treatments. This subtype is often associated with lower rates of early recurrence and a greater likelihood of responding to treatments over the long term. However, ER^+ cancers also have a risk of late recurrence, sometimes occurring several years or even decades after initial treatment, which underscores the importance of extended hormonal therapy in many patients (Thomas and Gustafsson, 2011).

1.2.2 Human Epidermal Growth Factor Receptor 2-Positive (HER^{2+})

HER^{2+} breast cancer is defined by an overproduction of the human epidermal growth factor receptor 2 (HER2) protein, a receptor tyrosine kinase that plays a key role in cell growth and survival. This subtype accounts for about 15%–20% of breast cancers and is frequently linked to more aggressive tumor characteristics and a faster growth rate compared to ER^+ breast cancer. The overexpression of HER2 results in constant growth signals, promoting rapid cell division and tumor progression (Elshazly and Gewirtz, 2022).

Targeted therapies have transformed the management of HER^{2+} breast cancer by blocking HER2 signaling and inhibiting tumor growth. The cornerstone of HER2-targeted therapy is trastuzumab (Herceptin), a monoclonal antibody that binds to the HER2 protein, preventing it from activating growth pathways. Trastuzumab is often combined with chemotherapy and significantly improves survival rates and reduces the risk of recurrence in patients with HER^{2+} breast cancer (Omidi et al., 2022). Other HER2-targeted therapies include pertuzumab, ado-trastuzumab emtansine (T-DM1), and lapatinib, each with distinct mechanisms to disrupt HER2 signaling (Xu et al., 2024).

Despite these advancements, HER^{2+} breast cancer can develop resistance to targeted therapies over time. This resistance often necessitates alternative or combination therapies, and ongoing research aims to better understand the mechanisms of resistance and develop new treatment options.

1.2.3 Triple-Negative Breast Cancer (TNBC)

TNBC is an aggressive subtype defined by the absence of estrogen receptors, progesterone receptors, and HER2 expression, representing about 10%–15% of breast cancers. The lack of these receptors makes TNBC unresponsive to hormonal therapies (like tamoxifen or AIs) and HER2-targeted therapies (like trastuzumab). Without these receptor targets, treatment options for TNBC are limited primarily to chemotherapy and, more recently, immunotherapy and PARP inhibitors (Kumar and Aggarwal, 2016).

TNBC has a poorer prognosis than other subtypes due to its aggressive nature, high likelihood of early recurrence, and propensity for distant metastasis, particularly to the brain, liver, and lungs. TNBC is also more common in younger women and those with BRCA1 gene mutations, which further complicates its management and necessitates a tailored approach.

Due to the limited efficacy of traditional treatments, novel therapeutic strategies are under investigation for TNBC. Immunotherapy, particularly checkpoint inhibitors like pembrolizumab, has shown promise in enhancing the immune system's ability to target TNBC cells. In addition, PARP inhibitors, which exploit DNA repair weaknesses in cancer cells, are being studied as targeted therapies, especially in BRCA-mutated TNBC cases (Jain et al., 2024). These emerging options represent important steps toward improving outcomes for patients with this challenging subtype.

Understanding these subtypes is essential for creating and utilizing models that accurately replicate each type's unique biology, thereby enabling the development of effective and personalized treatments.

1.3 TYPES OF BREAST CANCER MODELS

Researchers employ various types of models to study breast cancer:

- *In vitro* (cell-based) models
- *In vivo* (animal-based) models
- *In silico* (computational) models

Each type of model provides unique advantages, contributing essential insights to breast cancer research.

1.4 *IN VITRO* MODELS

Developing effective treatments for cancer remains a significant challenge due to its complexity, which stems from the diversity of the disease and the variations observed between patients as well as within individual tumors. Traditional preclinical cancer drug studies have relied heavily on two-dimensional (2D) *in vitro* models to assess drug toxicity. However, these 2D models fail to accurately replicate the 3D TME, leading to a disconnect between laboratory results and real-world tumor responses (Bağci et al., 2022). This lack of correlation contributes significantly to the high failure rate of drug development, as 95% of drugs that show efficacy in preclinical trials fail in clinical settings. Furthermore, only around 7.5% of drugs successfully advance from Phase 1 trials to market approval (Kunnumakkara et al., 2019).

While 2D models are valuable for isolating specific variables, they fall short in capturing the complex interactions between cells and the materials they interact with in the body. In a 2D environment, cells often adapt in unnatural ways, losing key phenotypic traits that are essential for realistic tumor modeling.

Conversely, *in vivo* models, though offering insight into tumor development, progression, and drug metabolism, are costly, time-consuming, and may not accurately mimic human responses due to species differences or variations in immune system function.

Ethical concerns also arise from the extensive use of animal subjects in such studies (Li et al., 2023). There is a pressing need for drug-testing platforms that more effectively simulate human biology. *In vitro* 3D models that replicate the TME provide a promising solution, offering researchers the opportunity to investigate the molecular mechanisms of tumor growth, metastasis, and drug efficacy in a more realistic context. Notably, results from 3D models have shown strong alignment with both *in vivo* studies and clinical outcomes (Bağci et al., 2022).

Recent progress in biomaterials and tissue engineering (TE) has made it more possible to create 3D models that closely mimic the human TME, especially for particular diseases or physiological conditions. This chapter will examine the components of the breast TME and their contributions to cancer progression, followed by a review of methods for designing 3D models that accurately replicate the TME (Curvello et al., 2023).

In contrast, *in vitro* models utilizing primary human cells and cell lines offer a more controlled environment to investigate the molecular mechanisms involved in tumor behavior. These models simplify the biological system, allowing for the study of tumor dynamics and the prediction of clinical drug efficacy, though the pharmacokinetics and pharmacodynamics may differ from *in vivo* conditions (Vaira et al., 2010).

1.4.1 Two-Dimensional (2D) Models

For many years, 2D cell culture systems have served as the foundation for biological research, including breast cancer studies. Most research begins with simple monolayer cultures or transwell systems before progressing to more complex 3D and *in vivo* models. The advantages of 2D cultures include ease of handling, straightforward measurement of cellular changes, cost-effectiveness, and reduced experimental time. They also facilitate manipulation of the culture substrate to explore the impact of various physical or chemical cues on cell behavior. Cancer cell lines are derived from human breast tumors and are among the most widely used tools in breast cancer research. Cell lines like MCF-7 (ER^{+}), SK-BR-3 (HER^{2+}), and MDA-MB-231 (TNBC) represent specific subtypes and allow for consistent, reproducible experimentation. However, cell lines have limitations, including genetic drift over time and the lack of a complex tumor microenvironment, which can influence tumor behavior and drug response (Fröhlich, 2023).

However, 2D models present several challenges in breast cancer research. Many findings from 2D cultures contradict results from *in vivo* studies, highlighting the random morphology of cells in this system, which fails to replicate the signaling related to cell shape and orientation. Furthermore, studies have shown that cancer cells grown in 2D do not exhibit their typical signaling responses in key biological processes such as growth, morphology, metabolism, and differentiation. Thus, while 2D cultures can be useful for preliminary investigations, they are increasingly regarded as inadequate for comprehensive studies (Bahcecioglu et al., 2020).

1.4.2 Three-Dimensional (3D) Models in Cancer Research

3D models are increasingly recognized for their ability to replicate the complex architecture of tissues, offering significant advantages over traditional 2D systems. Unlike 2D cultures, 3D models allow cells to interact with one another and the extracellular matrix (ECM) in a manner that closely mimics *in vivo* conditions. Ex vivo models, which utilize thin slices of tumors from humans or animals, are another promising approach. These models preserve the native ECM structure of the TME, supporting the cellular phenotype and maintaining the heterogeneity found in the original tissue. While the use of *ex vivo* models has traditionally been limited by the availability of explants, the development of explant banks has made them more accessible for drug screening (Randall-Demllo et al., 2024). Furthermore, advancements in culture techniques have enabled the long-term maintenance of explants without significant loss of phenotype. Despite these improvements, variability between patient-derived tumors can introduce challenges when comparing results across studies.

Spheroids, a type of 3D model, allow cancer cells to aggregate into rounded, tumor-like structures, providing a more accurate representation of gene expression, cell proliferation, migration, and heterogeneity compared to 2D cultures (Boix-Montesinos et al., 2021). These models are highly valuable for cancer research, particularly in drug testing. Tumor cells within 3D structures often exhibit increased resistance to treatments, a characteristic that mirrors clinical drug resistance observed in solid tumors. In some cases, specific drugs may demonstrate reduced resistance under hypoxic conditions within 3D cultures. Research has shown that tumor cells in 3D models display distinct gene expression profiles, which can predict how different breast cancer subtypes will respond to various therapies. Additionally, 3D models are well-suited for investigating interactions between tumor cells, immune cells, and stromal components, as well as for studying the signaling pathways involved in tumor progression (Fontana et al., 2021).

3D models are particularly advantageous for drug testing because they often exhibit treatment resistance patterns more similar to those in actual tumors. Some tumor cells in 3D conditions even show unique gene expression profiles, which may predict drug responsiveness across different cancer subtypes. Additionally, these models allow for studying interactions among cancer cells, immune cells, and stromal cells, along with important tumor signaling pathways.

Figure 1.1 illustrates four cutting-edge 3D modeling technologies used in biomedical research, particularly in breast cancer studies. Table 1.1 presents a comparative analysis of various breast cancer models, highlighting their sources, advantages, disadvantages, and applications.

1.4.3 Spheroids

Spheroids, also known as multicellular tumor spheroids, are aggregates of tightly packed cancer cells typically ranging in size from 100 to 600 microns. These cell clusters promote interactions between individual cells and replicate mechanical properties, oxygen gradients, and gene expression profiles found in tumors *in vivo*, though some differences persist. Several methods exist for creating spheroids, such as the hanging-drop technique, spinner flasks, basement membrane extracts, and magnetic levitation (Zhao, 2018).

Smaller spheroids (100–200 μm) are particularly useful for studying cellular interactions and drug responses. Larger spheroids (400–600 μm) develop hypoxic cores, making them ideal for investigating the effects of conditions like hypoxia, which significantly influence cancer cell behavior. The size of a spheroid can also correlate with its aggressiveness, impacting cell migration and biomarker expression. Additionally, spheroids are beneficial for evaluating drug penetration, as the size of the spheroid can influence the effectiveness of drug delivery systems (Zhao, 2018).

TABLE 1.1
Comparison Between Different Breast Cancer Models

Model Type	Source	Advantage	Disadvantage	Application
Two-dimensional cell-culture model	Often derived from an immortalized cell line with known properties (e.g., ER Positive, BRCA1 mutant)	Low price, rapid hypothesis testing	Simplest model with least clinical representation	Used for drug screening, toxicology studies, and basic cell biology research
In vivo animal model	May model pathogenesis by genetic knockout, carcinogenic exposure, or by xenograft	Demonstrates multisystem effects, well standardized	Time-consuming and does not take immune system into account in most cases	Studying complex biological processes, disease mechanisms, and testing therapies in a whole organism context
Three-Dimensional (3D) spheroids model	Derived from Multiple cell types, Primary Cells, Cell lines, tumor cells and tissues	Easy and standardized production, includes many cells	Limited control over formation	Tumor features, Cellular interactions, angiogenesis, drug testing, and metabolic profile of cancer cells
3D organoid model	Derived from at least one epithelial and one mesenchymal cell type, tumor cells and tissues, ESCs, iPSCs, or ASCs	Can quickly replicate multiple cell types *in vitro* from patient biopsy, allows for native cell arrangement	Poorly standardized	Mimicking organ-specific architecture, Cellular interactions, and drug screening for personalized medicine
Microfluidic devices organ-on-chip model	May be derived from ESCs, iPSCs, or ASCs	Replicates a multi-organ system from biopsy, predicts distal involvement	Inhibits native cell organization	organ-level functions for drug testing and physiological studies. Cellular interactions, angiogenesis, chemotaxis, and intravasation/extravasation
3D printing—natural matrices	Bioinks like alginate, collagen, and gelatin	High biocompatibility, strong cell adhesion, modifiable by cells	Batch variability and uncontrolled degradation	Studies on stiffness, migration, invasion, angiogenesis, and hypoxia
3D printing—synthetic matrices	Polymers like PCL, PLA, and PEG	Highly tunable	Poor cell adhesion and limited biocompatibility	Research on stiffness, migration, invasion, angiogenesis, and hypoxia
3D printing—composite material	Combinations like hydroxyapatite with collagen or PCL	Adjustable biophysical and biochemical properties	Complex, potential cytotoxicity, and batch variability	Studies on stiffness, porosity, migration, invasion, angiogenesis, and hypoxia
Decellularized scaffolds	Organs/tissues decellularized for 3D scaffolds	Similar properties and composition to *in vivo* tissues	Potential alteration by treatment, residual treatment agents	Cellular interaction

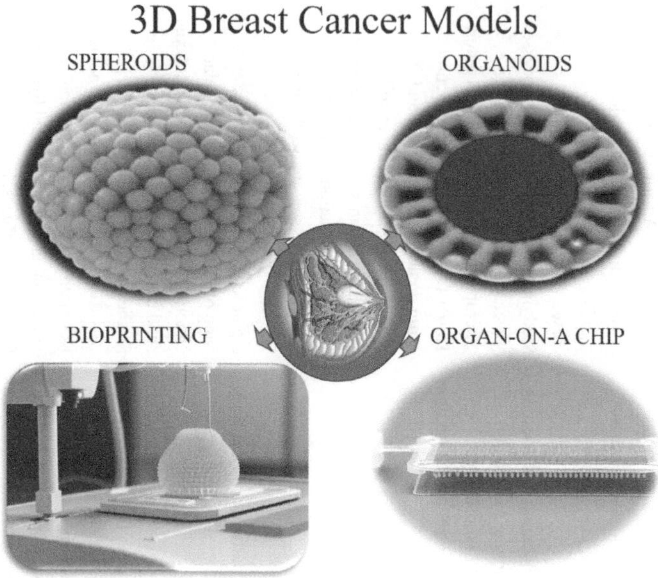

FIGURE 1.1 Types of 3D breast cancer models.

Spheroids are increasingly recognized as valuable models for drug screening and cancer research due to their ability to mimic tumor dynamics more accurately than traditional 2D cultures. These models display more physiologically relevant responses to chemotherapy and radiation, including heightened resistance to treatments, which mirrors the behavior of tumor cells in clinical settings (El Harane et al., 2023). Furthermore, spheroids offer a reproducible platform for evaluating drug efficacy and predicting clinical outcomes, making them particularly useful in developing precision medicine strategies, especially for breast cancer (Jadhav et al., 2024).

The selection of appropriate cell lines for generating 3D breast cancer models is critical, as different breast cancer cell lines exhibit varied responses to therapeutic agents. This variability emphasizes the importance of using models that capture the heterogeneity of tumor responses to treatment. Spheroids effectively replicate the structure of solid tumors and provide an ideal system for studying drug penetration, cellular interactions, and tumor hypoxia, offering advantages over 2D cell cultures in preclinical drug testing.

1.4.4 Patient-Derived Organoids (PDOs)

Organoids are self-organizing 3D cultures derived directly from patient tumors, preserving the heterogeneity and architecture of the original cancer. They are derived from at least one epithelial and one mesenchymal cell type, tumor cells, and tissues. They can also be generated from embryonic stem cells (ESCs), induced pluripotent stem cells (iPSCs), or adult stem cells (ASCs). They are valuable for personalized medicine and high-throughput drug screening, as they retain the genetic and phenotypic characteristics of patient tumors. Patient-derived organoids (PDOs) can replicate the cellular diversity and genetic mutations of breast cancer more accurately than traditional cell lines, making them highly valuable for personalized medicine approaches and drug response testing. These models enable high-throughput screening of potential drugs while closely mimicking *in vivo* tumor characteristics. Breast cancer models using PDOs have become a powerful tool for cancer research. By creating 3D cultures from patient breast cancer tissue or needle biopsies from metastatic lesions, these organoids successfully capture both the histological and genetic characteristics of the original tumors (Ebrahimi et al., 2022). By incorporating factors like Neuregulin 1, researchers have managed to overcome challenges in both the initial formation and long-term maintenance of breast cancer organoids (Tzeng et al., 2023). These models occupy a unique position in preclinical research, offering a middle ground between traditional 2D cell cultures and patient-derived xenografts (PDXs) used *in vivo*.

Breast cancer organoids exhibit a striking diversity that mirrors breast cancer histology. Ductal breast cancers typically form solid organoid structures, while lobular breast cancers often create less cohesive, discohesive organoids. This discohesive phenotype is particularly valuable as an experimental model, as it may provide insights into the mechanisms underlying lobular breast cancer's clinical behavior and survival outcomes. For example, loss of E-cadherin function—a key factor in cell adhesion—has been identified in some discohesive organoids, suggesting a potential avenue for exploring molecular targets. Additionally, mammary gland organoids are suitable for gene-editing techniques, enabling researchers to conduct phenotypic screens and test specific hypotheses regarding tumor biology (Boudreault, 2024).

1.4.5 Scaffold-Based 3D Models for Breast Cancer Tissue Engineering

TE has emerged as a promising approach for breast cancer treatment, offering innovative techniques for regenerating damaged or diseased tissues (Teixeira and Martins, 2023). A central aspect of this field involves the creation of scaffolds that mimic the ECM, providing a supportive microenvironment for cell growth and tissue repair (Sebestyén et al., 2022). As a branch of regenerative medicine, TE combines biological sciences and engineering principles to develop artificial tissues and organs, potentially addressing the shortage of donor organs. This approach involves the integration of cells within biological structures to restore tissue function. Key components of TE include scaffolds, cells, and signaling molecules, with scaffolds offering structural support, facilitating cell attachment, and delivering critical growth factors (Whitaker et al., 2001).

Scaffolds can be derived from natural biological tissues obtained from human or animal sources or synthesized using biomaterials. Natural scaffolds exhibit superior biocompatibility, degradability, and the ability to closely replicate native tissue microenvironments, promoting effective cellular interactions (Zhang and King, 2020). 3D scaffold-based models are crucial in cancer TE, as they create a biomimetic environment for studying tumor development, drug responses, and cell interactions. Several types of scaffolds have been developed to more closely mimic the tumor microenvironment compared to traditional 2D cultures. In breast cancer TE, a variety of scaffold-based 3D models have been designed, including hydrogels, 3D bioprinted scaffolds, decellularized scaffolds, nanoparticle-incorporated scaffolds, microspheres, and fibrous scaffolds, each offering distinct advantages and applications (Figure 1.2).

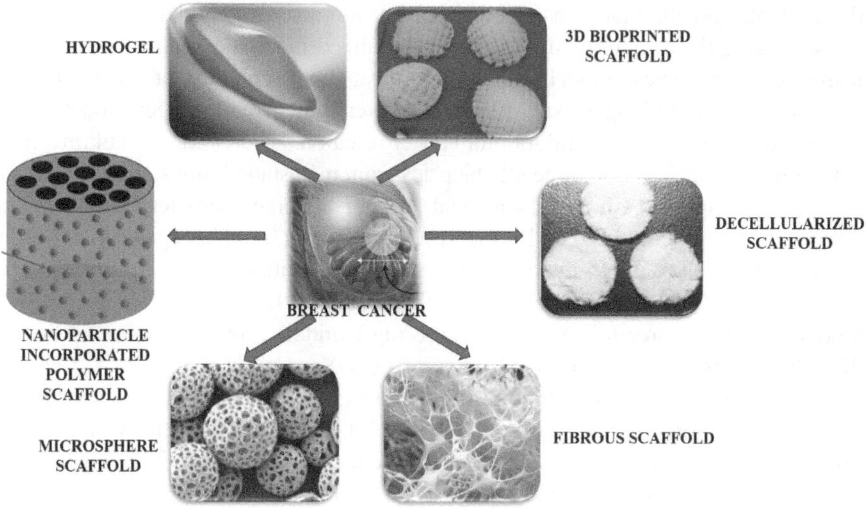

FIGURE 1.2 3D based scaffold for breast cancer TE.

1.4.5.1 Hydrogels

Hydrogels are hydrophilic, soft scaffolds capable of encapsulating cells and simulating the ECM environment. Due to their high biocompatibility, these materials can be engineered to match the mechanical properties of target tissues. Hydrogels can also be infused with bioactive molecules like growth factors to facilitate cell adhesion, proliferation, and differentiation (Ganeson et al., 2023). Morales et al. (2021) used 3D hydrogel scaffolds to study the mechanobiology of cancer cell migration. In a similar vein, Pradhan et al. showed that polyethylene glycol (PEG)-fibrinogen hydrogel models serve as a functional ECM scaffold for breast cancer cells (BCCs), with cell morphology and colony size being affected by the stiffness of the scaffold (Pradhan et al., 2017). Additionally, PEG hydrogels have been used to encapsulate estrogen-driven BCCs, resulting in enhanced cell proliferation without negatively impacting cell viability (Livingston et al., 2019).

1.4.5.2 Bioprinted Scaffolds

3D bioprinting technologies have enabled the creation of custom-designed scaffolds with precise control over cell placement and tissue architecture. These scaffolds can be engineered to recapitulate the complex structure and composition of the tumor microenvironment, allowing researchers to study cancer cell behavior in a more physiologically relevant context (Morales et al., 2021).

Recent advancements have led to the creation of complex 3D- and 4D-printed scaffolds specifically designed for breast cancer research. These structures facilitate personalized cancer modeling and targeted therapies by preserving key tumor characteristics (Moroni et al., 2022). For instance, bioprinted tumor models incorporating curcumin nanoparticles have been developed to enhance breast cancer treatment and serve as reliable platforms for preclinical drug testing (Su et al., 2022). Additionally, cellulose nanocrystal-based hydrogels have been integrated into thermosensitive scaffolds to allow controlled drug release for breast cancer therapy (Phan et al., 2022).

Hydrogels are widely used in bioprinting due to their biocompatibility and ability to simulate the ECM, supporting cell growth and interaction. Research has demonstrated that 3D-printed hydrogel scaffolds can be applied in breast cancer photothermal therapy, providing potential applications in targeted tumor ablation and tissue repair (Luo et al., 2019). These advanced biomaterials also contribute to personalized medicine by enabling high-throughput drug screening and combination therapy evaluation, bridging the gap between in vitro studies and clinical applications (Sharifi et al., 2021). Furthermore, additive-manufactured metastatic breast cancer models allow researchers to study tumor invasion and progression in a controlled 3D environment, offering deeper insights into the mechanisms underlying cancer metastasis (Nanou et al., 2022).

Different types of bioprinted scaffolds have been developed to optimize research applications. Natural matrices, such as alginate, collagen, and gelatin, are biocompatible and mimic the ECM, promoting cell adhesion, proliferation, and differentiation. Synthetic matrices, including polycaprolactone (PCL), polylactic acid (PLA), and PEG, offer tunable mechanical properties and controlled degradation rates, making them ideal for studying cancer cell behavior and drug delivery. Composite materials, such as hydroxyapatite (HAp) combined with collagen or PCL, integrate the advantages of both natural and synthetic matrices, enhancing mechanical stability and bioactivity to create biomimetic tumor environments.

Bioprinted scaffolds have become a crucial tool in cancer research, enabling more precise modeling of tumor biology and drug responses. With ongoing advancements in bioprinting technologies, these engineered platforms offer great potential for creating enhanced treatment strategies and personalized therapies in oncology.

1.4.5.3 Decellularized Scaffolds

Decellularized scaffolds are derived from natural tissues through a process that removes cellular components while preserving the ECM. This approach maintains the biochemical and structural properties of native tissues, making these scaffolds valuable for breast cancer research and TE. By retaining essential ECM components, they provide a physiologically relevant microenvironment for studying tumor-stroma interactions and evaluating therapeutic responses (Zhang et al., 2022).

The decellularization process enhances biological recognition, promotes cell adhesion, and supports tissue regeneration while minimizing immunogenicity. These properties make decellularized scaffolds particularly useful for drug testing and personalized medicine applications. Studies have shown that breast cancer-derived decellularized scaffolds can influence tumor progression by promoting epithelial-to-mesenchymal transition and enhancing cancer cell stemness, demonstrating their role in modeling disease dynamics (Liu et al., 2019).

Engineered decellularized adipose tissue scaffolds have been used as models to support breast cancer cell growth and assess drug efficacy. These biomaterials replicate the biochemical and mechanical properties of breast tissue, providing an advanced platform for testing anticancer therapies in a controlled environment (Dunne et al., 2014). Additionally, recent advancements have led to the development of bioengineered 3D scaffolds based on decellularized ECM, which facilitate personalized therapy and high-throughput drug screening, bridging the gap between traditional *in vitro* models and clinical applications (Bhattacharya et al., 2024).

Innovations in bioprinting technology have further enhanced the use of decellularized breast tissue in creating 3D cancer models. These printed scaffolds closely mimic the native tumor microenvironment, allowing researchers to explore cancer cell behavior and treatment responses with greater accuracy (Blanco-Fernandez et al., 2022). With their ability to preserve essential tissue characteristics, decellularized scaffolds continue to be a crucial tool in breast cancer research, offering new possibilities for therapeutic advancements and precision medicine.

1.4.5.4 Nanoparticle-Incorporated Scaffolds

Among the various 3D *in vitro* culture scaffolds explored in research, nanofibrous scaffolds, particularly those fabricated using the electrospinning technique, have gained prominence due to their high surface-to-volume ratio, significant porosity, and structural similarity to the native ECM (Flores-Rojas et al., 2023). These scaffolds, composed of biodegradable and biocompatible nanofibers, have been widely utilized in TE and regenerative medicine because they closely mimic the size and architecture of collagen fibrils within the ECM (Ding et al., 2019). The nanoscale structure of these fibers fosters an optimal environment for cellular functions such as adhesion, proliferation, differentiation, and biomineralization, making them highly effective in biomedical applications (Feng et al., 2023)

Beyond their role in normal TE, electrospun nanofibers have also been employed in 3D tumor models for cancer research, particularly in drug development and *in vitro* cancer studies (Keirouz et al., 2020). Among biodegradable polymers used in scaffold fabrication, poly(lactic-co-glycolic acid) (PLGA) has garnered significant attention due to its excellent biocompatibility, biodegradability, and favorable mechanical properties (Lu et al., 2023). Approved by regulatory agencies such as the U.S. Food and Drug Administration (FDA) and the European Medicines Agency (EMA), PLGA is extensively used in drug delivery and scaffold fabrication for both normal and cancerous cell cultures. HAp-incorporated PLGA scaffolds represent a significant advancement in the development of *in vitro* cancer models. By combining the beneficial properties of HAp and PLGA, these scaffolds offer a biomimetic and versatile platform for cancer research, enabling more accurate and reproducible studies of cancer cell behavior and therapeutic responses. Kargozar et al. (2022) highlighted the potential of HAp nanoparticles in cancer theranostics, emphasizing their role in improving cancer diagnosis and treatment.

In this study, electrospinning was employed to fabricate nanofibrous PLGA-HAp composite scaffolds for breast cancer research. The incorporation of HAp, a bioactive phase known for its mechanical strength and biocompatibility, was hypothesized to enhance the mechanical stability and cancer cell compatibility of PLGA scaffolds. This study represents the first attempt to evaluate the impact of HAp incorporation within nanofibrous PLGA scaffolds on breast cancer cell behavior, providing a novel approach to developing improved 3D *in vitro* cancer models.

Nanoparticle-enhanced scaffolds integrate nanomaterials to improve scaffold functionality, such as drug delivery or imaging applications. Studies show that

BCCs cultured on nanohydroxyapatite (nHA)/PLGA scaffolds exhibit enhanced cell viability and growth, highlighting their potential for cancer research and bone metastasis studies (Pradhan et al., 2017).

1.4.5.5 Microspheres

Microspheres are spherical scaffolds that can be used as carriers for cells and bioactive molecules. Their uniform shape and size allow for precise control over drug release kinetics. Microspheres have been used in breast cancer models to target the delivery of anticancer drugs directly to tumor sites, while preserving a 3D environment that supports cell growth. Study proposed a modular approach to addressing this gap using microsphere-enabled technology to construct *in vitro* breast tumor models that integrate spatial organization and various cell types.

PLGA microspheres (sized 150–200 µm) with customizable pore sizes and surface topographies serve as carriers for tumor-associated cells. These cell-laden microspheres were cultured in a custom microfluidic chamber, allowing for the formation of 3D tumor models with controlled spatial cell distribution. Furthermore, introducing endothelial cell-laden microspheres into cancer-cell-laden microspheres at varying ratios facilitated angiogenesis, enabling the creation of vascularized tumor models (Lu et al., 2023).

1.4.5.6 Fibrous Scaffolds

Fibrous scaffolds are composed of interconnected fibers that mimic the fibrous structure of the natural ECM. These scaffolds provide a high surface area for cell attachment and can be engineered to include specific mechanical and biochemical properties. Fibrous scaffolds have been used to study cell-matrix interactions and tumor invasion dynamics.

Fibrous scaffolds are a vital component in breast cancer TE, as they provide a biomimetic framework that supports cell adhesion, penetration, and proliferation. Scaffolds with tunable architectural features and material properties are particularly desirable for replicating the complex tumor microenvironment. Since fibers represent the fundamental structural elements of the ECM), engineered nonwoven fibrous scaffolds are widely used to study and promote the growth of various cell and tissue types (Keirouz et al., 2020). Electrospinning, a commonly employed fabrication technique, produces fibrous scaffolds with nano- to microscale diameters and allows for random or aligned fiber configurations, mimicking the structural characteristics of native ECM (Ding et al., 2019). While electrospun scaffolds have been extensively utilized in normal TE, limited research has been conducted on their role in supporting tumor cell behavior (Flores-Rojas et al., 2023). This gap highlights the need to investigate BCCs, particularly chemoresistant subpopulations, using fibrous scaffolds as *in vitro* models.

In a recent study, electrospun 3D scaffolds were fabricated with micron-sized fibers arranged randomly or in alignment to simulate the size and orientation of collagen fibers in the ECM. The ECM surrounding mammary epithelial and tumor cells consists of collagen fibrils (~67 nm in diameter) bundled into micron-sized fibers. BCCs tend to align along these fibers, but the functional implications of fiber orientation on tumor progression remain poorly understood (Ashworth and Cox, 2024).

The scaffolds were constructed using PCL, a biocompatible polymer with favorable mechanical properties, slow degradation rates, and a low glass transition temperature, making it suitable for long-term *in vitro* studies (Dash and Konkimalla, 2012).

These electrospun scaffolds were thoroughly characterized for their morphology, fiber diameter, interfiber spacing, porosity, and mechanical properties. BCCs were cultured on these scaffolds and compared to traditional 2D monolayer cultures. Chemoresistant BCCs were identified via cell viability assays and further analyzed for stem cell marker expression. Both treated (chemoresistant) and untreated BCC populations were evaluated on the scaffolds to assess cell attachment, growth, cell cycle progression, apoptosis-related protein expression, and stem cell characteristics. The findings aimed to provide insights into how the 3D fibrous scaffold environment influences breast cancer cell behavior, particularly for chemoresistant populations, thereby advancing the development of *in vitro* models for breast cancer research (de Abreu Queirós Osório, 2024).

1.4.5.7 Applications in Breast Cancer Tissue Engineering

Scaffold-based 3D models have broad applications in breast cancer research and treatment development. These models are used to:

- **Investigate Tumor Biology**: Scaffold-based models replicate the tumor microenvironment, enabling researchers to investigate cancer cell behaviors such as proliferation, migration, and drug resistance.
- **Drug Testing and Development**: 3D scaffolds provide a more physiologically relevant platform for evaluating the efficacy and toxicity of anticancer drugs compared to traditional 2D cultures.
- **Personalized Medicine**: Customizable scaffold-based models can be developed using patient-derived cells, enabling the design of personalized treatment strategies.

Scaffold-based 3D models represent a significant advancement in breast cancer TE, offering new opportunities for understanding tumor biology and developing effective therapies. The diverse range of scaffolds, including hydrogels, 3D bioprinted scaffolds, decellularized scaffolds, nanoparticle-incorporated scaffolds, microspheres, and fibrous scaffolds, highlights the potential of this technology to transform breast cancer research and treatment. Future efforts should focus on improving the scalability, reproducibility, and clinical translation of these models to maximize their impact. Organoids are 3D cellular clusters derived from patient tissues that can self-organize to mimic organ structure and function. Zhou et al. (2021) established a living biobank of breast cancer organoids that retain the genomic features of the original tumors, providing a platform for personalized drug testing.

1.4.6 Microfluidic Devices ("Organs-on-a-Chip")

These devices simulate the dynamic microenvironment of breast tissue, including fluid flow and mechanical forces. Scientists developed a breast cancer-on-a-chip model that recapitulates tumor-stromal interactions and can be used for drug screening (Brooks et al., 2024). Microfluidic organ-on-a-chip models have emerged as revolutionary tools in cancer research, offering a bridge between traditional *in vitro* models and *in vivo* studies (Regmi et al., 2022). These advanced platforms provide a more physiologically relevant environment for studying cancer biology, drug responses, and metastatic processes. Various microfluidic devices have been designed to study the origins and early progression of breast cancer. For instance, one platform screens patient blood plasma for BRCA1/2 mutations to detect ovarian and breast cancer-related mutations. Other devices model early tumor conditions, including a system replicating patient-like low-perfusion vasculature, where restricted blood flow promoted tumor growth (Huang et al., 2024). Unlike traditional dish-based culture methods, microfluidics manipulates fluids within submillimeter-scale geometries, offering benefits such as cost efficiency, reduced reagent use, high sensitivity, and better spatiotemporal resolution with laminar flow. This makes microfluidics a promising tool for studying cell migration, enhancing our understanding of cancer invasion and its response to microenvironmental changes. Collaborative efforts between biologists and engineers continue to advance microfluidic applications in cancer research (Huang et al., 2011). Models also test the enhanced permeation and retention (EPR) effect and create breast cancer spheroids to aid in early tumor studies and drug discovery (Frankman et al., 2022).

1.4.7 Engineered 3D Models

Engineered 3D tumor models have emerged as powerful tools for studying breast cancer. These models better recapitulate the tumor microenvironment compared to traditional 2D cell cultures. González-Callejo et al. (2023) developed a 3D printed scaffold that mimics the extracellular matrix of breast tissue, allowing for more accurate drug screening. TE combines principles from material science and biology to create functional tissues or tissue-like structures. A critical component of TE is the in vitro creation of tissues using natural or synthetic polymer scaffolds that encase cells. These scaffolds deliver vital physical, mechanical, and biochemical signals, allowing for a more comprehensive investigation of tumor interactions with the ECM and the immune system. In comparison to traditional research techniques, these models provide a more predictive and physiologically relevant approach.

1.4.7.1 Tumor Microenvironment (TME) in Breast Cancer Models

The TME is a highly intricate and dynamic environment that significantly impacts the onset, progression, metastasis, and therapy resistance of breast cancer. While traditional research has largely relied on 2D cell cultures or animal models, these approaches fail to accurately replicate the complexity and heterogeneity of human tumors. Consequently, there is a growing need for engineering advanced models that simulate *in vivo* conditions more effectively. These engineered TME models are pivotal for understanding cancer biology, studying tumor progression, and assessing therapeutic interventions. Recent advancements have emphasized creating 3D breast cancer models that integrate ECM proteins and stromal cells to mimic *in vivo* conditions more faithfully, thereby enhancing research accuracy and reproducibility (Esposito et al., 2024).

1.4.7.2 The Role of the Tumor Microenvironment in Breast Cancer

The TME consists of cancer cells, stromal cells (such as fibroblasts, immune cells, and endothelial cells), ECM components, and signaling molecules. These elements interact to promote tumor growth and metastasis. In breast cancer, the TME plays a crucial role in processes like angiogenesis, immune evasion, invasion, and drug resistance (Yousefi et al., 2021). ECM proteins such as collagen, laminin, and fibronectin provide both structural support and act as platforms for signaling molecules that influence cancer cell behavior (Wu and Chu, 2021). Stromal cells, including cancer-associated fibroblasts (CAFs) and macrophages, further modify the tumor niche by secreting cytokines and growth factors that enhance tumor proliferation and suppress immune responses (Du et al., 2024).

1.4.7.3 Advances in Engineering the TME

Engineered TME models now incorporate both cancer cells and critical TME elements, such as ECM proteins and stromal cells, for enhanced replication of *in vivo* conditions. ECM stiffness, for example, has been linked to tumor cell invasiveness in breast cancer (Schedin and Keely, 2011). Models engineered with appropriate ECM characteristics allow researchers to study tumor progression and therapeutic responses more effectively.

Stromal cells, including fibroblasts and immune cells, are integral to these models. Studies have demonstrated that CAFs contribute to tumor invasion by secreting ECM proteins and cytokines (Joshi et al., 2021). Immune cells, depending on the tumor's immune evasion strategies, can either promote or inhibit cancer growth, making them critical for studying tumor progression (Wegiel et al., 2018). Organotypic culture systems and bioprinting are among the advanced techniques that enable the construction of complex 3D tumor models with precise architecture and cellular heterogeneity (Bağci et al., 2022).

1.4.7.4 Drug Response and Metastasis Research

Engineered TME models enhance the predictive accuracy of drug response studies. Traditional 2D cultures often fail to replicate the tumor's complexity, leading to inconsistent preclinical and clinical outcomes. By integrating ECM and stromal components, 3D models provide a more realistic environment for testing therapies, including chemotherapies and immunotherapies. These models also facilitate the study of hypoxic and nutrient-deprived conditions commonly found in tumors, improving the relevance of preclinical drug testing (Balachander et al., 2021).

For metastasis and dormancy research, engineered TME models offer unparalleled insights. Tumor cells in 3D environments exhibit realistic migration and invasion patterns, reflecting complex cancer-ECM-stromal interactions (Bahcecioglu et al., 2020). Additionally, these models help study dormancy, a critical phase where disseminated tumor cells remain quiescent before reactivating to form secondary tumors. Understanding these interactions in engineered TME models is crucial for developing therapies targeting dormant cells and preventing relapse (Bissell and LaBarge, 2005).

1.4.7.5 Biomaterials for TME Engineering

Biomaterials play a critical role in engineering the TME for 3D cancer models, as they provide the structural and biochemical cues necessary to mimic the native tumor niche. The choice of biomaterials is crucial to accurately replicate the complex interactions between cancer cells, stromal cells, and the ECM in the TME. Natural polymers, such as collagen and gelatin, are widely used due to their inherent bioactivity, biocompatibility, and ability to promote cell adhesion and proliferation. These materials closely resemble the native ECM, facilitating cell-matrix interactions and supporting the growth of tumor cells. However, natural polymers often exhibit variability in mechanical properties and degradation rates, which can limit their reproducibility and applicability in standardized models. Synthetic polymers offer controlled mechanical properties and degradation rates but lack the bioactivity of natural ECM components.

Hybrid approaches combining natural and synthetic materials allow researchers to study specific ECM effects on tumor biology. Farzamfar et al. (2020) demonstrated the potential of PCL/gelatin nanofibrous scaffolds in TE, highlighting their ability to support cell proliferation and differentiation.

Recent advancements in biomaterials have also focused on enhancing immunotherapy for breast cancer. Santerre et al. (2024) highlighted the role of biomaterials in targeting functional cells within the TME to improve immunotherapeutic outcomes. Biomaterial-based platforms, such as hydrogels and nanofibrous scaffolds, can be engineered to deliver immune-modulating agents, such as checkpoint inhibitors or cytokines, directly to the tumor site. These platforms not only enhance the efficacy of immunotherapies but also reduce off-target effects, offering a promising strategy for breast cancer treatment.

HAp nanoparticles have also emerged as a valuable component in TME engineering. Kargozar et al. (2022) demonstrated the potential of HAp nanoparticles in cancer theranostics, emphasizing their ability to improve drug delivery and imaging. When incorporated into biomaterial scaffolds, HAp nanoparticles can enhance the mechanical properties and bioactivity of the constructs, making them suitable for studying bone metastasis in breast cancer. These nanoparticles can also be functionalized with therapeutic agents, enabling targeted delivery and controlled release within the TME.

1.4.7.6 Microfluidic Devices and Breast Cancer Research

Microfluidic devices, or organ-on-chip models, recreate the breast TME on small-scale chips equipped with microchannels. These systems enable the study of tumor cell behavior under conditions such as flow-induced shear stress and matrix stiffness. These are used to investigate breast cancer cell clustering and migration, revealing significant differences in migration rates between cell types. Additionally, these devices have been employed to simulate breast-to-bone metastasis, incorporating osteo-differentiated cells to study cancer cell extravasation and microtumor formation (Sung, 2023).

Microfluidic models offer real-time analysis of tumor progression and therapy responses while enabling precise control of the TME. Future improvements may include integrating advanced sensors and analytical tools to maximize their utility in breast cancer research. Organ-on-a-chip models excel at recreating the complex 3D structure and dynamic conditions of the TME. This capability allows researchers to study cell-cell interactions, extracellular matrix influences, and the impact of mechanical forces on tumor progression (El Harane et al., 2023). Microfluidic devices can simulate different stages of the metastatic cascade, including intravasation and extravasation. This feature is particularly useful for studying organ-specific metastasis patterns (Sigdel et al., 2021).

Organ-on-a-chip platforms can be combined with tumor spheroids or organoids for enhanced physiological relevance. This integration allows for more comprehensive studies of tumor biology and drug responses (Li et al., 2023). These platforms enable more accurate predictions of drug efficacy and toxicity compared to traditional 2D cultures. The ability to incorporate patient-derived cells makes them valuable tools for personalized medicine approaches.

For instance, Hwangbo et al. (2024) discuss the combination of tumor-on-a-chip models with mini-tissues or organoids to engineer more accurate tumor tissues. This approach allows for better modeling of the heterogeneous nature of tumors and their interactions with surrounding tissues. Subia et al. (2021) highlight the potential of breast tumor-on-chip models in personalized drug screening, emphasizing the need for further development to fully realize their potential in clinical applications. Microfluidic organ-on-a-chip models represent a significant advancement in cancer research tools. By providing a more physiologically relevant platform for studying tumor biology, drug responses, and metastatic processes, these models have the potential to accelerate drug discovery and development of personalized treatment strategies for cancer patients.

1.4.7.7 Advantages over Traditional Models

- **Physiological Relevance**: Better recapitulation of *in vivo* conditions compared to 2D cultures.
- **Controlled Microenvironment**: Ability to precisely manipulate environmental factors.
- **Reduced Animal Testing**: Potential to decrease reliance on animal models in preclinical studies.
- **Real-Time Monitoring**: Many platforms allow for real-time imaging and analysis of cellular behaviors.

1.4.7.8 Challenges and Future Directions

Despite their potential, organ-on-a-chip models face several challenges:

- **Standardization**: There is a need for standardized protocols to ensure reproducibility across different laboratories.
- **Vascularization**: Improving the integration of functional blood vessels remains a challenge.
- **Immune Component**: Incorporating immune cells is crucial for studying immunotherapies and tumor-immune interactions.
- **Scaling Up**: Adapting these models for high-throughput applications is an ongoing area of research.

1.5 *IN VIVO* MODELS

In vivo models utilize live organisms, such as mice, to study breast cancer within a whole-body context, providing insights into tumor development, metastasis, and response to therapy (Holen et al., 2017). *In vivo* breast cancer models are considered the gold standard, serving as the final assessment prior to advancing any treatment to clinical trials. Both xenograft and syngeneic models are employed to investigate breast cancer, each presenting distinct advantages and disadvantages.

Xenografts utilize immunocompromised mice to examine the behavior and response of human cells, particularly those derived from patients, in a living organism. PDXs are especially valuable as they allow for the study of specific cancer lines that cannot be represented by existing animal models. However, xenograft models face challenges, primarily the absence of immune cells, which significantly influence breast cancer cell behavior and are vital for tumor development and progression.

In vivo models are essential tools in breast cancer research, providing critical insights into tumor biology, progression, and therapeutic responses. These models, which include genetically engineered mice, xenografts, and zebrafish, allow researchers to study cancer in a living organism, closely mimicking the complex interactions between tumors and their microenvironment. According to Singhal et al. (2023), recent advancements in mouse and zebrafish models have significantly enhanced our understanding of breast cancer mechanisms, including metastasis, drug resistance, and tumor heterogeneity. Mouse models, particularly genetically engineered mice (GEMs), have been widely used due to their ability to recapitulate human breast cancer subtypes and their suitability for studying gene function and signaling pathways (Singhal et al., 2023).

Kim et al. (2003) emphasize the importance of integrating human and animal models to bridge the gap between preclinical research and clinical applications. They argue that xenograft models, where human tumor cells are implanted into immunodeficient mice, provide a valuable platform for testing targeted therapies and understanding tumor-host interactions. However, these models often lack the immune component, which is crucial for studying immunotherapy. To address this limitation, humanized mouse models, which incorporate human immune cells, have been developed to better represent the tumor microenvironment and immune responses (Kim et al., 2003).

Silva et al. (2025) provide a comprehensive overview of mammary cancer models, tracing their evolution from early carcinogen-induced models to modern genetically engineered and PDX models. They highlight the role of zebrafish as an emerging model organism in breast cancer research due to their transparency, rapid development, and genetic tractability. Zebrafish models are particularly useful for high-throughput drug screening and studying metastatic processes in real-time (Silva et al., 2025).

Each model has its strengths and limitations, the integration of these systems continues to provide valuable insights into breast cancer progression and treatment. These models remain indispensable for advancing our understanding of breast cancer and developing effective therapies.

Syngeneic mouse models have been widely utilized for decades to replicate breast tumors through reductionist cell lines, genetic manipulation, and environmental induction. Unlike xenografts, these models are primarily employed to explore the fundamental biology of breast cancer rather than drug responses (Silva et al., 2025). They are particularly advantageous for studying metastasis comprehensively within a single organism, avoiding cross-species interactions. However, the lack of human cells limits the direct applicability of results to clinical settings.

Both *in vivo* model types share notable drawbacks that have spurred the development of *in vitro* methods for breast cancer research. Although whole-organism studies are beneficial, they often make it challenging to isolate and analyze specific mechanisms involved in tumor progression. Additionally, controlling specific variables within an entire organism can be difficult. Furthermore, species variability can hinder the translation of findings from animal models to clinical trials.

1.5.1 Patient-Derived Xenografts (PDXs)

PDX models involve transplanting tumor samples from patients into immunodeficient mice, maintaining the genetic and molecular features of the original tumor. In contrast to PDOs, which are *ex vivo* 3D cultures, PDX models offer an in vivo system for investigating tumor biology and treatment responses. Byrne et al. (2017) and Gao et al. (2015) demonstrated the utility of PDX models in predicting drug responses across breast cancer subtypes, reinforcing their role in personalized medicine. Recently, these models have been increasingly incorporated into co-clinical trials, where treatments are tested concurrently in patients and preclinical models. This approach has been particularly effective in identifying responsive subgroups and optimizing therapeutic strategies. In upper gastrointestinal cancers, PDX models have provided crucial insights into tumor behavior and drug efficacy, though challenges such as model establishment and scalability persist (Gao et al., 2023).

Since immunodeficient mice lack functional immune systems, PDXs are not rejected, allowing them to maintain tumor heterogeneity and faithfully replicate human cancer characteristics (Dobrolecki et al., 2016; Tang and Liu, 2024). Their ability to retain the tumor microenvironment enables researchers to investigate cancer progression and treatment resistance, particularly in aggressive subtypes like TNBC, which has limited targeted therapies. Furthermore, PDX models facilitate high-throughput drug screening and combination therapy evaluations, making them an essential tool for translational research (Tang and Liu, 2024).

Despite their limitations in immunotherapy studies due to the absence of an active immune system, Liu et al. (2023) discuss the application of PDX models in co-clinical trials, where drugs are tested simultaneously in patients and preclinical models. This strategy has proven effective in predicting clinical responses and accelerating the development of targeted therapies. Their ability to preserve tumor complexity ensures that findings closely align with human tumor behavior, advancing the field of precision oncology.

1.5.2 Genetically Engineered Mouse Models (GEMMs)

Genetically Engineered Mouse Models (GEMMs) are created by introducing specific genetic mutations into the mouse genome that are associated with human cancers, such as breast cancer. Unlike xenograft models, where tumors are transplanted into mice, GEMMs enable tumors to develop spontaneously within the mouse, closely mimicking the natural progression of cancer in humans. This unique feature makes GEMMs particularly valuable for studying the early stages of tumor initiation, progression, and metastasis, as well as the complex interactions between tumors and their microenvironment (Rottenberg et al., 2010). GEMMs have transformed cancer research by providing a robust platform to explore tumor biology, drug resistance, and the molecular pathways driving cancer progression. By replicating specific genetic alterations found in human cancers, these models allow researchers to investigate the functional roles of individual genes and signaling pathways in tumor development and therapeutic responses. One of the key strengths of GEMMs lies in their ability to recapitulate the complexity of human cancers, including tumor heterogeneity and the tumor microenvironment. Pfefferle et al. (2013) demonstrated that GEMMs can be classified based on their transcriptomic profiles, which closely resemble human breast cancer subtypes. This classification allows researchers to identify mouse models that accurately reflect specific human cancers, facilitating the study of subtype-specific mechanisms and therapeutic responses. By bridging the gap between preclinical models and clinical applications, GEMMs have become invaluable tools for translational cancer research (Pfefferle et al., 2013).

GEMMs are particularly useful for studying drug resistance, a major challenge in cancer treatment. Rottenberg et al. (2010) highlighted the role of GEMMs in understanding the mechanisms of resistance to chemotherapy and targeted therapies. By introducing specific genetic alterations, researchers can model the development of resistance and identify potential strategies to overcome it. For example, GEMMs have been instrumental in studying the PI3K signaling pathway, which is frequently dysregulated in breast cancer. Klarenbeek et al. (2013) used GEMMs to investigate the role of PI3K mutations in tumor initiation, progression, and response to PI3K inhibitors, providing critical insights into the development of targeted therapies (Klarenbeek et al., 2013).

In addition to their use in basic research, GEMMs have significant applications in translational cancer medicine. Kersten et al. (2017) emphasized the utility of GEMMs in preclinical drug testing and the evaluation of combination therapies. These models enable researchers to study the effects of therapeutic interventions in a physiologically relevant context, improving the predictive value of preclinical studies. Furthermore, GEMMs have been used to model cancer evolution, a process driven by genetic and epigenetic changes that contribute to tumor heterogeneity and treatment resistance. Hill et al. (2021) discussed how GEMMs can capture the dynamic nature of cancer evolution, providing insights into the mechanisms of tumor adaptation and progression (Hill et al., 2021).

Non-germline GEMMs, which involve somatic gene editing, offer additional flexibility for cancer research. Heyer et al. (2010) highlighted the advantages of these models, including the ability to introduce specific genetic alterations in a spatially and temporally controlled manner. This approach allows researchers to study the effects of oncogenic mutations in specific cell types and tissues, providing a more nuanced understanding of tumor biology and therapeutic responses (Heyer et al., 2010).

One significant advantage of GEMMs over PDX models is the presence of a fully functional immune system. This aspect allows for the exploration of immune-tumor interactions, providing insight into how genetic mutations might influence the immune response to tumors. GEMMs are also instrumental for testing therapies that involve an intact immune response, such as immunotherapies. Although creating and maintaining GEMMs can be costly and time-consuming, their ability to provide a more comprehensive view of cancer development and progression makes them invaluable in breast cancer research.

1.5.3 Humanized Mouse Models

Humanized mouse models are advanced preclinical tools created by engrafting human immune cells or entire human immune systems into immunodeficient mice. These models are designed to replicate a partially or fully humanized immune response, making them invaluable for studying the complex interactions between breast tumors and human immune cells. By incorporating human immune components, these models provide a more accurate representation of human biology, enabling researchers to investigate tumor-immune dynamics and evaluate the efficacy of immunotherapies in a controlled setting (Guil-Luna et al., 2021; Chuprin et al., 2023).

One of the primary applications of humanized mouse models is in the testing of immunotherapies, such as immune checkpoint inhibitors, CAR-T cell therapies, and other immune-modulating treatments. These models allow researchers to observe how human immune cells respond to breast tumors, providing critical insights into immune cell behavior, tumor-immune interactions, and the mechanisms of therapeutic resistance. For example, Rosato et al. (2018) used humanized mouse models to evaluate anti-PD-1-based therapies against TNBC, demonstrating their utility in predicting immunotherapy responses and guiding treatment strategies (Rosato et al., 2018).

Humanized mouse models are particularly advantageous for studying the TME and its role in cancer progression and treatment. By incorporating human immune cells, these models enable researchers to explore how immune cells influence tumor growth, metastasis, and response to therapy. Wege et al. (2011) highlighted the use of humanized tumor mice to manipulate and study immune responses in advanced cancer therapy, providing a platform to investigate the effects of immunotherapeutic agents on tumor progression and immune evasion (Wege et al., 2011).

The development of humanized mouse models has also advanced the field of personalized medicine. By engrafting patient-derived tumors into humanized mice, researchers can test individualized treatment regimens and identify the most effective therapies for specific patients. This approach has been particularly useful in breast cancer research, where tumor heterogeneity and immune responses vary significantly among patients. Walsh et al. (2017) emphasized the role of humanized mouse models in studying clinical diseases, including cancer, and their potential to bridge the gap between preclinical research and clinical applications (Walsh et al., 2017).

Despite their advantages, humanized mouse models face challenges, such as the complexity of establishing and maintaining these models and the variability in immune cell engraftment. Advancements in genetic engineering and the development of immunodeficient mouse strains have enhanced the reliability and reproducibility of these models, broadening their use in immuno-oncology research (Chuprin et al., 2023). Humanized mouse models bridge a critical gap in cancer research by combining human immune responses with a living organism, enabling a more accurate assessment of immune-based treatments for breast cancer. As these models continue to evolve, they will play an increasingly important role in advancing immunotherapy research and improving outcomes for breast cancer patients. While the development of humanized mice is complex and resource-intensive, these models are essential for advancing immunotherapy research and understanding how immune cells and tumors interact in ways that could inform new therapeutic strategies for patients.

1.6 *IN SILICO* MODELS

In silico models utilize computational simulations to analyze extensive datasets, predict treatment responses, and uncover insights into the complex biological mechanisms of breast cancer.

1.6.1 Multiomics Approaches in Breast Cancer Management

Multiomics, also known as integrative omics, combines various "omics" disciplines—such as genomics, proteomics, metabolomics, and transcriptomics—within computational models to develop a comprehensive understanding of breast cancer biology (Figure 1.3). By integrating molecular datasets, multiomics enables researchers to identify critical pathways, molecular interactions, and biomarkers that drive tumor progression and drug resistance (Hasin et al., 2017).

Genomic data provide insights into driver mutations responsible for cancer initiation, while proteomic analysis reveals how these mutations affect protein function and cellular signaling pathways (Simpson et al., 2019). Similarly, metabolomic profiling highlights alterations in energy metabolism that fuel rapid cancer cell proliferation (Chakraborty et al., 2024). By combining these data layers, multiomics approaches elucidate the interconnected mechanisms of tumor development and progression, offering potential therapeutic targets.

In addition to uncovering disease mechanisms, multiomics facilitates systems-level modeling, allowing researchers to simulate tumor behavior under various conditions, such as drug exposure. These integrative models assist in predicting patient responses to therapy by identifying biomarkers across multiple molecular layers (Lan et al., 2024). Multiomics also facilitates the design of combination therapies targeting

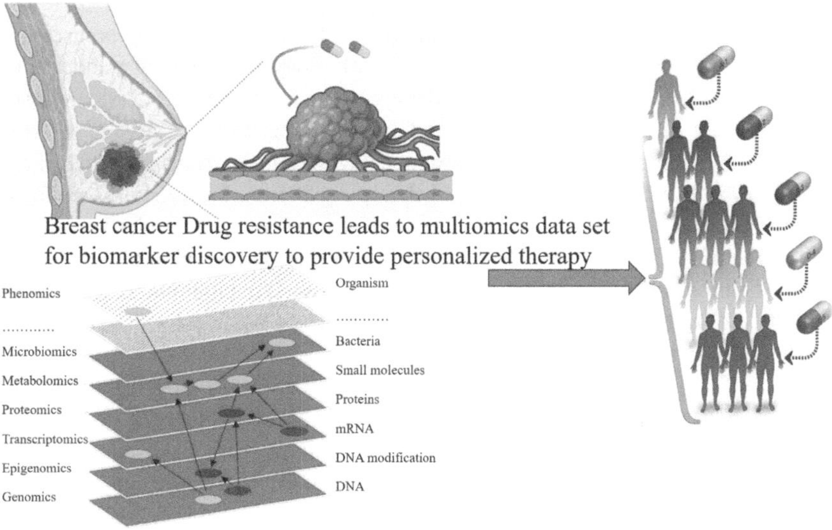

FIGURE 1.3 Multiomics approaches in breast cancer management.

multiple pathways, thereby reducing the likelihood of drug resistance and improving treatment efficacy (Otálora-Otálora et al., 2023).

Advancements in computational modeling, artificial intelligence (AI), and machine learning (ML) have further enhanced multiomics applications in breast cancer research. AI-driven multiomics models provide a predictive framework for understanding tumor biology, guiding treatment selection, and personalizing therapeutic strategies (Wu and Xie, 2025). By integrating diverse molecular datasets, multiomics not only advances precision oncology but also improves patient outcomes by offering a holistic approach to breast cancer management.

1.6.2 Genomics in Breast Cancer Management

Next-generation sequencing (NGS) has revolutionized genomic research by enabling faster DNA sequencing and improving the classification of breast cancer subtypes. Genomic studies have identified important mutations and single nucleotide polymorphisms (SNPs) that serve as biomarkers in breast cancer treatment. The discovery of molecular markers like the estrogen receptor (ER), progesterone receptor (PR), and HER2 has facilitated the development of targeted therapies. Patients with ER^+ and PR^+ breast cancer benefit from endocrine therapy, while those with HER2-positive tumors are effectively treated with trastuzumab (Rakha et al., 2006). However, TNBC, which lacks ER, PR, and HER2 expression, presents treatment challenges due to its aggressive nature and the limited availability of targeted therapies (Obidiro et al., 2023). Genomic screening tools like BRACAnalysis® help detect BRCA1 and BRCA2 mutations, facilitating early detection and management of hereditary breast cancer (Ellsworth et al., 2010).

1.6.3 Transcriptomics in Breast Cancer Management

Transcriptomics, the study of RNA transcripts, has revolutionized breast cancer research by providing deeper insights into tumor biology, heterogeneity, and potential therapeutic targets. By analyzing gene expression patterns, researchers can develop biomarkers for diagnosis, prognosis, and treatment response assessment (Albitar et al., 2024). Techniques such as RNA sequencing and microarrays help identify biomarkers for breast cancer prognosis and therapy selection. Transcriptomics focuses on analyzing RNA molecules to study gene expression patterns under specific conditions. Techniques such as RNA sequencing and microarrays play a crucial role in identifying biomarkers for breast cancer prognosis and treatment selection (Castillo et al., 2017). Transcriptome-wide association studies (TWAS) have established connections between particular genes, signaling pathways, and breast cancer development (Li et al., 2025). One of the significant applications of transcriptomics is in TNBC, a subtype lacking estrogen, progesterone, and HER2 receptors. Given its aggressive nature, transcriptomic and proteomic studies have been instrumental in identifying molecular subtypes and prognostic markers that could aid in personalized treatment approaches (Li et al., 2022).

Commercial transcriptomic tests, including Oncotype DX, MammaPrint, and EndoPredict, utilize gene expression data to assess disease prognosis and aid in treatment planning (Jamshidii et al., 2020). MammaPrint, for example, analyzes the expression of 70 genes associated with cell cycle regulation and angiogenesis to determine the recurrence risk in early-stage breast cancer (Cardoso et al., 2016). These advancements in transcriptomics contribute significantly to personalized breast cancer management by enabling more precise diagnostic and therapeutic strategies (Zhang et al., 2024).

Advancements in transcriptomics, alongside genomic and proteomic profiling, have also been crucial in understanding metastatic breast cancer. By analyzing tumor progression at the molecular level, researchers can identify key genetic and transcriptomic alterations that drive metastasis, leading to improved targeted therapies (Akcakanat et al., 2021).

Furthermore, the integration of single-cell transcriptomics has enabled a more refined analysis of breast cancer heterogeneity. This approach provides detailed insights into the tumor microenvironment, immune interactions, and drug resistance mechanisms, ultimately facilitating better therapeutic strategies (Zhu et al., 2023).

1.6.4 Proteomics in Breast Cancer Research

Proteomics plays a pivotal role in breast cancer research by analyzing protein expression patterns to uncover biomarkers essential for diagnosis, prognosis, and treatment strategies. The application of mass spectrometry has greatly enhanced the ability to detect specific protein signatures, contributing to the early identification of breast cancer (Korkola and Gray, 2010). Key biomarkers, including Ki-67, HER2, and cyclin D1, have been widely studied for their significance in guiding clinical decision-making and personalized therapies (Kędzierska and Bańkosz, 2024).

Recent research has emphasized the involvement of certain proteins in tumor progression. For example, α-B-crystallin has been identified as a crucial factor in cancer development, particularly through its role in activating the MAPK/ERK signaling pathway, which influences tumor growth and aggressiveness (Moyano, 2006). The integration of proteomics with other omics approaches, such as genomics and transcriptomics, provides a more comprehensive understanding of breast cancer heterogeneity, ultimately aiding in the advancement of precision medicine (Li et al., 2022; Zhu et al., 2023).

Proteomic profiling has also proven beneficial in identifying molecular signatures linked to metastatic breast cancer, helping to predict disease progression and treatment responses (Akcakanat et al., 2021). As research continues to evolve, proteomics remains a powerful tool in refining breast cancer classification, enabling targeted therapeutic approaches tailored to individual patients.

Proteomics has emerged as a crucial tool in breast cancer research, offering valuable insights into disease mechanisms, biomarker discovery, and therapeutic targets. By analyzing protein expression and modifications, researchers can identify molecular signatures associated with different breast cancer subtypes, aiding in early diagnosis and treatment planning (Hondermarck et al., 2001).

One of the primary applications of proteomics in breast cancer is the identification of protein biomarkers that help classify tumors based on their molecular characteristics. These biomarkers play a key role in distinguishing between aggressive and less aggressive forms of the disease, ultimately guiding personalized treatment strategies

(Mueller et al., 2018). Advanced proteomic techniques, such as mass spectrometry and protein microarrays, have facilitated the detection of specific protein signatures linked to tumor progression and resistance to therapy.

Additionally, integrating proteomics with other omics approaches, such as genomics and transcriptomics, has enhanced the understanding of breast cancer heterogeneity. This multiomics approach enables the identification of novel therapeutic targets and personalized treatment options, improving patient outcomes (Mehmood et al., 2022). Through continued advancements in proteomic technologies, breast cancer research is making significant strides toward precision medicine, leading to more effective and tailored therapeutic interventions.

1.6.5 Computational Modeling and Simulation

Computational modeling applies mathematical algorithms and data-driven simulations to study tumor dynamics, growth patterns, and metastasis. These models integrate patient-specific genomic, transcriptomic, and clinical data to predict tumor behavior and treatment responses.

Agent-based modeling (ABM) simulates the interactions of individual cancer cells within the TME, considering factors such as oxygen levels, drug diffusion, and immune responses. These simulations enable researchers to test treatment strategies and predict therapy outcomes before clinical application. Computational models facilitate *in silico* testing of various therapeutic interventions, optimizing drug development and personalized treatment plans.

1.6.6 Artificial Intelligence and Machine Learning in Breast Cancer Risk Prediction and Screening

AI and its subset, ML, have transformed breast cancer research by facilitating the analysis of extensive and complex datasets, uncovering hidden patterns, and generating predictive models that aid in diagnosis and treatment planning (Yala et al., 2022). Deep Learning (DL), a specialized form of ML that simulates neural networks in the human brain, processes unstructured data from diverse sources—including medical imaging, genetic sequences, and clinical records—into actionable insights for breast cancer management (Balkenende et al., 2022).

AI-powered algorithms significantly enhance breast cancer imaging analysis, improving early detection by identifying minute features in mammograms, MRI scans, and histopathological images that may be imperceptible to the human eye (Uzun Ozsahin et al., 2022). Additionally, ML models facilitate the identification of genetic mutations, molecular markers, and disease subtypes from large genomic datasets, revealing potential therapeutic targets and biomarkers for precision oncology (Adam et al., 2023). AI can also predict genomic alterations associated with drug resistance, enabling clinicians to anticipate treatment failure and adjust therapeutic strategies accordingly (Mema and McGinty, 2020).

Beyond diagnostics, ML algorithms effectively categorize patients into molecularly or genetically similar subgroups, optimizing personalized treatment approaches. This stratification is fundamental to precision medicine, ensuring that patients receive therapies tailored to their unique tumor profiles, thereby improving treatment efficacy and clinical outcomes (Manikis et al., 2023).

Furthermore, AI's predictive potential extends to prognosis. By analyzing vast datasets, AI can identify patterns correlated with patient survival, recurrence risks, and therapeutic responses. For example, AI models trained on historical patient records can predict which individuals are at higher risk of post-surgical recurrence, allowing oncologists to recommend additional adjuvant therapies proactively (Romanov et al., 2023).

1.6.7 AI in Breast Cancer Risk Prediction

AI has significantly advanced breast cancer risk prediction by analyzing diverse demographic data to provide personalized care. Romanov et al. developed a model that achieved an area under the curve (AUC) of 0.747 for predicting cancer-free mammograms of women who later developed breast cancer. The Mirai model, on the other hand, demonstrated consistent accuracy across populations in five countries, highlighting AI's potential in addressing demographic variability and promoting equity in healthcare (Romanov et al., 2023; Yala et al., 2022). AI-driven tools are also instrumental in motivating individuals to adhere to screening protocols and adopt preventive measures, such as hormone therapy or chemoprevention. Additionally, AI applications extend to psychological support for patients, identifying those at risk of poor resilience post-diagnosis and enabling early interventions to improve quality of life (Manikis et al., 2023). These tools are cost-effective and have shown promise in improving healthcare access for underserved populations (Mema and McGinty, 2020).

1.6.8 Challenges in AI Implementation

Despite its potential, AI adoption in healthcare faces several barriers. A major challenge is the limited generalizability of AI models due to reliance on non-representative or small datasets. This issue undermines model efficacy in diverse populations, necessitating the collection of extensive datasets and integration of social determinants of health (SDOH) into AI algorithms. SDOH encompasses socioeconomic status, education, and access to healthcare, influencing health outcomes and ensuring comprehensive risk assessments (Segar et al., 2022).

The underrepresentation of minority groups in datasets further exacerbates healthcare inequities. Federated learning, which enables collaborative AI model development while preserving data privacy, has emerged as a solution to enhance model diversity and inclusivity (Shams et al., 2023; Rieke et al., 2020). Addressing healthcare literacy gaps is another critical aspect. AI-powered educational tools and chatbots can help bridge these gaps by providing tailored health information, thereby improving understanding and participation in breast cancer screening (Levy and Janke, 2016).

1.6.9 Deep Learning in Breast Cancer Imaging

Deep learning techniques have transformed breast cancer imaging by effectively analyzing large datasets from mammograms, ultrasounds, and MRI scans to detect early-stage malignancies. Balkenende et al. (2022) emphasized its transformative role in enhancing diagnostic accuracy and reducing false positives. Similarly, Adam et al. (2023) highlighted the effectiveness of deep learning applications in MRI-based breast cancer detection. Uzun Ozsahin et al. (2022) provided insights into how AI technologies automate diagnostic processes, improve detection rates, and enable personalized treatment plans, particularly for high-risk patients.

1.6.10 AI in Screening Protocols and Healthcare Systems

AI models have optimized breast cancer screening by enhancing the efficiency of population-based programs. Marinovich et al. (2022) proposed a protocol for AI-driven cancer detection in large-scale screening, emphasizing resource optimization. Furthermore, Ng et al. (2023) reported a 45% reduction in diagnostic workloads for healthcare professionals while improving screening accuracy. These applications demonstrate the potential of AI to streamline workflows and improve healthcare delivery. The generalizability of AI models across diverse healthcare settings remains a challenge. Studies by Yang et al. (2023) and Yang et al. (2022) highlighted performance variability across healthcare infrastructures, emphasizing the need for adaptable and inclusive AI systems.

1.7 FUTURE DIRECTIONS IN BREAST CANCER MODELING

To fully leverage AI in breast cancer care, addressing data diversity, refining model generalizability, and incorporating SDOH are crucial. Investments in federated learning, algorithm development, and clinician training will further enhance AI's role, particularly in resource-limited settings. These advancements have the potential to improve diagnostic accuracy, reduce healthcare disparities, and lower clinical workloads, ultimately contributing to more equitable breast cancer management.

A key frontier in breast cancer modeling is the integration of diverse systems. Combining *in vitro* organoid cultures with *in silico* data analytics offers a more comprehensive understanding of tumor biology and treatment responses. By analyzing PDOs alongside computational predictions, researchers can identify optimal therapeutic strategies tailored to individual patients.

Advancements in humanized *in vivo* models, where human immune cells are engrafted into immunocompromised mice, provide a more physiologically relevant platform for studying breast cancer and evaluating immunotherapies. This approach enhances our ability to understand immune-tumor interactions and optimize immunotherapeutic strategies.

Emerging technologies, such as organ-on-a-chip systems and 3D bioprinting, further improve the fidelity of breast cancer models. Organ-on-a-chip platforms replicate the tumor microenvironment, enabling real-time observation of tumor behavior and drug responses. Similarly, 3D bioprinting allows for the creation of complex tissue structures that mimic breast tumors, providing more accurate models for drug discovery and testing.

The integration of multiomics data within computational frameworks is expected to deepen our understanding of breast cancer at a systems level. The PREDICT breast cancer prognostic model, originally developed to estimate survival outcomes and guide treatment decisions, has been recently updated to incorporate the benefits and risks of radiotherapy (Grootes et al., 2024). By incorporating genomics, transcriptomics, proteomics, and metabolomics data, researchers can uncover novel biomarkers and therapeutic targets that reflect the intricate molecular landscape of individual tumors. This holistic approach paves the way for more precise and effective treatment strategies.

1.8 CONCLUSION

In summary, breast cancer models are indispensable tools that bridge the gap between basic scientific research and clinical applications. The ongoing refinement and expansion of these models are essential for the development of more effective, personalized therapies. As we advance our understanding of breast cancer biology and integrate innovative technologies and modeling approaches, we can expect to see significant improvements in patient outcomes and a reduction in the global burden of breast cancer. The commitment to personalized medicine in breast cancer will ultimately lead to more targeted and effective treatment options for patients, improving their quality of life and survival rates worldwide.

The integration of these diverse breast cancer models promises to enhance our understanding of the disease's complexities, improve treatment response predictions, and contribute to the development of personalized therapies. Emerging technologies like organ-on-a-chip, advanced 3D bioprinting, and integrative AI-driven multi-omics are pushing the boundaries of what can be achieved with each model type. As our capacity to combine and refine these models grows, so will our ability to develop precise, effective, and patient-tailored interventions, ultimately reducing breast cancer's global burden and improving patient outcomes.

ACKNOWLEDGEMENT

Authors are thankful to the Management and Department of Biotechnology, PSGR Krishnammal College for Women, Coimbatore, and Karpagam Academy of Higher Education, Coimbatore, for providing necessary infrastructure facilities and encouragement to carry out this research work. Dr. VB thanks DBT Builder for providing necessary infrastructural facilities. Dr. VB also acknowledges the DST FIST, New Delhi, for the infrastructure provided to PSGR Krishnammal College for Women. Dr. RA thanks the DST FIST, New Delhi, for the infrastructure provided to Karpagam Academy of Higher Education.

CONFLICT OF INTEREST

The authors declare that they have no conflict of interests.

REFERENCES

Adam, R., Dell'Aquila, K., Hodges, L., Maldjian, T., & Duong, T.Q. (2023). Deep learning applications to breast cancer detection by magnetic resonance imaging: A literature review. *Breast Cancer Research*, 25, 87. https://doi.org/10.1186/s13058-023-01687-4

Akcakanat, A., Zheng, X., Cruz Pico, C.X., Kim, T.B., Chen, K., Korkut, A., Sahin, A., Holla, V., Tarco, E., Singh, G., & Damodaran, S. (2021). Genomic, transcriptomic, and proteomic profiling of metastatic breast cancer. *Clinical Cancer Research*, 27(11), 3243–3252. https://doi.org/10.1158/1078-0432.ccr-20-4048

Albitar, M., Goy, A., Pecora, A., Graham, D., McNamara, D., Charifa, A., Andrew, I.P., Ma, W., & Waintraub, S. (2024). The use of transcriptomic data in developing biomarkers in breast cancer. *ImmunoMedicine*, 4(1), e1051. https://doi.org/10.1002/imed.1051

Almeida, C.F., Oliveira, A., Ramos, M.J., Fernandes, P.A., Teixeira, N., & Amaral, C. (2020). Estrogen receptor-positive (ER+) breast cancer treatment: Are multi-target compounds the next promising approach?. *Biochemical Pharmacology*, 177, 113989. https://doi.org/10.1016/j.bcp.2020.113989

Ashworth, J.C., & Cox, T.R. (2024). The importance of 3D fibre architecture in cancer and implications for biomaterial model design. *Nature Reviews Cancer*, 24(7), 461–479. https://doi.org/10.1038/s41568-024-00704-8

Bağci, G., Ximenes-Carballo, C., Perez-Amodio, S., Castaño, O., Engel, E., & Blanco-Fernandez, B. (2022). Advanced 3D *in vitro* models to recapitulate the breast tumor microenvironment. In Rezaei, N. (Eds.), *Breast Cancer Pathophysiology: An Interdisciplinary Approach* (pp. 193–232). Springer, Cham. https://doi.org/10.1007/16833_2022_56

Bahcecioglu, G., Basara, G., Ellis, B.W., Ren, X., & Zorlutuna, P. (2020). Breast cancer models: Engineering the tumor microenvironment. *Acta Biomaterialia*, 106, 1–21. https://doi.org/10.1016/j.actbio.2020.02.006

Balachander, G.M., Kotcherlakota, R., Nayak, B., Kedaria, D., Rangarajan, A., & Chatterjee, K. (2021). 3D tumor models for breast cancer: Whither we are and what we need. *ACS Biomaterials Science and Engineering*, 7(8), 3470–3486. https://doi.org/10.1021/acsbiomaterials.1c00230

Balkenende, L., Teuwen, J., & Mann, R.M. (2022). Application of deep learning in breast cancer imaging. *Seminars in Nuclear Medicine*, 52, 584–596. https://doi.org/10.1053/j.semnuclmed.2022.02.003

Bhattacharya, T., Kumari, M., Kaur, K., Kaity, S., Arumugam, S., Ravichandiran, V., & Roy, S. (2024). Decellularized extracellular matrix-based bioengineered 3D breast cancer scaffolds for personalized therapy and drug screening. *Journal of Materials Chemistry B*, 29, 101280. https://doi.org/10.1039/d4tb00680a

Bissell, M.J., & LaBarge, M.A. (2005). Context, tissue plasticity, and cancer: Are tumor stem cells also regulated by the microenvironment?. *Cancer Cell*, 7(1), 17–23. https://doi.org/10.1016/j.ccr.2004.12.013

Blanco-Fernandez, B., Rey-Vinolas, S., Bağcı, G., Rubi-Sans, G., Otero, J., Navajas, D., Perez-Amodio, S., & Engel, E. (2022). Bioprinting decellularized breast tissue for the development of three-dimensional breast cancer models. *ACS Applied Materials and Interfaces*, 14(26), 29467–29482. https://doi.org/10.1021/acsami.2c00920

Boix-Montesinos, P., Soriano-Teruel, P.M., Arminan, A., Orzaez, M., & Vicent, M.J. (2021). The past, present, and future of breast cancer models for nanomedicine development. *Advanced Drug Delivery Reviews*, 173, 306–330. https://doi.org/10.1016/j.addr.2021.03.018

Boudreault, J. (2024). *Crispr Genome-Editing Studies Reveal Potential Vulnerabilities in Targeting Tumor-Suppressors and Oncogenes of Aggressive Cancers* [Doctoral dissertation, McGill University]. McGill University (Canada) ProQuest Dissertations & Theses. https://www.proquest.com/openview/27a68eac32745763def550c67cd73f1a/

Brooks, A., Zhang, Y., Chen, J., & Zhao, C.X. (2024). Cancer metastasis-on-a-chip for modeling metastatic cascade and drug screening. *Advanced Healthcare Materials*, 13(21), e2302436. https://doi.org/10.1002/adhm.202302436

Byrne, A.T., Alférez, D.G., Amant, F., Annibali, D., Arribas, J., Biankin, A.V., Bruna, A., Budinská, E., Caldas, C., Chang, D.K., Clarke, R.B., Clevers, H., Coukos, G., Dangles-Marie, V., Eckhardt, S.G., Gonzalez-Suarez, E., Hermans, E., Hidalgo, M.,

Jarzabek, M.A., de Jong, S., Jonkers, J., … & Trusolino, L. (2017). Interrogating open issues in cancer precision medicine with patient-derived xenografts. *Nature Reviews Cancer*, 17(4), 254–268. https://doi.org/10.1038/nrc.2016.140

Cardoso, F., van't Veer, L.J., Bogaerts, J., Slaets, L., Viale, G., Delaloge, S., Pierga, J.-Y., Brain, E., Causeret, S., DeLorenzi, M., Glas, A.M., Golfinopoulos, V., Goulioti, T., Knox, S., Matos, E., Meulemans, B., Neijenhuis, P.A., Nitz, U., Passalacqua, R., Ravdin, P., Rubio, I.T. Cardoso, F., van't Veer, L.J., Bogaerts, J., Slaets, L., Viale, G., Delaloge, S., Pierga, J.Y., … & Piccart, M. (2016). 70-gene signature as an aid to treatment decisions in early-stage breast cancer. *New England Journal of Medicine*, 375(8), 717–729. https://doi.org/10.1056/nejmoa1602253

Castillo, D., Gálvez, J.M., Herrera, L.J., Román, B.S., Rojas, F., & Rojas, I. (2017). Integration of RNA-Seq data with heterogeneous microarray data for breast cancer profiling. *BMC Bioinformatics*, 18, 1–15. https://doi.org/10.1186/s12859-017-1925-0

Chakraborty, S., Sharma, G., Karmakar, S., & Banerjee, S. (2024). Multi-OMICS approaches in cancer biology: New era in cancer therapy. *Biochimica et Biophysica Acta (BBA)-Molecular Basis of Disease*, 1870(5), 167120. https://doi.org/10.1016/j.bbadis.2024.167120

Chuprin, J., Buettner, H., Seedhom, M.O., Greiner, D.L., Keck, J.G., Ishikawa, F., Shultz, L.D., & Brehm, M.A. (2023). Humanized mouse models for immuno-oncology research. *Nature Reviews Clinical Oncology*, 20(3), 192–206. https://doi.org/10.1038/s41571-022-00721-2

Curvello, R., Kast, V., Ordóñez-Morán, P., Mata, A., & Loessner, D. (2023). Biomaterial-based platforms for tumour tissue engineering. *Nature Reviews Materials*, 8(5), 314–330. https://doi.org/10.1038/s41578-023-00535-3

Dash, T.K., & Konkimalla, V.B. (2012). Polymeric modification and its implication in drug delivery: Poly-ε-caprolactone (PCL) as a model polymer. *Molecular Pharmaceutics*, 9(9), 2365–2379. https://doi.org/10.1021/mp3001952

de Abreu Queirós Osório, L.A. (2024). *Can Electrospun Scaffolds be Used for the Support of Breast Tissue Culture in a Breast-on-a-Chip Model?* [Doctoral dissertation, Brunel University]. Brunel University Research Archive. https://bura.brunel.ac.uk/handle/2438/30116

Ding, Y., Li, W., Zhang, F., Liu, Z., Zanjanizadeh Ezazi, N., Liu, D., & Santos, H.A. (2019). Electrospun fibrous architectures for drug delivery, tissue engineering and cancer therapy. *Advanced Functional Materials*, 29(2), 1802852. https://doi.org/10.1002/adfm.201802852

Dobrolecki, L.E., Airhart, S.D., Alferez, D.G., Aparicio, S., Behbod, F., Bentires-Alj, M., Brisken, C., Bult, C.J., Cai, S., Clarke, R.B., Dowst, H., Ellis, M.J., Gonzalez-Suarez, E., Iggo, R.D., Kabos, P., Li, S., Lindeman, G.J., Marangoni, E., McCoy, A., Meric-Bernstam, F., Piwnica-Worms, H., … & Lewis, M.T. (2016). Patient-derived xenograft (PDX) models in basic and translational breast cancer research. *Cancer and Metastasis Reviews*, 35, 547–573.

Du, W., Xia, X., Hu, F., & Yu, J. (2024). Extracellular matrix remodeling in the tumor immunity. *Frontiers in Immunology*, 14, 1340634. https://doi.org/10.3389/fimmu.2023.1340634

Dunne, L.W., Huang, Z., Meng, W., Fan, X., Zhang, N., Zhang, Q., & An, Z. (2014). Human decellularized adipose tissue scaffold as a model for breast cancer cell growth and drug treatments. *Biomaterials*, 35(18), 4940–4949. https://doi.org/10.1016/j.biomaterials.2014.03.003

Ebrahimi, N., Nasr Esfahani, A., Samizade, S., Mansouri, A., Ghanaatian, M., Adelian, S., Shadman Manesh, V., & Hamblin, M.R. (2022). The potential application of organoids in breast cancer research and treatment. *Human Genetics*, 141(2), 1–16. https://doi.org/10.1007/s00439-021-02390-0

El Harane, S., Zidi, B., El Harane, N., Krause, K.H., Matthes, T., & Preynat-Seauve, O. (2023). Cancer spheroids and organoids as novel tools for research and therapy: State of the art and challenges to guide precision medicine. *Cells*, 12(7), 1001. https://doi.org/10.3390/cells12071001

Ellsworth, R.E, Decewicz, D.J, Shriver, C.D, & Ellsworth, D.L (2010). Breast cancer in the personal genomics era. *Current Genomics*, 11(3), 146–161. https://doi.org/10.2174/138920210791110951

Elshazly, A.M., & Gewirtz, D.A. (2022). An overview of resistance to human epidermal growth factor receptor 2 (HER2)-targeted therapies in breast cancer. *Cancer Drug Resistance*, 5(2), 472. https://doi.org/10.20517/cdr.2022.09

Esposito, A., Ferraresi, A., Vallino, L., Garavaglia, B., Dhanasekaran, D.N., & Isidoro, C. (2024). Three-dimensional in vitro cell cultures as a feasible and promising alternative to two-dimensional and animal models in cancer research. *International Journal of Biological Sciences*, 20(13), 5293. https://doi.org/10.7150/ijbs.96469

Farzamfar, S., Aleahmad, M., Gholamreza, S.K., Majid, S., & Niloofar, N. (2020). Polycaprolactone/gelatin nanofibrous scaffolds for tissue engineering. *Biointerface* Research in Applied Chemistry, 11(4), 11104–11115. https://doi.org/10.33263/BRIAC114.1110411115

Feng, Y., Cölfen, H., & Xiong, R. (2023). Organized mineralized cellulose nanostructures for biomedical applications. *Journal of Materials Chemistry B*, 11(24), 5321–5349. https://doi.org/10.1039/d2tb02611b

Flores-Rojas, G.G., Gómez-Lazaro, B., López-Saucedo, F., Vera-Graziano, R., Bucio, E., & Mendizábal, E. (2023). Electrospun scaffolds for tissue engineering: A review. *Macromol*, 3(3), 524–553. https://doi.org/10.3390/macromol3030031

Fontana, F., Marzagalli, M., Sommariva, M., Gagliano, N., & Limonta, P. (2021). *In vitro* 3D cultures to model the tumor microenvironment. *Cancers*, 13(12), 2970. https://doi.org/10.3390/cancers13122970

Frankman, Z.D., Jiang, L., Schroeder, J.A., & Zohar, Y. (2022). Application of microfluidic systems for breast cancer research. *Micromachines*, 13(2), 152. https://doi.org/10.3390/mi13020152

Fröhlich, E. (2023). The variety of 3D breast cancer models for the study of tumor physiology and drug screening. *International Journal of Molecular Sciences*, 24(8), 7116. https://doi.org/10.3390/ijms24087116

Ganeson, K., Tan Xue May, C., Abdullah, A.A.A., Ramakrishna, S., & Vigneswari, S. (2023). Advantages and prospective implications of smart materials in tissue engineering: Piezoelectric, shape memory, and hydrogels. *Pharmaceutics*, 15(9), 2356. https://doi.org/10.3390/pharmaceutics15092356

Gao, H., Korn, J.M., Ferretti, S., Monahan, J.E., Wang, Y., Singh, M., Zhang, C., Schnell, C., Yang, G., Zhang, Y., & Balbin, O.A. (2015). High-throughput screening using patient-derived tumor xenografts to predict clinical trial drug response. *Nature Medicine*, 21(11), 1318–1325. https://doi.org/10.1038/nm.3954

Gao, J., Lan, J., Liao, H., Yang, F., Qiu, P., Jin, F., Wang, S., Shen, L., Chao, T., Zhang, C., & Zhu, Y. (2023). Promising preclinical patient-derived organoid (PDO) and xenograft (PDX) models in upper gastrointestinal cancers: Progress and challenges. *BMC Cancer*, 23(1), 1205. https://doi.org/10.1186/s12885-023-11434-9

González-Callejo, P., Vázquez-Aristizabal, P., García-Astrain, C., de Aberasturi, D.J., Henriksen-Lacey, M., Izeta, A., & Liz-Marzán, L.M. (2023). 3D bioprinted breast tumor-stroma models for pre-clinical drug testing. *Materials Today Bio*, 23, 100826. https://doi.org/10.1016/j.mtbio.2023.100826

Grootes, I., Wishart, G.C., & Pharoah, P.D.P. (2024). An updated PREDICT breast cancer prognostic model including the benefits and harms of radiotherapy. *NPJ Breast Cancer*, 10(1), 6. https://doi.org/10.1038/s41523-024-00612-y

Guil-Luna, S., Sedlik, C., & Piaggio, E. (2021). Humanized mouse models to evaluate cancer immunotherapeutics. *Annual Review of Cancer Biology*, 5(1), 119–136. https://doi.org/10.1146/annurev-cancerbio-050520-100526

Hasin, Y., Seldin, M., & Lusis, A. (2017). Multi-omics approaches to disease. *Genome Biology*, 18, 1–15. https://doi.org/10.1186/s13059-017-1215-1

Heyer, J., Kwong, L.N., Lowe, S.W., & Chin, L. (2010). Non-germline genetically engineered mouse models for translational cancer research. *Nature Reviews Cancer*, 10(7), 470–480. https://doi.org/10.1038/nrc2877

Hill, W., Caswell, D.R., & Swanton, C. (2021). Capturing cancer evolution using genetically engineered mouse models (GEMMs). *Trends in Cell Biology*, 31(12), 1007–1018. https://doi.org/10.1016/j.tcb.2021.07.003

Holen, I., Speirs, V., Morrissey, B., & Blyth, K. (2017). *In vivo* models in breast cancer research: Progress, challenges and future directions. *Disease Models and Mechanisms*, 10(4), 359–371. https://doi.org/10.1242/dmm.028274

Hondermarck, H., Vercoutter-Edouart, A.S., Révillion, F., Lemoine, J., El-Yazidi-Belkoura, I., Nurcombe, V., & Peyrat, J.P. (2001). Proteomics of breast cancer for marker discovery and signal pathway profiling. *Proteomics: International Edition*, 1(10), 1216–1232. https://doi.org/10.1002/1615-9861(200110)1:10%3C1216::aid-prot1216%3E3.0.co;2-p

Huang, Y., Agrawal, B., Sun, D., Kuo, J.S., & Williams, J.C. (2011). Microfluidics-based devices: New tools for studying cancer and cancer stem cell migration. *Biomicrofluidics*, 5(1), 013412. https://doi.org/10.1063/1.3555195

Huang, Y., Liu, T., Huang, Q., & Wang, Y. (2024). From organ-on-a-chip to human-on-a-chip: A review of research progress and latest applications. *ACS Sensors*, 9(7), 3466–3488. https://doi.org/10.1021/acssensors.4c00004

Huber-Keener, K.J. (2022). Cancer genetics and breast cancer. *Best Practice and Research Clinical Obstetrics and Gynaecology*, 82, 3–11. https://doi.org/10.1016/j.bpobgyn.2022.01.007

Hwangbo, H., Chae, S., Kim, W., Jo, S., & Kim, G.H. (2024). Tumor-on-a-chip models combined with mini-tissues or organoids for engineering tumor tissues. *Theranostics*, 14(1), 33. https://doi.org/10.7150/thno.90093

Irigaray, P., Newby, J.A., Clapp, R., Hardell, L., Howard, V., Montagnier, L., Epstein, S., & Belpomme, D. (2007). Lifestyle-related factors and environmental agents causing cancer: An overview. *Biomedicine and Pharmacotherapy*, 61(10), 640–658. https://doi.org/10.1016/j.biopha.2007.10.006

Jadhav, S., Rath, S.N., & Roopavath, U.K. (2024). A review on multicellular spheroids and organoids for breast cancer diagnosis and therapy. *Biomedical Materials and Devices*, 3, 1–23. https://dx. doi.org/10.1007/s44174-024-00225-w

Jain, A., Barge, A., & Parris, C.N. (2024). Combination strategies with PARP inhibitors in BRCA-mutated triple-negative breast cancer: Overcoming resistance mechanisms. *Oncogene*, 44, 1–15. https://doi.org/10.1038/s41388-024-03227-6

Jamshidii, N., Chang, J., Mock, K., Nguyen, B., Dauphine, C., & Kuo, M.D. (2020). Evaluation of the predictive ability of ultrasound-based assessment of breast cancer using BI-RADS natural language reporting against commercial transcriptome-based tests. *PLoS One*, 15(1), e0226634. https://doi.org/10.1371/journal.pone.0226634

Joshi, R.S., Kanugula, S.S., Sudhir, S., Pereira, M.P., Jain, S., & Aghi, M.K. (2021). The role of cancer-associated fibroblasts in tumor progression. *Cancers*, 13(6), 1399. https://doi.org/10.3390/cancers13061399

Kargozar, S., Mollazadeh, S., Kermani, F., Webster, T.J., Nazarnezhad, S., Hamzehlou, S., & Baino, F. (2022). Hydroxyapatite nanoparticles for improved cancer theranostics. *Journal of Functional Biomaterials*, 13(3), 100. https://doi.org/10.3390/jfb13030100

Kędzierska, M., & Bańkosz, M. (2024). Role of proteins in oncology: Advances in cancer diagnosis, prognosis, and targeted therapy—A narrative review. *Journal of Clinical Medicine*, 13(23), 7131. https://doi.org/10.3390/jcm13237131

Keirouz, A., Chung, M., Kwon, J., Fortunato, G., & Radacsi, N. (2020). 2D and 3D electrospinning technologies for the fabrication of nanofibrous scaffolds for skin tissue engineering: A review. *Wiley Interdisciplinary Reviews: Nanomedicine and Nanobiotechnology*, 12(4), e1626. https://doi.org/10.1002/wnan.1626

Kersten, K., de Visser, K.E., van Miltenburg, M.H., & Jonkers, J. (2017). Genetically engineered mouse models in oncology research and cancer medicine. *EMBO Molecular Medicine*, 9(2), 137–153. https://doi.org/10.15252/emmm.201606857

Kim, J.B., O'Hare, M.J., & Stein, R. (2003). Models of breast cancer: Is merging human and animal models the future?. *Breast Cancer Research*, 6, 1–9. https://doi.org/10.1186/bcr645

Klarenbeek, S., van Miltenburg, M.H., & Jonkers, J. (2013). Genetically engineered mouse models of PI3K signaling in breast cancer. *Molecular Oncology*, 7(2), 146–164. https://doi.org/10.1016/j.molonc.2013.02.003

Korkola, J., & Gray, J.W. (2010). Breast cancer genomes—form and function. *Current Opinion in Genetics & Development*, 20(1), 4–14. https://doi.org/10.1016/j.gde.2009.11.005

Kumar, P., & Aggarwal, R. (2016). An overview of triple-negative breast cancer. *Archives of Gynecology and Obstetrics*, 293, 247–269. https://doi.org/10.1007/s00404-015-3859-y

Kunnumakkara, A.B., Bordoloi, D., Sailo, B.L., Roy, N.K., Thakur, K.K., Banik, K., Shakibaei, M., Gupta, S.C., & Aggarwal, B.B. (2019). Cancer drug development: The missing links. *Experimental Biology and Medicine*, 244(8), 663–689. https://doi.org/10.1177/1535370219839163

Lan, W., Liao, H., Chen, Q., Zhu, L., Pan, Y., & Chen, Y.P.P. (2024). DeepKEGG: A multi-omics data integration framework with biological insights for cancer recurrence prediction and biomarker discovery. *Briefings in Bioinformatics*, 25(3), bbae185. https://doi.org/10.1093/bib/bbae185

Levy, H., & Janke, A. (2016). Health literacy and access to care. *Journal of Health Communication*, 21(Suppl. S1), 43–50. https://doi.org/10.1080/10810730.2015.1131776

Li, S., Gui, J., Karagas, M.R., & Passarelli, M.N. (2025). Transcriptome-wide association study identifies genes associated with bladder cancer risk. *Scientific Reports*, 15(1), 1390. https://doi.org/10.1038/s41598-025-85565-3

Li, W., Zhou, Z., Zhou, X., Khoo, B.L., Gunawan, R., Chin, Y.R., Zhang, L., Yi, C., Guan, X., & Yang, M. (2023). 3D biomimetic models to reconstitute tumor microenvironment in vitro: Spheroids, organoids, and tumor-on-a-chip. *Advanced Healthcare Materials*, 12(18), 2202609. https://doi.org/10.1002/adhm.202202609

Li, Y., Kong, X., Wang, Z., & Xuan, L. (2022). Recent advances of transcriptomics and proteomics in triple-negative breast cancer prognosis assessment. *Journal of Cellular and Molecular Medicine*, 26(5), 1351–1362. https://doi.org/10.1111/jcmm.17124

Liu, G., Wang, B., Li, S., Jin, Q., & Dai, Y. (2019). Human breast cancer decellularized scaffolds promote epithelial-to-mesenchymal transitions and stemness of breast cancer cells in vitro. *Journal of Cellular Physiology*, 234(6), 9447–9456. https://doi.org/10.1002/jcp.27630

Liu, Y., Wu, W., Cai, C., Zhang, H., Shen, H., & Han, Y. (2023). Patient-derived xenograft models in cancer therapy: Technologies and applications. *Signal Transduction and Targeted Therapy*, 8(1), 160. https://doi.org/10.1038/s41392-023-01419-2

Livingston, M.K., Morgan, M.M., Daly, W.T., Murphy, W.L., Johnson, B.P., Beebe, D.J., & Virumbrales-Muñoz, M. (2019). Evaluation of PEG-based hydrogel influence on estrogen-receptor-driven responses in MCF7 breast cancer cells. *ACS Biomaterials Science and Engineering*, 5(11), 6089–6098. https://doi.org/10.1021/acsbiomaterials.9b00480

Lu, Y., Cheng, D., Niu, B., Wang, X., Wu, X., & Wang, A. (2023). Properties of poly (lactic-co-glycolic acid) and progress of poly (lactic-co-glycolic acid)-based biodegradable materials in biomedical research. *Pharmaceuticals*, 16(3), 454. https://doi.org/10.3390/ph16030454

Lui, A.L. (2017). *The Role of IFITM1 in Promoting Breast Cancer Aggression and Aromatase-Inhibitor Resistance* [Doctoral dissertation, University of Kansas]. KU ScholarWorks. https://kuscholarworks.ku.edu/bitstream/handle/1808/25887/Lui_ku_0099D_15348_DATA_1.pdf?sequence=1

Luo, Y., Wei, X., Wan, Y., Lin, X., Wang, Z., & Huang, P. (2019). 3D printing of hydrogel scaffolds for future application in photothermal therapy of breast cancer and tissue repair. *Acta Biomaterialia*, 92, 37–47. https://doi.org/10.1016/j.actbio.2019.05.039

Manikis, G.C., Simos, N.J., Kourou, K., Kondylakis, H., Poikonen-Saksela, P., Mazzocco, K., Pat-Horenczyk, R., Sousa, B., Oliveira-Maia, A.J., Mattson, J., et al. (2023). Personalized risk analysis to improve the psychological resilience of women undergoing treatment for breast cancer: Development of a machine learning–driven clinical decision support tool. *Journal of Medical Internet Research*, 25, e43838. https://doi.org/10.2196/43838

Marinovich, M.L., Wylie, E., Lotter, W., Pearce, A., Carter, S.M., Lund, H., Waddell, A., Kim, J.G., Pereira, G.F., Lee, C.I., et al. (2022). Artificial intelligence (AI) to enhance breast cancer screening: Protocol for population-based cohort study of cancer detection. *BMJ Open*, 12, e054005. https://doi.org/10.1136/bmjopen-2021-054005

Mehmood, S., Faheem, M., Ismail, H., Farhat, S.M., Ali, M., Younis, S., & Asghar, M.N. (2022). Breast cancer resistance likelihood and personalized treatment through integrated multiomics. *Frontiers in Molecular Biosciences*, 9, 783494. https://doi.org/10.3389/fmolb.2022.783494

Mema, E., & McGinty, G. (2020). The role of artificial intelligence in understanding and addressing disparities in breast cancer outcomes. *Current Breast Cancer Reports*, 12, 168–174. https://link.springer.com/article/10.1007/s12609-020-00368-x

Momenimovahed, Z., & Salehiniya, H. (2019). Epidemiological characteristics of and risk factors for breast cancer in the world. *Breast Cancer: Targets and Therapy*, 11, 151–164. https://doi.org/10.2147/bctt.s176070

Morales, X., Cortes-Dominguez, I., & Ortiz-de-Solorzano, C. (2021). Modeling the mechanobiology of cancer cell migration using 3D biomimetic hydrogels. *Gels*, 7(1), 17. https://doi.org/10.3390/gels7010017

Moroni, S., Casettari, L., & Lamprou, D.A. (2022). 3D and 4D printing in the fight against breast cancer. *Biosensors*, 12(8), 568. https://doi.org/10.3390/bios12080568

Moyano, J.V. (2006). AlphaB-crystallin is a novel oncoprotein that predicts poor clinical outcome in breast cancer. *Journal of Clinical Investigation*, 116(1), 30–32. https://doi.org/10.1172/jci25888

Mueller, C., Haymond, A., Davis, J.B., Williams, A., & Espina, V. (2018). Protein biomarkers for subtyping breast cancer and implications for future research. *Expert Review of Proteomics*, 15(2), 131–152. https://doi.org/10.1080/14789450.2018.1421071

Nanou, A., Lorenzo-Moldero, I., Gazouleas, K.D., Cortese, B., & Moroni, L. (2022). 3D culture modeling of metastatic breast cancer cells in additive manufactured scaffolds. *ACS Applied Materials & Interfaces*, 14(24), 28389–28402. https://doi.org/10.1021/acsami.2c07492

Ng, A.Y., Oberije, C.J.G., Ambrózay, É., Szabó, E., Serfőző, O., Karpati, E., Fox, G., Glocker, B., Morris, E.A., Forrai, G., et al. (2023). Prospective implementation of AI-assisted screen reading to improve early detection of breast cancer. *Nature Medicine*, 29, 3044–3049. https://doi.org/10.1038/s41591-023-02625-9

O'Regan, R.M., & Jordan, V.C. (2002). The evolution of tamoxifen therapy in breast cancer: Selective oestrogen-receptor modulators and downregulators. *The Lancet Oncology*, 3(4), 207–214. https://doi.org/10.1016/s1470-2045(02)00711-8

Obidiro, O., Battogtokh, G., & Akala, E.O. (2023). Triple negative breast cancer treatment options and limitations: Future outlook. *Pharmaceutics*, 15(7), 1796. https://doi.org/10.3390/pharmaceutics15071796

Omidi, Y., Mobasher, M., Castejon, A.M., & Mahmoudi, M. (2022). Recent advances in nanoscale targeted therapy of HER2-positive breast cancer. *Journal of Drug Targeting*, 30(7), 687–708. https://doi.org/10.1080/1061186x.2022.2055045

Otálora-Otálora, B.A., López-Rivera, J.J., Aristizábal-Guzmán, C., Isaza-Ruget, M.A., & Álvarez-Moreno, C.A. (2023). Host transcriptional regulatory genes and microbiome networks crosstalk through immune receptors establishing normal and tumor multiomics metafirm of the oral-gut-lung axis. *International Journal of Molecular Sciences*, 24(23), 16638. https://doi.org/10.3390/ijms242316638

Parise, C.A., & Caggiano, V. (2019). Differences in clinicopathologic characteristics and risk of mortality between the triple positive and ER+/PR+/HER2− breast cancer subtypes. *Cancer Causes and Control*, 30, 417–424. https://doi.org/10.1007/s10552-019-01152-8

Pfefferle, A.D., Herschkowitz, J.I., Usary, J., Harrell, J.C., Spike, B.T., Adams, J.R., Torres-Arzayus, M.I., Brown, M., Egan, S.E., Wahl, G.M., & Rosen, J.M. (2013). Transcriptomic classification of genetically engineered mouse models of breast cancer identifies human subtype counterparts. *Genome Biology*, 14, 1–16. https://doi.org/10.1186/gb-2013-14-11-r125

Phan, V.G., Murugesan, M., Huong, H., Le, T.T., Phan, T.H., Manivasagan, P., Mathiyalagan, R., Jang, E.S., Yang, D.C., Li, Y., & Thambi, T. (2022). Cellulose nanocrystals-incorporated thermosensitive hydrogel for controlled release, 3D printing, and breast cancer treatment applications. *ACS Applied Materials & Interfaces*, 14(38), 42812–42826. https://doi.org/10.1021/acsami.2c05864

Pistollato, F., Bernasconi, C., McCarthy, J., Campia, I., Desaintes, C., Wittwehr, C., Deceuninck, P., & Whelan, M. (2020). Alzheimer's disease, and breast and prostate cancer research: Translational failures and the importance to monitor outputs and impact of funded research. *Animals*, 10(7), 1194. https://doi.org/10.3390/ani10071194

Pradhan, S., Hassani, I., Seeto, W.J., & Lipke, E.A. (2017). PEG-fibrinogen hydrogels for three-dimensional breast cancer cell culture. *Journal of Biomedical Materials Research Part A*, 105(1), 236–252. https://doi.org/10.1002/jbm.a.35899

Rakha, E.A., Putti, T.C., Abd El-Rehim, D.M., Paish, C., Green, A.R., Powe, D.G., Lee, A.H., Robertson, J.F., & Ellis, I.O. (2006). Morphological and immunophenotypic analysis of breast carcinomas with basal and myoepithelial differentiation. *The Journal of Pathology: A Journal of the Pathological Society of Great Britain and Ireland*, 208(4), 495–506. https://doi.org/10.1002/path.1916

Randall-Demllo, S., Al-Qadami, G., Raposo, A.E., Ma, C., Priebe, I.K., Hor, M., Singh, R., & Fung, K.Y. (2024). Ex vivo intestinal organoid models: Current state-of-the-art and challenges in disease modelling and therapeutic testing for colorectal cancer. *Cancers*, 16(21), 3664. https://doi.org/10.3390/cancers16213664

Regmi, S., Poudel, C., Adhikari, R., & Luo, K.Q. (2022). Applications of microfluidics and organ-on-a-chip in cancer research. *Biosensors*, 12(7), 459. https://doi.org/10.3390/bios12070459

Rieke, N., Hancox, J., Li, W., Milletarì, F., Roth, H.R., Albarqouni, S., Bakas, S., Galtier, M.N., Landman, B.A., Maier-Hein, K., et al. (2020). The future of digital health with federated learning. *NPJ Digital Medicine*, 3, 119. https://doi.org/10.1038/s41746-020-00323-1

Romanov, S., Howell, S., Harkness, E., Bydder, M., Evans, D.G., Squires, S., Fergie, M., & Astley, S. (2023). Artificial intelligence for image-based breast cancer risk prediction using attention. *Tomography*, 9, 2103–2115. https://doi.org/10.1038/s41746-020-00323-1

Rosato, R.R., Dávila-González, D., Choi, D.S., Qian, W., Chen, W., Kozielski, A.J., Wong, H., Dave, B., & Chang, J.C. (2018). Evaluation of anti-PD-1-based therapy against triple-negative breast cancer patient-derived xenograft tumors engrafted in humanized mouse models. *Breast Cancer Research*, 20, 1–16. https://doi.org/10.1186/s13058-018-1037-4

Rottenberg, S., Pajic, M., & Jonkers, J. (2010). Studying drug resistance using genetically engineered mouse models for breast cancer. *Multi-Drug Resistance in Cancer*, 596, 33–45. https://doi.org/10.1007/978-1-60761-416-6_3

Santerre, J.P., Yang, Y., Du, Z., Wang, W., & Zhang, X. (2024). Biomaterials' enhancement of immunotherapy for breast cancer by targeting functional cells in the tumor micro-environment. *Frontiers in Immunology*, 15, 1492323. https://doi.org/10.3389/fimmu.2024.1492323

Schedin, P., & Keely, P.J. (2011). Mammary gland ECM remodeling, stiffness, and mechanosignaling in normal development and tumor progression. *Cold Spring Harbor Perspectives in Biology*, 3(1), a003228. https://doi.org/10.1101/cshperspect.a003228

Sebestyén, A., Dankó, T., Sztankovics, D., Moldvai, D., Krencz, I., Raffay, R., & Petővári, G. (2022). Extracellular matrix as a metabolic niche in cancer. In Kovalszky, I., Franchi, M., & Alaniz, L.D. (Eds.), *The Extracellular Matrix and the Tumor Microenvironment* (pp. 345–373). Springer International Publishing, Cham. https://doi.org/10.1007/978-3-030-99708-3_13

Segar, M.W., Hall, J.L., Jhund, P.S., Powell-Wiley, T.M., Morris, A.A., Kao, D., Fonarow, G.C., Hernandez, R., Ibrahim, N.E., Rutan, C., et al. (2022). Machine learning-based models incorporating social determinants of health vs traditional models for predicting in-hospital mortality in patients with heart failure. *JAMA Cardiology*, 7, 844–854. https://doi.org/10.1001/jamacardio.2022.1900

Shams, R.A., Zowghi, D., & Bano, M. (2023). AI and the quest for diversity and inclusion: A systematic literature review. *AI Ethics*, 5, 411–438. https://doi.org/10.1007/s43681-023-00362-w

Sharifi, M., Bai, Q., Babadaei, M.M.N., Chowdhury, F., Hassan, M., Taghizadeh, A., Derakhshankhah, H., Khan, S., Hasan, A., & Falahati, M. (2021). 3D bioprinting of engineered breast cancer constructs for personalized and targeted cancer therapy. *Journal of Controlled Release*, 333, 91–106. https://doi.org/10.1016/j.jconrel.2021.03.026

Sigdel, I., Gupta, N., Faizee, F., Khare, V.M., Tiwari, A.K., & Tang, Y. (2021). Biomimetic microfluidic platforms for the assessment of breast cancer metastasis. *Frontiers in Bioengineering and Biotechnology*, 9, 633671. https://doi.org/10.3389/fbioe.2021.633671

Silva, J., Oliveira, P.A., Duarte, J.A., & Faustino-Rocha, A.I. (2025). Mammary cancer models: An overview from the past to the future. *In Vivo*, 39(1), 1–16. https://doi.org/10.21873/invivo.13800

Simpson, C.M., Zhang, B., Hornbeck, P.V., & Gnad, F. (2019). Systematic analysis of the intersection of disease mutations with protein modifications. *BMC Medical Genomics*, 12, 1–10. https://doi.org/10.1186/s12920-019-0543-2

Singhal, S.S., Garg, R., Mohanty, A., Garg, P., Ramisetty, S.K., Mirzapoiazova, T., Soldi, R., Sharma, S., Kulkarni, P., & Salgia, R. (2023). Recent advancement in breast cancer research: Insights from model organisms—Mouse models to zebrafish. *Cancers*, 15(11), 2961. https://doi.org/10.3390/cancers15112961

Su, Y., Hu, X., Kang, Y., Zhang, C., Cheng, Y.Y., Jiao, Z., Nie, Y., & Song, K. (2022). Curcumin nanoparticles combined with 3D printed bionic tumor models for breast cancer treatment. *Biofabrication*, 15(1), 014105. https://doi.org/10.1088/1758-5090/aca5b8

Subia, B., Dahiya, U.R., Mishra, S., Ayache, J., Casquillas, G.V., Caballero, D., Reis, R.L., & Kundu, S.C. (2021). Breast tumor-on-chip models: From disease modeling to personalized drug screening. *Journal of Controlled Release*, 331, 103–120. https://doi.org/10.1016/j.jconrel.2020.12.057

Sung, B. (2023). Biomaterials and microfluidics systems used for modeling pathological tissues. In Kumar, A., Dhinasekaran, D., Savina, I., & Han, S.S. (Eds.), *Functional Biomaterials* (pp. 289–308). CRC Press, Boca Raton, FL. https://doi.org/10.1201/9781003251767-14

Tang, R., & Liu, X.Q. (2024). Modeling development of breast cancer: From tumor microenvironment to preclinical applications. *Frontiers in Pharmacology*, 15, 1466017. https://doi.org/10.3389/fphar.2024.1466017

Tao, Z., Shi, A., Lu, C., Song, T., Zhang, Z., & Zhao, J. 2015. Breast cancer: Epidemiology and etiology. *Cell Biochemistry and Biophysics*, 72, 333–338. https://doi.org/10.1007/s12013-014-0459-6

Teixeira, A.M., & Martins, P. (2023). A review of bioengineering techniques applied to breast tissue: Mechanical properties, tissue engineering and finite element analysis. *Frontiers in Bioengineering and Biotechnology*, 11, 1161815. https://doi.org/10.3389/fbioe.2023.1161815

Thakuri, P.S., Liu, C., Luker, G.D., & Tavana, H. (2018). Biomaterials-based approaches to tumor spheroid and organoid modeling. *Advanced Healthcare Materials*, 7(6), 1700980. https://doi.org/10.1002/adhm.201700980

Thomas, C., & Gustafsson, J.Å. (2011). The different roles of ER subtypes in cancer biology and therapy. *Nature Reviews Cancer*, 11(8), 597–608. https://doi.org/10.1038/nrc3093

Torre, L.A., Islami, F., Siegel, R.L., Ward, E.M., & Jemal, A. (2017). Global cancer in women: Burden and trends. *Cancer Epidemiology, Biomarkers and Prevention*, 26(4), 444–457. https://doi.org/10.1158/1055-9965.epi-16-0858

Tzeng, Y.D.T., Hsiao, J.H., Tseng, L.M., Hou, M.F., & Li, C.J. (2023). Breast cancer organoids derived from patients: A platform for tailored drug screening. *Biochemical Pharmacology*, 217, 115803. https://doi.org/10.1016/j.bcp.2023.115803

Uzun Ozsahin, D., Ikechukwu Emegano, D., Uzun, B., & Ozsahin, I. (2022). The systematic review of artificial intelligence applications in breast cancer diagnosis. *Diagnostics*, 13, 45. https://doi.org/10.3390/diagnostics13010045

Vaira, V., Fedele, G., Pyne, S., Fasoli, E., Zadra, G., Bailey, D., Snyder, E., Faversani, A., Coggi, G., Flavin, R., & Bosari, S. (2010). Preclinical model of organotypic culture for pharmacodynamic profiling of human tumors. *Proceedings of the National Academy of Sciences*, 107(18), 8352–8356. https://doi.org/10.1073/pnas.0907676107

Walsh, N.C., Kenney, L.L., Jangalwe, S., Aryee, K.E., Greiner, D.L., Brehm, M.A., & Shultz, L.D. (2017). Humanized mouse models of clinical disease. *Annual Review of Pathology: Mechanisms of Disease*, 12(1), 187–215. https://doi.org/10.1146/annurev-pathol-052016-100332

Wege, A.K., Ernst, W., Eckl, J., Frankenberger, B., Vollmann-Zwerenz, A., Männel, D.N., Ortmann, O., Kroemer, A., & Brockhoff, G. (2011). Humanized tumor mice—a new model to study and manipulate the immune response in advanced cancer therapy. *International Journal of Cancer*, 129(9), 2194–2206. https://doi.org/10.1002/ijc.26159

Wegiel, B., Vuerich, M., Daneshmandi, S., & Seth, P. (2018). Metabolic switch in the tumor microenvironment determines immune responses to anti-cancer therapy. *Frontiers in Oncology*, 8, 284. https://doi.org/10.3389/fonc.2018.00284

Whitaker, M.J., Quirk, R.A., Howdle, S.M., & Shakesheff, K.M. (2001). Growth factor release from tissue engineering scaffolds. *Journal of Pharmacy and Pharmacology*, 53(11), 1427–1437. https://doi.org/10.1211/0022357011777963

Wu, H.J., & Chu, P.Y. (2021). Recent discoveries of macromolecule-and cell-based biomarkers and therapeutic implications in breast cancer. *International Journal of Molecular Sciences*, 22(2), 636. https://doi.org/10.3390/ijms22020636

Wu, Y., & Xie, L. (2025). AI-driven multi-omics integration for multi-scale predictive modeling of genotype-environment-phenotype relationships. *Computational and Structural Biotechnology Journal*, 27, 265–277. https://doi: 10.1016/j.csbj.2024.12.030

Xu, L., Xie, Y., Gou, Q., Cai, R., Bao, R., Huang, Y., & Tang, R. (2024). HER2-targeted therapies for HER2-positive early-stage breast cancer: Present and future. *Frontiers in Pharmacology*, 15, 1446414. https://doi.org/10.3389/fphar.2024.1446414

Yala, A., Mikhael, P.G., Strand, F., Lin, G., Satuluru, S., Kim, T., Banerjee, I., Gichoya, J., Trivedi, H., Lehman, C.D., et al. (2022). Multi-institutional validation of a mammography-based breast cancer risk model. *Journal of Clinical Oncology*, 40, 1732–1740. https://doi.org/10.1200/jco.21.01337

Yang, J., Dung, N.T., Thach, P.N., Phong, N.T., Phu, V.D., Phu, K.D., Yen, L.M., Xuan Thy, D.B., Soltan, A.A., Thwaites, L., et al. (2023). Generalizability assessment of AI models across hospitals: A comparative study in low-middle income and high-income countries. *bioRxiv*, 15(1), 8270. https://doi.org/10.1038/s41467-024-52618-6

Yang, J., Soltan, A.A.S., & Clifton, D.A. (2022). Machine learning generalizability across healthcare settings: Insights from multi-site COVID-19 screening. *NPJ Digital Medicine*, 5, 69. https://doi.org/10.1038/s41746-022-00614-9

Yersal, O., & Barutca, S. (2014). Biological subtypes of breast cancer: Prognostic and therapeutic implications. *World Journal of Clinical Oncology*, 5(3), 412. https://doi.org/10.5306/wjco.v5.i3.412

Yousefi, H., Vatanmakanian, M., Mahdiannasser, M., Mashouri, L., Alahari, N.V., Monjezi, M.R., Ilbeigi, S., & Alahari, S.K. (2021). Understanding the role of integrins in breast cancer invasion, metastasis, angiogenesis, and drug resistance. *Oncogene*, 40(6), 1043–1063. https://doi.org/10.1038/s41388-020-01588-2

Zhang, F., & King, M.W. (2020). Biodegradable polymers as the pivotal player in the design of tissue engineering scaffolds. *Advanced Healthcare Materials*, 9(13), 1901358. https://doi.org/10.1002/adhm.201901358

Zhang, X., Chen, X., Hong, H., Hu, R., Liu, J., & Liu, C. (2022). Decellularized extracellular matrix scaffolds: Recent trends and emerging strategies in tissue engineering. *Bioactive Materials*, 10, 15–31. https://doi.org/10.1016/j.bioactmat.2021.09.014

Zhang, Y., Gong, S., & Liu, X. (2024). Spatial transcriptomics: A new frontier in accurate localization of breast cancer diagnosis and treatment. *Frontiers in Immunology*, 15, 1483595. https://doi.org/10.3389/fimmu.2024.1483595

Zhao, L. (2018). *Versatile Cradle-to-Grave Culture Platform for 3D Spheroid Formation and Analysis*. [Doctoral dissertation, McGill University]. eScholarship@McGill University (Canada). https://escholarship.mcgill.ca/collections/theses?locale=en&view=gallery

Zhou, Z., Cong, L., & Cong, X. (2021). Patient-derived organoids in precision medicine: Drug screening, organoid-on-a-chip and living organoid biobank. *Frontiers in Oncology*, 11, 762184. https://doi.org/10.3389/fonc.2021.762184

Zhu, Z., Jiang, L., & Ding, X. (2023). Advancing breast cancer heterogeneity analysis: Insights from genomics, transcriptomics and proteomics at bulk and single-cell levels. *Cancers*, 15(16), 4164. https://doi.org/10.3390/cancers15164164

2 Lung Cancer Models

Roma Pandey, Tanya Aggarwal, Madhav Sharan, Anjali Singh, Anamika Kushwaha, Sorabh Chaudhary, Avinash Singh, Santosh Kumar Mishra, and Pankaj Verma

2.1 INTRODUCTION

Lung cancer is one of the leading causes of death in the world. It is reported that in the year 2022 nearly 20 million new cases of lung cancers were registered and 9.7 million deaths (Bray et al., 2024). Men are majorly diagnosed with cancer in comparison to women due to tobacco smoking, as tobacco smoking is the reason for 90% of lung cancer cases (Chaitanya Thandra et al., 2021). Lung cancer progression is majorly divided into four different stages: Stage-1, Stage-2, Stage-3, and Stage-4, shown in Figure 2.1. Small cell lung cancer (SCLC) and non-small cell lung cancer (NSCLC) are the two subtypes of lung cancer that are separated according to the cell of origin. According to WHO in 2015, prevalent forms of lung cancer are squamous cell carcinoma (SCC), adenocarcinoma (cancer of glandular cells), and neuroendocrine tumors such as SCLC, large cell neuroendocrine carcinoma (LCNEC), and carcinoid (Travis et al., 2015). Poorly differentiated neuroendocrine cells are another source of SCLC, which leads to quick metastases, poor response to treatment, and a dismal prognosis. A history of smoking is linked to a higher likelihood of centrally situated squamous cell and small cell malignancies, particularly in men. Adenocarcinoma is more common in women and people who have never smoked. Adenocarcinoma appears in the periphery and has positive test results for targetable driver mutations, including those in the ROS1, BRAF, anaplastic lymphoma kinase (ALK), and epidermal growth factor receptor (EGFR). In recent years, immunotherapies such as cytotoxic T-lymphocyte-associated protein 4 (CTLA-4) inhibitors and receptor tyrosine kinase small molecule inhibitors have either replaced or enhanced chemotherapy in patients who qualify (Denisenko et al., 2018).

2.2 EPIDEMIOLOGY

Lung cancer is majorly diagnosed globally among all the cancer types. It accounts for about 12.4% of all cancer cases, with breast cancer coming in second at 11.4%. Lung cancer is responsible for the death of approximately 1,817,172 peoples in 2022 (Bray et al., 2024). It is reported by the National Cancer Institute's Surveillance that in 2018, approximately 229,000 cases of lung cancer were reported in the USA, which is 12.4% of all cancer cases. SEER reports that the most recent 5-year survival rate for lung cancer in the United States, spanning from 2010 to 2016, was 20.5%.

 DOI: 10.1201/9781003583349-3

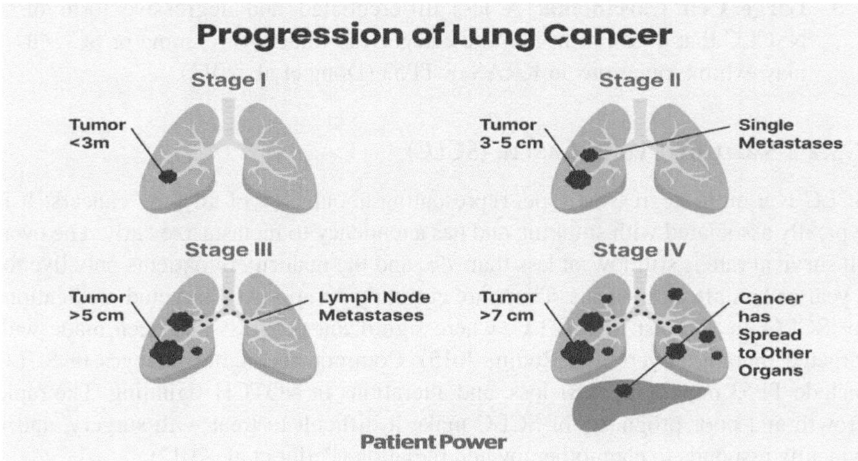

FIGURE 2.1 Different stages of lung cancer (https://www.patientpower.info/lung-cancer/lung-cancer-staging).

In 1975, the 5-year survival rate was 11.5%, the earliest recorded (Barta et al., 2019; Chaitanya Thandra et al., 2021). India is the world's third-largest producer and second-largest user of tobacco. In India, tobacco products are used by about 28.6% of people (42.4% of males and 14.2% of females). Lung cancer accounts for 5.9% of all cancer cases in India, with 1.16 lakh cases reported, according to GLOBOCAN India statistics 2018 (Singh et al., 2021).

2.3 MOLECULAR SUBTYPES AND DIAGNOSIS OF LUNG CANCER

Lung cancer is a complex disease that can be classified into molecular subtypes based on genetic and molecular alterations. These subtypes play a significant role in diagnosis, treatment decisions, and prognosis. The two primary types of lung cancer are NSCLC and SCLC, with NSCLC further divided into various subtypes. Understanding the molecular and genetic basis of these subtypes helps guide targeted therapies and immunotherapy treatments (Fukuda et al., 2024).

2.3.1 Non-Small Cell Lung Cancer (NSCLC) Subtypes

NSCLC accounts for about 85% of all lung cancers and is further subdivided into several subtypes:

- **Adenocarcinoma**: The most common subtype, particularly in nonsmokers. It typically originates in the peripheral lung tissue (close to the pleura). Common mutations include EGFR, ALK, ROS1, and KRAS (Sholl, 2017).
- **Squamous Cell Carcinoma**: Often associated with smoking and originating in the central parts of the lungs, near the bronchi. Genetic alterations include TP53 mutations, CDKN2A loss, and FGFR1 amplifications (Ruiz-Cordero et al., 2020).

- **Large Cell Carcinoma**: A less differentiated and aggressive form of NSCLC that does not fit into the categories of adenocarcinoma or SCC. It may exhibit mutations in KRAS or TP53 (Dong et al., 2017).

2.3.2 Small Cell Lung Cancer (SCLC)

SCLC is a more aggressive type, representing about 15% of all lung cancers. It is typically associated with smoking and has a tendency to metastasize early. The overall survival rate is still low, at less than 7%, and the majority of patients only live for a year or less after diagnosis. There are currently no approved targeted medications for SCLC, in contrast to NSCLC, where significant progress has been made with targeted therapy (Byers and Rudin, 2015). Common molecular changes in SCLC include TP53 mutations, RB1 loss, and alterations in NOTCH signaling. The rapid growth and poor prognosis of SCLC make it difficult to treat with surgery, and it typically responds to chemotherapy and radiation (Peifer et al., 2012).

2.4 MOLECULAR SUBTYPES AND GENETIC ALTERATIONS IN LUNG CANCER

Lung cancer is a heterogeneous disease with distinct molecular subtypes, primarily classified into NSCLC and SCLC. NSCLC, which includes adenocarcinoma, SCC, and large cell carcinoma, is the most common form. Genetic alterations play a critical role in lung cancer development and progression, with mutations in key oncogenes like *EGFR*, *KRAS*, *ALK*, and *ROS1* influencing tumor behavior and treatment responses. Additionally, tumor suppressor genes such as *TP53* and *RB1* are frequently altered in both NSCLC and SCLC. These genetic alterations guide targeted therapies and immunotherapies, making molecular profiling crucial for personalized treatment strategies. Understanding the specific mutations and molecular subtypes allows for more accurate prognosis, improved treatment options, and better patient outcomes in lung cancer management. Table 2.1 shows key genetic mutations responsible for lung cancer.

2.4.1 Key Genetic Mutations in NSCLC

- **EGFR (Epidermal Growth Factor Receptor)**: EGFR belongs to the erbB family gene, which includes erbB1 (also named as EGFR), erbB2, erbB3, and erbB4 genes. Mutation in EGFR is commonly found in adenocarcinoma and particularly in nonsmokers. Overexpression and mutation are observed in 43%–89% of cases of NSCLC (Gupta et al., 2009). In another report, 75% of NSCLC cases had elevated receptor expression linked to mutations in the EGFR tyrosine kinase domain, which were present in 25% of cases (Shigematsu and Gazdar, 2006; Suzuki et al., 2005).
- **ALK (Anaplastic Lymphoma Kinase)**: Oncologic drivers such as ALK and EGFR fusions are more commonly detected at an advanced stage (discovered in 19% of stage IV tumors compared to 2%–7% of early-stage malignancies) (Shi et al., 2020). Rearrangements involving the ALK gene,

TABLE 2.1
Key Genetic Mutations in Lung Cancer

Molecular Subtype	Description	Key Genetic Alterations	References
Non-small cell lung cancer (NSCLC)	The most common form of lung cancer, making up ~85% of all lung cancer cases.	• EGFR mutations • ALK	Shigematsu and Gazdar (2006), Chen and Chaft (2023)
Squamous cell carcinoma (SCC)	A subtype of NSCLC, often associated with smoking.	• TP53 mutations SOX_2 amplification PTEN mutation • FGFR1 amplification	Perez-Moreno et al. (2012), Joshi et al. (2021)
Adenocarcinoma	The most common NSCLC subtype, more frequent in nonsmokers	• EGFR mutations • ALK rearrangements • ROS1 rearrangements • KRAS mutations	Warrier et al. (2022), Soda et al. (2007)
Large cell carcinoma	A subtype of NSCLC with a poor prognosis; often lacks specific molecular alterations	• TP53 mutations • Lack of identifiable genetic markers	Larsen and Minna (2011), Rodriguez-Canales et al. (2016)
Small cell lung cancer	A highly aggressive form of lung cancer, commonly associated with smoking	• TP53 mutations • RB1 mutations • MYC amplification	Salgia and Skarin (1998), Rudin et al. (2021)
KRAS-mutated lung cancer	A major driver in lung cancer, especially in smokers	• KRAS mutations (G12C, G12V, etc.)	Mao et al. (2010), Wankhede et al. (2023)

such as EML4-ALK, are most commonly found in younger, non-smoking patients with adenocarcinoma. ALK inhibitors like Crizotinib and Alectinib are used for targeted therapy (Chen and Chaft, 2023; Chaft et al., 2018).

- **KRAS**: An oncogenic event that frequently occurs in lung cancer and other kinds of epithelial carcinoma is the mutational activation of KRAS. Mutations in KRAS, particularly in codon 12, are frequent in lung adenocarcinoma, especially among smokers. This mutation is challenging to target, but the development of inhibitors, such as Sotorasib, has shown promise (Westcott and To, 2013). Lung adenocarcinomas, the most prevalent histological subtype of NSCLC, are the primary site of KRAS mutations, occurring in 16%–40% of the examined samples. KRAS mutations are rare in squamous cell carcinoma, another form of NSCLC, but they are never found in SCLC (Nelson et al., 1996; Suzuki et al., 1990; Mitsudomi et al., 1991).
- **ROS1**: ROS1 gene fusions are a known carcinogen that accounts for 1%–2% of NSCLC, it is about 10,000–15,000 cases per year. ALK, ROS1, RET, and NTRK genomic rearrangements or mutations in KRAS, BRAF, EGFR, MET, and ERBB2 are among the main genetic changes seen in two out of three patients with advanced NSCLC (D'Angelo et al., 2020; Takeuchi et al., 2012).

2.4.1.1 *In Vitro* and *In Vivo* Models for Lung Cancer

Both *in vitro* and *in vivo* models play a crucial role in studying lung cancer. *In vitro* models allow for the isolated study of cellular behavior, pharmacological responses, and molecular pathways by cultivating cancer cells in a controlled laboratory environment. Despite being helpful for learning basic biology and screening treatments, they are not as complex as living things. *In vivo* models, on the other hand, reproduce the tumor microenvironment using real animals, usually mice, and allow for a more comprehensive, physiological view of cancer progression, metastasis, and treatment outcomes. *In vivo* models may not fully capture human-specific responses and are more resource-intensive, but they offer a more accurate depiction of human disease. These models work well together to assist researchers.

2.4.1.2 *In Vitro* Models for Lung Cancer

The interaction between cells and the extracellular matrix surrounding them during the invasion process cannot be seen using the conventional 2-D culture method, and animal models are typically too complicated to provide a detailed explanation of the mechanisms. Thus, a 3-D *in vitro* model that is accurate and effective is ideal for drug screening testing and cell invasion research. 3-D cell culture models are divided into two types: scaffold-supported (hydrogel) and non-scaffold-supported (spheroids, bioreactor) on the basis of the requirement of supporting material (Han and Wu, 2022)

1. **Hydrogel Based Model**: Biopolymer models like collagen and fibrin, together with other synthetic polymer materials, are used to produce 3D hydrogel models, which are then used to construct extracellular matrix components as cell culture supporting materials. Hydrogel-based models contain alginate or gelatin, which are sensitive to temperatures, so they are printed under the low temperatures. The metabolic rate of cells slows down at low temperatures and increases the survival capacity of the cells and the success rate of the model development (Zou et al., 2023).
2. **Spheroids**: Spheroids are scaffold-independent 3D culture models that exhibit characteristics of the corresponding *in vivo* tissue, including spatial growth and cell-cell interaction, and are distinguished by the clustering of cultured cells due to their high adhesiveness (Zanoni et al., 2016; Rolver et al., 2019). Spheroids can be formed using a variety of techniques, including rotating wall vessel bioreactors, hanging-drop plates, and rounded-bottom microplates (Zhang et al., 2024).
3. **Organoids**: Organoid-based models are implemented on both alveolar and bronchial tissues successfully. Pluripotent stem cells can be used for the development of organoid models. Lung cancer organoid is one of the recent cultures systems, but it also has some limitations, such as growth limitations, and its properties are not similar to the original tumors.

2.4.1.3 *In Vitro* Models for Lung Cancers

Lung cancer *in vivo* models have advantages over *in vitro* models for a number of reasons, including the fact that they more accurately resemble the complexity of human tumors in terms of metastasis, immune system interactions, tumor microenvironment, and overall disease development. *In vivo* models—usually involving mice—are essential for comprehending how tumors grow, spread, and react to treatment in a whole organism. They enable the investigation of intricate relationships within the tumor microenvironment. Different types of models are given below.

1. **Xenograft Models**: Involve transplanting human lung cancer cells (from cell lines or patient-derived tumors into immunodeficient mice). Cells isolated from the patients are injected into the mice having a compromised immune system and then transferred *in vivo* from mouse to mouse directly. Some different types of Xenografts are
    - Patient-Derived Xenografts: In this, human tumor cells are transferred to immunocompromised mice.
    - Human tumor tissue is implanted into immunodeficient mice. These models closely replicate human tumor biology and are useful for personalized medicine (Hutchinson and Kirk, 2011).
2. **Carcinogen-Induced Mouse Model**: This type of mouse model is developed by exposure to radiation, viruses, chemicals, and different physical factors. Cancer is developed in the mice models by the exposure to 4-(methylnitrosamino)-1-(3-pyridyl)-1-butanone (NNK), and this is an excellent way to investigate how smoking cigarettes causes lung cancer. For creating chemically induced lung cancer mice models, other substances including polyaromatic hydrocarbons like benzo(a)pyrene (BaP), N-nitroso-tris-chloroethylurea (NTCU), diethylnitrosamine, and 2-methylcholanthrene (MCA) are also frequently utilized (Kasala et al., 2015).
3. **Syngeneic Mouse Models**: Syngeneic mouse models are effective tools that enable immunocompetent mice to develop malignancies. They are developed by injecting compatible cancer cells into normal mice. Lewis lung carcinoma (LLC) is the only reproducible model present in syngeneic models (Sakai et al., 2006). Unlike the popular xenograft models, which involve implanting human cells into mouse tissue, the LLC model has the benefit that the implanted cells are immunologically compatible with the murine system. Even though the LLC model is the best animal model for lung cancer, it has a number of drawbacks. Responses tested in a whole murine system might not be applicable to human circumstances because it is a syngeneic model (Kellar et al., 2015).
4. **New generation PDX model**: This type of mouse model is generated by implanting stem cells, mononuclear cells, liver or thymus cells into immunocompromised mice. NSG and MISTRG mice are mainly used for this purpose because of their three key features: (a) Hemolytic complement deficit and a loss-of-function allele Sirpa, they also come from the NOD background, which lowers phagocytosis. (b) Reduction in T and B cells due to mutation in Prkdc gene which results in severe immunodeficiency. (c) Natural killer (NK) cell abnormalities and immunodeficiency are exacerbated by a mutation in the IL-2 receptor (Ishikawa et al., 2005). MISTRG mice are $Rag^{2-/-}\gamma c^{-/-}$ immunosuppressed mice that have had human genes supporting myeloid cell function knocked into their mouse

locus. These genes are responsible for the coding of Interleukin-3 (IL-3), thrombopoietin, granulocyte macrophage CSF (GM-CSF), human macrophage colony-stimulating factor (M-CSF), and a transgenic encoding human SIRPA (Walsh et al., 2017; Allen et al., 2019).

2.5 COMPUTATIONAL MODELS OF LUNG CANCER PROGRESSION

Computational models for lung cancer progression simulate the dynamic behavior of tumor cells and their interactions with the microenvironment. These models help understand how genetic mutations, tumor growth, and metastasis occur over time.

1. **Mathematical and Agent-Based Models (ABMs)**: These models simulate the interactions between individual cancer cells, immune cells, and the extracellular matrix. ABMs stimulate the spatial distribution of cells, allowing researchers to investigate the effects of mutations (e.g., KRAS, EGFR) on tumor growth and spread. These models can predict tumor dynamics, such as growth rates and the emergence of metastatic sites (Smieja, 2023).
2. **Network Models**: Modeling studies have focused on signaling pathways and regulatory networks linked to cancer progression, metastasis, response to treatment, and resistance to therapy. Gene expression networks or signaling pathway models are used to understand how mutations in key oncogenes (e.g., EGFR, BRAF) and tumor suppressors (e.g., TP53) drive cancer progression. These models simulate the effect of various mutations on cellular behavior and help identify potential therapeutic targets (Tong et al., 2022; Smieja, 2023) (Figure 2.2).

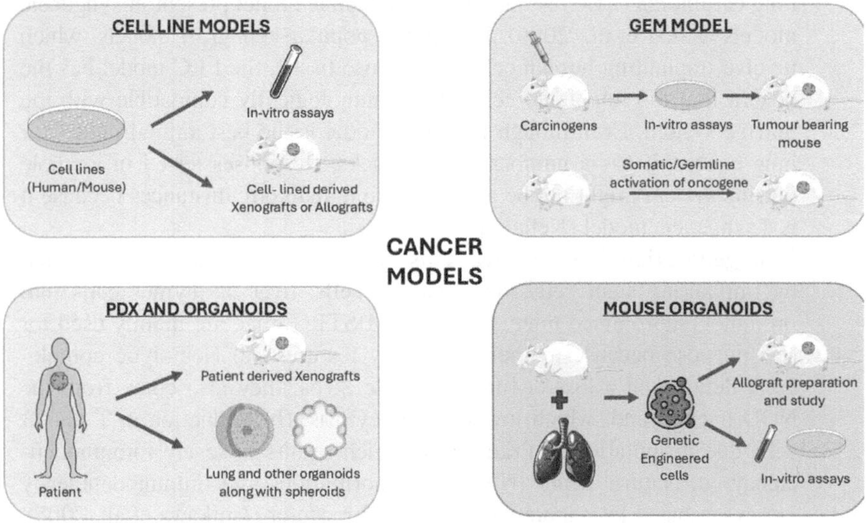

FIGURE 2.2 In vitro and in vivo models of lung cancer.

2.6 CONCLUSION

Models of lung cancer are essential for expanding our knowledge of the condition, enhancing early detection, and creating efficient treatment plans. These models, which range from intricate animal models and computational simulations to *in vitro* cell cultures, offer important insights into the molecular mechanisms underlying the progression and metastasis of lung cancer. They work in tandem to help researchers close the gap between laboratory results and clinical applications. Lung cancer models also support the development of personalized therapy by facilitating the testing of new therapeutic methods and the discovery of biomarkers for early diagnosis. But issues like tumor complexity, model heterogeneity, and the requirement for more realistic depictions of human illness still exist. These models must be continuously developed and improved in order to increase their clinical relevance and, eventually, patient outcomes.

REFERENCES

Allen, T. M., Brehm, M. A., Bridges, S., Ferguson, S., Kumar, P., Mirochnitchenko, O., Palucka, K., Pelanda, R., Sanders-Beer, B., Shultz, L. D., Su, L., & PrabhuDas, M. (2019). Humanized immune system mouse models: Progress, challenges and opportunities. *Nature Immunology*, 20(7), 770–774. https://doi.org/10.1038/s41590-019-0416-z

Barta, J. A., Powell, C. A., & Wisnivesky, J. P. (2019). Global epidemiology of lung cancer. *Annals of Global Health*, 85(1), 8. https://doi.org/10.5334/aogh.2419

Bray, F., Laversanne, M., Sung, H., Ferlay, J., Siegel, R. L., Soerjomataram, I., & Jemal, A. (2024). Global cancer statistics 2022: GLOBOCAN estimates of incidence and mortality worldwide for 36 cancers in 185 countries. *CA: A Cancer Journal for Clinicians*, 74(3), 229–263. https://doi.org/10.3322/caac.21834

Byers, L. A., & Rudin, C. M. (2015). Small cell lung cancer: Where do we go from here? *Cancer*, 121(5), 664–672. https://doi.org/10.1002/cncr.29098

Chaft, J. E., Dagogo-Jack, I., Santini, F. C., Eng, J., Yeap, B. Y., Izar, B., Chin, E., Jones, D. R., Kris, M. G., Shaw, A. T., & Gainor, J. F. (2018). Clinical outcomes of patients with resected, early-stage ALK-positive lung cancer. *Lung Cancer*, 122, 67–71. https://doi.org/10.1016/j.lungcan.2018.05.020

Chaitanya Thandra, K., Barsouk, A., Saginala, K., Sukumar Aluru, J., & Barsouk, A. (2021). Epidemiology of lung cancer. *Współczesna Onkologia*, 25(1), 45–52. https://doi.org/10.5114/wo.2021.103829

Chen, M. F., & Chaft, J. E. (2023). Early-stage ALK-positive lung cancer: A narrative review. *Translational Lung Cancer Research*, 12(2), 337–345. https://doi.org/10.21037/tlcr-22-631

D'Angelo, A., Sobhani, N., Chapman, R., Bagby, S., Bortoletti, C., Traversini, M., Ferrari, K., Voltolini, L., Darlow, J., & Roviello, G. (2020). Focus on ROS1-Positive non-small cell lung cancer (NSCLC): Crizotinib, resistance mechanisms and the newer generation of targeted therapies. *Cancers*, 12(11), 3293. https://doi.org/10.3390/cancers12113293

Denisenko, T. V., Budkevich, I. N., & Zhivotovsky, B. (2018). Cell death-based treatment of lung adenocarcinoma. *Cell Death & Disease*, 9(2), 117. https://doi.org/10.1038/s41419-017-0063-y

Dong, Z.-Y., Zhong, W.-Z., Zhang, X.-C., Su, J., Xie, Z., Liu, S.-Y., Tu, H.-Y., Chen, H.-J., Sun, Y.-L., Zhou, Q., Yang, J.-J., Yang, X.-N., Lin, J.-X., Yan, H.-H., Zhai, H.-R., Yan, L.-X., Liao, R.-Q., Wu, S.-P., & Wu, Y.-L. (2017). Potential predictive value of *TP53* and *KRAS* mutation status for response to PD-1 blockade immunotherapy in lung adenocarcinoma. *Clinical Cancer Research*, 23(12), 3012–3024. https://doi.org/10.1158/1078-0432.CCR-16-2554

Fukuda, H., Arai, K., Mizuno, H., Nishito, Y., Motoi, N., Arai, Y., Hiraoka, N., Shibata, T., Sonobe, Y., Kayukawa, Y., Hashimoto, E., Takahashi, M., Fujii, E., Maruyama, T., Kuwabara, K., Nishizawa, T., Mizoguchi, Y., Yoshida, Y., Watanabe, S., …, & Aoki, K. (2024). Molecular subtypes of lung adenocarcinoma present distinct immune tumor microenvironments. *Cancer Science*, 115(6), 1763–1777. https://doi.org/10.1111/cas.16154

Gupta, R., Dastane, A. M., Forozan, F., Riley-Portuguez, A., Chung, F., Lopategui, J., & Marchevsky, A. M. (2009). Evaluation of EGFR abnormalities in patients with pulmonary adenocarcinoma: The need to test neoplasms with more than one method. *Modern Pathology*, 22(1), 128–133. https://doi.org/10.1038/modpathol.2008.182

Han, S., & Wu, J. (2022). Three-dimensional (3D) scaffolds as powerful weapons for tumor immunotherapy. *Bioactive Materials*, 17, 300–319. https://doi.org/10.1016/j.bioactmat.2022.01.020

Hutchinson, L., & Kirk, R. (2011). High drug attrition rates—Where are we going wrong? *Nature Reviews Clinical Oncology*, 8(4), 189–190. https://doi.org/10.1038/nrclinonc.2011.34

Ishikawa, F., Yasukawa, M., Lyons, B., Yoshida, S., Miyamoto, T., Yoshimoto, G., Watanabe, T., Akashi, K., Shultz, L. D., & Harada, M. (2005). Development of functional human blood and immune systems in NOD/SCID/IL2 receptor γ chainnull mice. *Blood*, 106(5), 1565–1573. https://doi.org/10.1182/blood-2005-02-0516

Joshi, A., Mishra, R., Desai, S., Chandrani, P., Kore, H., Sunder, R., Hait, S., Iyer, P., Trivedi, V., Choughule, A., Noronha, V., Joshi, A., Patil, V., Menon, N., Kumar, R., Prabhash, K., & Dutt, A. (2021). Molecular characterization of lung squamous cell carcinoma tumors reveals therapeutically relevant alterations. *Oncotarget*, 12(6), 578–588. https://doi.org/10.18632/oncotarget.27905

Kasala, E. R., Bodduluru, L. N., Barua, C. C., Sriram, C. S., & Gogoi, R. (2015). Benzo(a) pyrene induced lung cancer: Role of dietary phytochemicals in chemoprevention. *Pharmacological Reports*, 67(5), 996–1009. https://doi.org/10.1016/j.pharep.2015.03.004

Kellar, A., Egan, C., & Morris, D. (2015). Preclinical murine models for lung cancer: Clinical trial applications. *BioMed Research International*, 2015, 1–17. https://doi.org/10.1155/2015/621324

Larsen, J. E., & Minna, J. D. (2011). Molecular biology of lung cancer: Clinical implications. *Clinics in Chest Medicine*, 32(4), 703–740. https://doi.org/10.1016/j.ccm.2011.08.003

Mao, C., Qiu, L.-X., Liao, R.-Y., Du, F.-B., Ding, H., Yang, W.-C., Li, J., & Chen, Q. (2010). KRAS mutations and resistance to EGFR-TKIs treatment in patients with non-small cell lung cancer: A meta-analysis of 22 studies. *Lung Cancer*, 69(3), 272–278. https://doi.org/10.1016/j.lungcan.2009.11.020

Mitsudomi, T., Viallet, J., Mulshine, J. L., Linnoila, R. I., Minna, J. D., & Gazdar, A. F. (1991). Mutations of ras genes distinguish a subset of non-small-cell lung cancer cell lines from small-cell lung cancer cell lines. *Oncogene*, 6(8), 1353–1362.

Nelson, M. A., Wymer, J., & Clements, N. (1996). Detection of K-ras gene mutations in non-neoplastic lung tissue and lung cancers. *Cancer Letters*, 103(1), 115–121. https://doi.org/10.1016/0304-3835(96)04202-4

Peifer, M., Fernández-Cuesta, L., Sos, M. L., George, J., Seidel, D., Kasper, L. H., Plenker, D., Leenders, F., Sun, R., Zander, T., Menon, R., Koker, M., Dahmen, I., Müller, C., Di Cerbo, V., Schildhaus, H.-U., Altmüller, J., Baessmann, I., Becker, C., …, & Thomas, R. K. (2012). Integrative genome analyses identify key somatic driver mutations of small-cell lung cancer. *Nature Genetics*, 44(10), 1104–1110. https://doi.org/10.1038/ng.2396

Perez-Moreno, P., Brambilla, E., Thomas, R., & Soria, J.-C. (2012). Squamous cell carcinoma of the lung: Molecular subtypes and therapeutic opportunities. *Clinical Cancer Research*, 18(9), 2443–2451. https://doi.org/10.1158/1078-0432.CCR-11-2370

Rodriguez-Canales, J., Parra-Cuentas, E., & Wistuba, I. I. (2016). Diagnosis and molecular classification of lung cancer. In Reckamp, K. L. (Ed.), *Lung Cancer* (Vol. 170, pp. 25–46). Springer International Publishing, New York. https://doi.org/10.1007/978-3-319-40389-2_2

Rolver, M. G., Elingaard-Larsen, L. O., & Pedersen, S. F. (2019). Assessing cell viability and death in 3D spheroid cultures of cancer cells. *Journal of Visualized Experiments*, 148, 59714. https://doi.org/10.3791/59714

Rudin, C. M., Brambilla, E., Faivre-Finn, C., & Sage, J. (2021). Small-cell lung cancer. *Nature Reviews Disease Primers*, 7(1), 3. https://doi.org/10.1038/s41572-020-00235-0

Ruiz-Cordero, R., Ma, J., Khanna, A., Lyons, G., Rinsurongkawong, W., Bassett, R., Guo, M., Routbort, M. J., Zhang, J., Skoulidis, F., Heymach, J., Roarty, E. B., Tang, Z., Medeiros, L. J., Patel, K. P., Luthra, R., & Roy-Chowdhuri, S. (2020). Simplified molecular classification of lung adenocarcinomas based on EGFR, KRAS, and TP53 mutations. *BMC Cancer*, 20(1), 83. https://doi.org/10.1186/s12885-020-6579-z

Sakai, Y., Sasahira, T., Ohmori, H., Yoshida, K., & Kuniyasu, H. (2006). Conjugated linoleic acid reduced metastasized LL2 tumors in mouse peritoneum. *Virchows Archiv*, 449(3), 341–347. https://doi.org/10.1007/s00428-006-0249-7

Salgia, R., & Skarin, A. T. (1998). Molecular abnormalities in lung cancer. *Journal of Clinical Oncology*, 16(3), 1207–1217. https://doi.org/10.1200/JCO.1998.16.3.1207

Shi, J., Gu, W., Zhao, Y., Zhu, J., Jiang, G., Bao, M., & Shi, J. (2020). Clinicopathological and prognostic significance of EML4-ALK rearrangement in patients with surgically resected lung adenocarcinoma: A propensity score matching study. *Cancer Management and Research*, 12, 589–598. https://doi.org/10.2147/CMAR.S229217

Shigematsu, H., & Gazdar, A. F. (2006). Somatic mutations of epidermal growth factor receptor signaling pathway in lung cancers. *International Journal of Cancer*, 118(2), 257–262. https://doi.org/10.1002/ijc.21496

Sholl, L. (2017). Molecular diagnostics of lung cancer in the clinic. *Translational Lung Cancer Research*, 6(5), 560–569. https://doi.org/10.21037/tlcr.2017.08.03

Singh, N., Agrawal, S., Jiwnani, S., Khosla, D., Malik, P. S., Mohan, A., Penumadu, P., & Prasad, K. T. (2021). Lung cancer in India. *Journal of Thoracic Oncology*, 16(8), 1250–1266. https://doi.org/10.1016/j.jtho.2021.02.004

Smieja, J. (2023). Mathematical modeling support for lung cancer therapy—A short review. *International Journal of Molecular Sciences*, 24(19), 14516. https://doi.org/10.3390/ijms241914516

Soda, M., Choi, Y. L., Enomoto, M., Takada, S., Yamashita, Y., Ishikawa, S., Fujiwara, S., Watanabe, H., Kurashina, K., Hatanaka, H., Bando, M., Ohno, S., Ishikawa, Y., Aburatani, H., Niki, T., Sohara, Y., Sugiyama, Y., & Mano, H. (2007). Identification of the transforming EML4–ALK fusion gene in non-small-cell lung cancer. *Nature*, 448(7153), 561–566. https://doi.org/10.1038/nature05945

Suzuki, M., Shigematsu, H., Hiroshima, K., Iizasa, T., Nakatani, Y., Minna, J. D., Gazdar, A. F., & Fujisawa, T. (2005). Epidermal growth factor receptor expression status in lung cancer correlates with its mutation. *Human Pathology*, 36(10), 1127–1134. https://doi.org/10.1016/j.humpath.2005.08.007

Suzuki, Y., Orita, M., Shiraishi, M., Hayashi, K., & Sekiya, T. (1990). Detection of ras gene mutations in human lung cancers by single-strand conformation polymorphism analysis of polymerase chain reaction products. *Oncogene*, 5(7), 1037–1043.

Takeuchi, K., Soda, M., Togashi, Y., Suzuki, R., Sakata, S., Hatano, S., Asaka, R., Hamanaka, W., Ninomiya, H., Uehara, H., Lim Choi, Y., Satoh, Y., Okumura, S., Nakagawa, K., Mano, H., & Ishikawa, Y. (2012). RET, ROS1 and ALK fusions in lung cancer. *Nature Medicine*, 18(3), 378–381. https://doi.org/10.1038/nm.2658

Tong, X., Tang, R., Xiao, M., Xu, J., Wang, W., Zhang, B., Liu, J., Yu, X., & Shi, S. (2022). Targeting cell death pathways for cancer therapy: Recent developments in necroptosis, pyroptosis, ferroptosis, and cuproptosis research. *Journal of Hematology & Oncology*, 15(1), 174. https://doi.org/10.1186/s13045-022-01392-3

Travis, W. D., Brambilla, E., Nicholson, A. G., Yatabe, Y., Austin, J. H. M., Beasley, M. B., Chirieac, Lucian. R., Dacic, S., Duhig, E., Flieder, D. B., Geisinger, K., Hirsch, F. R., Ishikawa, Y., Kerr, K. M., Noguchi, M., Pelosi, G., Powell, C. A., Tsao, M. S., & Wistuba, I. (2015). The 2015 world health organization classification of lung tumors. *Journal of Thoracic Oncology*, 10(9), 1243–1260. https://doi.org/10.1097/JTO.0000000000000630

Walsh, N. C., Kenney, L. L., Jangalwe, S., Aryee, K.-E., Greiner, D. L., Brehm, M. A., & Shultz, L. D. (2017). Humanized mouse models of clinical disease. *Annual Review of Pathology: Mechanisms of Disease*, 12(1), 187–215. https://doi.org/10.1146/annurev-pathol-052016-100332

Wankhede, D., Bontoux, C., Grover, S., & Hofman, P. (2023). Prognostic role of KRAS G12C mutation in non-small cell lung cancer: A systematic review and meta-analysis. *Diagnostics*, 13(19), 3043. https://doi.org/10.3390/diagnostics13193043

Warrier, N. E., Vallathol, D. H., Sankar, U. V., Salahuddin, S., Sreedharan, P. S., Parameswaran, P. V., & Sajeevan, K. V. (2022). EGFR mutations in lung carcinomas: Mutation profile and survival analysis. *Journal of Clinical Oncology*, 40(16_suppl), e20518–e20518. https://doi.org/10.1200/JCO.2022.40.16_suppl.e20518

Westcott, P. M. K., & To, M. D. (2013). The genetics and biology of KRAS in lung cancer. *Chinese Journal of Cancer*, 32(2), 63–70. https://doi.org/10.5732/cjc.012.10098

Zanoni, M., Piccinini, F., Arienti, C., Zamagni, A., Santi, S., Polico, R., Bevilacqua, A., & Tesei, A. (2016). 3D tumor spheroid models for *in vitro* therapeutic screening: A systematic approach to enhance the biological relevance of data obtained. *Scientific Reports*, 6(1), 19103. https://doi.org/10.1038/srep19103

Zhang, C., Sui, Y., Liu, S., & Yang, M. (2024). *In vitro* and *in vivo* experimental models for cancer immunotherapy study. *Current Research in Biotechnology*, 7, 100210. https://doi.org/10.1016/j.crbiot.2024.100210

Zou, S., Ye, J., Wei, Y., & Xu, J. (2023). Characterization of 3D-bioprinted *in vitro* lung cancer models using RNA-sequencing techniques. *Bioengineering*, 10(6), 667. https://doi.org/10.3390/bioengineering10060667

3 Colorectal Cancer Models

Soumya Aggarwal and Khushboo Kumari

3.1 INTRODUCTION

Colorectal cancer (CRC) is one of the main causes of cancer-related morbidity and death globally, affecting both men and women across diverse populations. It arises from the colon or rectum and is often preceded by the formation of benign polyps, which, through a series of genetic mutations and environmental influences, may evolve into malignant tumors. As the third most commonly diagnosed cancer globally, CRC continues to represent a significant public health concern despite advances in early detection and treatment (Bray et al., 2018). This chapter provides an overview of colorectal cancer, examining the molecular, environmental, and genetic variables that contribute to its development. The pathogenesis of CRC is complex, involving multiple genetic mutations, most notably in key tumor suppressor genes (such as APC) and oncogenes (such as Kirsten Rat Sarcoma Viral Oncogene Homolog (KRAS)), which drive tumor initiation, progression, and metastasis (Fearon & Vogelstein, 1990; Q. Li et al., 2024). Environmental factors, such as diet, physical activity, and inflammatory bowel disease, furthermore play a pivotal role in modulating the risk of CRC (Song & Chan, 2019).

Early detection remains critical for improving survival outcomes, and screening strategies, such as colonoscopy and fecal occult blood tests, have been pivotal in reducing CRC-related mortality through the identification and removal of precancerous lesions (J. Li et al., 2024). Furthermore, treatment options for CRC have expanded significantly, with multimodal approaches that combine surgery, chemotherapy, radiation, and newer targeted therapies (Wang et al., 2021; Grothey et al., 2013). The use of immunotherapy, especially immune checkpoint inhibitors, has been an innovative treatment avenue for specific subgroups of CRC patients, demonstrating promising results in advanced stages (Le et al., 2015).

As the field of colorectal cancer research continues to evolve, new information about the molecular mechanisms, novel therapeutic targets, and personalized treatment strategies offers hope for improving patient outcomes. This chapter will explore the biological foundations of CRC, the clinical manifestations and diagnostic techniques, as well as current and emerging therapeutic approaches.

3.2 EPIDEMIOLOGY AND PREVALENCE OF COLORECTAL CANCER

CRC is one of the most prevalent and impactful cancers globally. It continues to pose significant public health challenges, with notable variations in incidence, mortality, and survival rates across different regions, populations, and risk groups. Understanding the

DOI: 10.1201/9781003583349-4

global and regional epidemiology of CRC is crucial for shaping prevention strategies, improving screening programs, and optimizing treatment and management approaches. This section gives an outline of the epidemiology and prevalence of CRC, focusing on global trends, regional variations, risk factors, and the demographic aspects that influence the disease burden (concise data are shown in Table 3.1).

3.2.1 Global Epidemiology of Colorectal Cancer

With around 1.9 million new cases and 935,000 fatalities recorded worldwide in 2020, colorectal cancer is one of the main causes of cancer-related deaths worldwide (Bray et al., 2018). The occurrence of CRC varies widely between countries and regions, largely due to differences in socioeconomic development, lifestyle, healthcare infrastructure, and cancer-screening practices. As of recent reports, the second most prevalent cause of cancer death globally and the third most often diagnosed malignancy is CRC (Siegel et al., 2023).

GLOBOCAN 2020 estimates that the age-standardized incidence rate (ASIR) of CRC is 19.7 cases per 100,000 individuals worldwide (Bray et al., 2018). However, CRC is disproportionately more prevalent in countries with higher incomes as opposed to those with lower and intermediate incomes, where the disease burden is increasing as a result of urbanization, changes in diet, and decreased physical activity (Arnold et al., 2017).

3.2.2 Regional Variations in Incidence and Mortality

The global distribution of colorectal cancer is marked by distinct regional differences. CRC is most prevalent in North America, Europe, and Oceania, with these regions accounting for the highest global incidence rates.

In North America, CRC is responsible for about 8% of cancer cases, with an age-adjusted incidence rate of about 38.4 per 100,000 for males and 27.7 per 100,000 for females (Siegel et al., 2024). Similarly, Europe—particularly countries such as France, Germany, and Denmark—shows high rates of CRC incidence, although mortality rates have been declining due to the widespread implementation of screening programs (Ait Ouakrim et al., 2015).

In contrast, many Asian and African nations have lower rates of colorectal cancer, where CRC remains less common, though the rates are gradually rising, particularly in urbanized areas. For example, Japan and South Korea have seen significant increases in CRC incidence over the last few decades due to the adoption of Western dietary patterns, including increased consumption of meat and processed foods, combined with decreased physical activity (Wong et al., 2019). Similarly, China has also experienced an increase in CRC incidence in urban areas, reflecting both lifestyle changes and aging populations (Li et al., 2022).

In many low- and middle-income countries (LMICs), the incidence of CRC is still relatively low, but the rates are increasing due to changes in diet, reduced physical activity, and an aging population. These countries face additional challenges in diagnosing and treating CRC due to limited access to screening programs and healthcare services (Wong et al., 2019).

TABLE 3.1
Epidemiology and Prevalence of Colorectal Cancer

Category	Details
Global burden	In 2020, there were 1.9 million new cases and 935,000 fatalities worldwide.
Incidence rate	19.7 cases per 100,000 people worldwide
Most affected regions	High incidence in North America, Europe, and Oceania
Regional variations	High rates in developed countries; rising rates in urban areas of Asia and Africa
Trends in high-income countries	Declining or stabilizing incidence due to screening programs (e.g., colonoscopy, FOBT)
Trends in low- and middle-income countries	Increasing incidence due to urbanization, dietary changes, and reduced physical activity
Early-onset CRC	2.2% annual increase in cases among individuals under 50, particularly in Western nations
Sex differences	CRC rates 30%–40% higher in men than women
Ethnic disparities	Higher risk and worse outcomes in African Americans in the U.S.
Key risk factors	High red/processed meat intake, low-fiber diets, obesity, smoking, inflammatory bowel disease, genetic mutations (APC, KRAS, TP53)

3.2.3 Trends in Incidence and Mortality

While the overall global incidence of CRC continues to rise, the trends in different regions and subgroups can be quite varied. In recent decades, the prevalence of CRC has decreased or stabilized in high-income nations. This is primarily attributed to the implementation of effective screening programs, such as colonoscopy and fecal occult blood tests (FOBT), which detect precancerous lesions and allow for early intervention (Siegel et al., 2024). For example, in the United States, the age-adjusted incidence rate of CRC has decreased by approximately 3% per year from 2007 to 2016, thanks to these screening efforts (Siegel et al., 2024).

However, previously there was a concerning increase in the prevalence of early-onset CRC (in individuals under 50), especially in Western countries like the United States and parts of Europe. Studies have shown a 2.2% increase per year in the incidence of CRC among younger adults (Siegel et al., 2020). It is believed that lifestyle issues, including poor nutrition, obesity, and inactivity, are responsible for this tendency, which is increasingly affecting younger populations.

Conversely, in low- and middle-income countries, the rising trend of CRC incidence is due to the adoption of Western lifestyles, urbanization, and a rise in risk variables, including obesity, smoking, and alcohol consumption (Wong et al., 2019). Despite the increase in incidence, the mortality rates in these regions remain high, mainly because of late-stage diagnoses and the lack of early-detection measures.

3.2.4 Demographic Factors

Several demographic factors influence the possibility of developing CRC, including age, sex, family history, and ethnicity. Most of the CRC cases are diagnosed in individuals over the age of 50, although there is an increasing trend of CRC cases among younger adults, particularly in developed countries (Arnold et al., 2017). This shift in age distribution has sparked concerns about the factors contributing to early-onset CRC, including increasing obesity rates and changing dietary habits.

In terms of sex differences, men generally have a greater frequency of CRC in contrast to women, with studies showing that CRC rates are 30%–40% higher in men across various populations (Bray et al., 2018). Additionally, ethnic disparities are evident in the incidence and outcomes of CRC. For instance, African Americans in the United States are more susceptible to developing CRC, and they also tend to present with more advanced disease stages, leading to worse outcomes (Siegel et al., 2023).

The risk of CRC is greatly increased by family history and the existence of hereditary disorders like Lynch syndrome and familial adenomatous polyposis (FAP). People are two to three times more likely to develop CRC if they have a first-degree relative who has the illness (Lynch & de la Chapelle, 2003).

3.2.5 Risk Factors for Colorectal Cancer

Environmental and genetic factors work together to impact the development of CRC. The most significant environmental risk factors for CRC include dietary habits, particularly a high intake of red meat, processed meats, and low-fiber diets, which have been consistently linked to increased CRC risk (Korn et al., 2022).

Obesity, physical inactivity, and smoking contribute significantly to the risk, as they lead to chronic inflammation and insulin resistance, both of which are implicated in CRC pathogenesis (Calle & Kaaks, 2004). IBD, or inflammatory bowel disorders, includes Crohn's disease and ulcerative colitis, are well-established risk factors, as chronic inflammation in the colon increases the likelihood of cancer development (Xie & Itzkowitz, 2008). Gene mutations caused by genetics, including APC, KRAS, and Tumor Protein p53 (TP53), are pivotal in both familial and sporadic forms of CRC (Fearon & Vogelstein, 1990).

3.3 MOLECULAR SUBTYPES OF COLORECTAL CANCER

CRC has been divided into several molecular subtypes by recent research using genomic, transcriptomic, and epigenetic characteristics. These subtypes vary in their prognosis, response to treatment, and genetic alterations.

3.3.1 Chromosomal Instability (CIN) Subtype

The most widespread subtype of CRC, accounting for 70%–85% of all cases, is chromosomal instability (CIN). The hallmarks of CIN cancers include aneuploidy and extensive chromosomal abnormalities, including translocations, amplifications, and deletions. Mutations in TP53, KRAS, and APC are frequently observed in these tumors, with APC being a crucial early event in tumor initiation.

- **APC Mutations**: APC is a major initiator of carcinogenesis and is mutated in around 80% of CIN-driven colorectal cancers. APC depletion triggers the Wnt/β-catenin signaling cascade, which builds up β-catenin in the nucleus and activates transcription factors like cyclin D1 and c-Myc that promote cell division (Morin et al., 1997). This results in benign adenomas that eventually evolve into carcinoma by accumulating more mutations, such as those in TP53 and KRAS (Fearon & Vogelstein, 1990).

- **KRAS Mutations**: KRAS mutations can be discovered in approximately 35%–45% of CIN CRCs. Usually occurring in codons 12, 13, and 61, these mutations cause constitutive activation of downstream signaling pathways, such as mitogen-activated protein kinase (MAPK)/ERK and PI3K/AKT which promote tumor cell survival and proliferation (Boughdady et al., 1992).
- **TP53 Mutations**: About 50%–60% of tumors, especially those in advanced stages, have mutations in TP53, the most often modified tumor suppressor gene in CRC (Kandoth et al., 2013). Tumor growth is aided by TP53 inactivation, which causes resistance to apoptosis and a loss of checkpoint regulation in the cell cycle.

3.3.2 Microsatellite Instability (MSI)

When repeating DNA sequences (microsatellites) accumulate mistakes owing to flawed DNA mismatch repair (MMR), this is known as microsatellite instability (MSI). MSI is linked to mutations in MLH1, MSH2, MSH6, and PMS2, which cause MMR, and it occurs in roughly 15%–20% of CRCs. In addition to having a high mutational burden, MSI tumors frequently have distinctive clinical characteristics, such as heightened immune checkpoint inhibitor (ICI) responsiveness.

- **BRAF Mutations**: MSI CRCs frequently have the BRAF V600E mutation (Davies et al., 2002). One serine/threonine kinase that triggers the MAPK/ERK signaling pathway is called BRAF. Constitutive activation of this mechanism due to the V600E mutation causes unchecked cell growth. BRAF V600E mutations are closely linked to poor prognosis and treatment resistance (Guerrero et al., 2022).
- **CIMP**: Widespread DNA hypermethylation at CpG islands, which silences tumor suppressor genes such as MLH1, CDKN2A, and MGMT, is a characteristic of CIMP-positive CRCs. A characteristic of BRAF V600E-mutant CRCs, CIMP is frequently linked to MSI tumors and represents a separate population with an unfavorable outlook and a distinctive response to treatment (Hinoue et al., 2009).
- **Immune Response**: MSI-H CRCs are especially sensitive to immune checkpoint blockade therapy because of their high mutation rate, which causes them to frequently show large quantities of neoantigens. In MSI-H CRCs, medications that target PD-1, like pembrolizumab and nivolumab, have demonstrated a substantial therapeutic benefit (Le et al., 2015).

3.3.3 CpG Island Methylator Phenotype (CIMP)

A molecular subtype referred to as the CpG island methylator phenotype (CIMP) is defined by the extensive hypermethylation of CpG islands in the promoter regions of tumor suppressor genes. Most frequently, CIMP is linked to the MSI subtype, especially in malignancies with a BRAF V600E mutation. Additionally, proximal tumor site, poor differentiation, and resistance to traditional therapy are commonly linked to these cancers (Ogino et al., 2009).

3.4 KEY ONCOGENES AND TUMOR SUPPRESSORS IN COLORECTAL CANCER

Colorectal cancer's growth is largely driven by key mutations in particular oncogenes and tumor suppressor genes. Important roles are played by APC, KRAS, BRAF, and TP53 among them (short and concise data are shown in Table 3.2).

3.4.1 APC (Adenomatous Polyposis Coli)

As a tumor suppressor, the Adenomatous Polyposis Coli (APC) gene controls the Wnt/β-catenin signaling cascade. Target genes that encourage cell proliferation are activated when APC mutations cause β-catenin to accumulate in the nucleus. Most CRCs begin with loss of APC, which causes colorectal adenomas to grow. Over time, these adenomas can develop into carcinoma (Fearon & Vogelstein, 1990). The conventional concept of the genesis of colorectal cancer, the adenoma-carcinoma sequence, is also related to the APC mutation.

Wnt/β-Catenin Pathway: When APC is lost, β-catenin signaling is triggered, which promotes the transcription of genes linked to proliferation, such as cyclin D1 and c-Myc. Nearly all CIN CRCs have dysregulation of this mechanism.

3.4.2 KRAS (Kirsten Rat Sarcoma Viral Oncogene Homolog)

An oncogene called KRAS encodes a little GTPase that modulates downstream signaling via the PI3K/AKT and MAPK/ERK pathways. A poor prognosis is linked

TABLE 3.2
The Key Oncogenes and Tumor Suppressors in Colorectal Cancer (CRC)

Gene	Type	Function	Mutation Effects	Clinical Impact
APC	Tumor Suppressor	Regulates Wnt/β-catenin signaling	Loss leads to β-catenin accumulation, promoting cell proliferation	Initiates adenoma formation; present in most CIN CRCs
KRAS	Oncogene	Activates MAPK/ERK & PI3K/AKT pathways	Constitutive activation leads to uncontrolled cell growth	Found in 35%–45% of CRCs; causes anti-EGFR therapy resistance
BRAF	Oncogene	Part of MAPK/ERK pathway	V600E mutation drives tumor progression	Present in 5%–10% of CRCs, mostly MSI; linked to poor prognosis & chemotherapy resistance
TP53	Tumor Suppressor	Controls cell cycle, DNA repair, apoptosis	Loss leads to genomic instability & resistance to apoptosis	Found in 50%–60% of CRCs, particularly in advanced stages

to KRAS mutations, which are present in 35%–45% of CRCs. These mutations cause signaling pathways to be constitutively activated, which encourages unchecked cell survival and proliferation (Zhu et al., 2021). KRAS mutations: Constitutive PI3K and MAPK pathway activation results from mutations in KRAS (exons 2, 3, and 4), which are most frequently detected at codons 12 and 13. Resistance to epidermal growth factor receptor (EGFR)-targeted treatments like cetuximab and panitumumab is also linked to these mutations (Porru et al., 2018).

3.4.3 BRAF (B-Raf Proto-Oncogene, Serine/Threonine Kinase)

The MAPK/ERK pathway includes the serine/threonine kinase BRAF. Five to ten percent of CRCs, primarily of the MSI subtype, have mutations in BRAF, namely the V600E substitution. The MAPK pathway is constitutively energized by the V600E mutation, promoting cell survival and proliferation (Davies et al., 2002).

BRAF V600E mutation: Chemotherapy resistance and a poor prognosis are commonly linked to colorectal cancers with this mutation. Nonetheless, in clinical studies, combination treatments involving BRAF inhibitors (e.g., vemurafenib, dabrafenib) and MEK inhibitors (e.g., trametinib) have displayed potential (Potocki & Wysocki, 2018).

3.4.4 TP53 (Tumor Protein p53)

One important tumor suppressor, the TP53 gene, controls apoptosis, DNA repair, and cell cycle checkpoints. About 50%–60% of CRCs have TP53 mutations, particularly those that are advanced in stage. Tumor growth is fueled by the accumulation of new mutations and resistance to apoptosis, which are linked to the loss of TP53 function (Kandoth et al., 2013).

3.5 SIGNALING PATHWAYS IMPLICATED IN COLORECTAL CANCER

The emergence of colorectal cancer is caused by several dysregulated signaling pathways, such as the TGF-β, MAPK/ERK, PI3K/AKT, and Wnt/β-catenin pathways. By encouraging uncontrolled cell division, survival, and metastasis, these pathways contribute to the growth of tumors.

3.5.1 Wnt/B-Catenin Signaling

Several physiological functions, tissue homeostasis, and embryonic development depend on the intricate and highly conserved Wnt signaling system. The transcriptional co-activator β-catenin, which is linked to the Wnt signaling pathway, is one of the genes whose expression levels are impacted by the presence or absence of the ligand Wnt (as shown in Figure 3.1a). When Wnt is absent, β-catenin is kept low because the ubiquitin-dependent proteasome breaks it down and the Axin/GSK3/APC/CK1 complex phosphorylates it. Nevertheless, Wnt activates the Wnt signaling pathway when it interacts with the Frizzled receptor (Lee et al., 2024). APC) and β-catenin in particular have been identified as typical core genes with common genetic changes in CRC within this signaling pathway, that have been demonstrated to have a major influence on the disease's genesis and progression. APC loss stabilizes β-catenin, which builds up in the nucleus and triggers transcription factors like cyclin D1 and c-Myc that support cell division. In the majority of CRCs, especially those with CIN, this pathway is dysregulated (He & Gan, 2023).

3.5.2 PI3K/AKT Signaling

Gene transcription, DNA repair, apoptosis, proliferation, and survival are just a few of the critical intracellular signaling cascades that are regulated by the PI3K/AKT/mTOR signaling pathway (as shown in Figure 3.1b). Based on their structure and particular substrates, the conserved lipid kinase family, which includes PI3K, is split into 3 categories (Class I, II, and III). Based on the kinds of catalytic and regulatory subunits, class I PI3K may be further separated into classes IA and IB. The catalytic subunits p110 (p110α, p110β, and p110δ) and the regulatory subunits p85 (p85α, p55α, p50α, p85β, and p55γ) make up class IA. However, class IB PI3K is composed of the regulatory subunit p101 and the catalytic subunit p110γ (Q. Li et al., 2024). Receptors include receptor tyrosine kinases and a variety of growth factors, such as fibroblast growth factor and vascular endothelial growth factor, can activate PI3K downstream. Activated PI3K phosphorylates phosphatidylinositol 4,5-bisphosphate to transform it into phosphatidylinositol 3,4,5-trisphosphate, which then triggers AKT (Xue et al., 2021). When active AKT phosphorylates additional substrates in the cytoplasm and nucleus, it promotes the activation of mTOR and activates the PI3K/AKT/mTOR signaling pathway. Specifically, this signaling pathway is essential for CRC in both its early and late phases, and its dysregulation alters angiogenesis, apoptosis, and proliferation, all of which contribute to carcinogenesis (Danielsen et al., 2015).

3.5.3 MAPK/ERK Signaling

KRAS, BRAF, and EGFR mutations frequently cause the MAPK/ERK pathway to become active in CRCs. Cell division, proliferation, survival, and death are among the many cellular processes that are regulated by the MAPK signaling pathway, an essential intracellular signaling system. A ligand's binding to receptor tyrosine kinases triggers this signaling pathway by phosphorylating and activating MAPK kinase kinase (MAP3K) (Q. Li et al., 2024) (as shown in Figure 3.1c). In order to influence gene expression and transfer signals from cell surface receptors to the nucleus, active MAP3K phosphorylates MAPK kinase (MAP2K), which phosphorylates MAPK via activating a number of kinases, enzymes, and transcription factors. The three primary families of MAPK are p38, c-Jun N-terminal kinase (JNK), and extracellular signal-regulated kinase 1/2 (ERK1/2). Through their individual or combined actions within the particular cellular environment and signaling needs of colorectal cancer, these families have been found to be important contributors to the formation, progression, and carcinogenic effects of CRC. Cell survival, differentiation, and division are all regulated by this route. BRAF V600E mutations in BRAF-mutant CRCs cause the MAPK/ERK cascade to be continuously activated, which encourages cell division and carcinogenesis (Peker et al., 2018).

3.5.4 TGF-B Signaling

Signaling by TGF-β is a complicated and intricate route that is essential for controlling immunological responses, homeostasis, cell development, and differentiation (as shown in Figure 3.1d). By attaching to its receptor, TGF-βR (TGF-βR I, TGF-βR II, and TGF-βR III), TGF-β (TGF-β I, TGF-β II, and TGF-β III) activates this pathway. TGF-β I and TGF-β III both directly link to TGF-βR II, however, TGF-β II can connect with TGF-βR II with the assistance of TGF-βR III (Q. Li et al., 2024). Cell differentiation and proliferation are largely controlled by the TGF-β pathway. TGF-βR

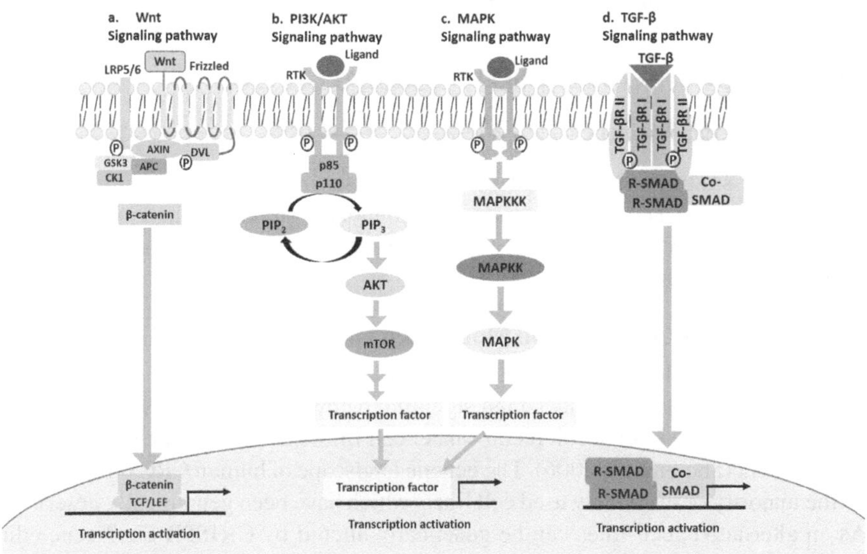

FIGURE 3.1 Activated signaling pathway schematic diagram: (a) The presence of Wnt activates the Wnt signaling pathway. The Axin/GSK3/APC/CK1 complex is attracted to the cell membrane when Wnt dimerizes with LRP5/6 after binding to the Frizzled receptor. This leads to the following transfer of β-catenin from the cytoplasm to the nucleus. β-catenin interacts with TCF/LEF transcription factors in the nucleus to cause downstream target genes to be expressed. (b) The binding of a ligand to receptors like RTK triggers the PI3K/AKT/mTOR signaling pathway. PIP2 is phosphorylated by activated PI3K, which then changes it into PIP3. Next, downstream genes like AKT and mTOR are activated by PIP3. (c) When a ligand attaches to RTK, MAP3K is phosphorylated and activated, starting the MAPK signaling cascade. When MAP3K is activated, it phosphorylates downstream MAP2K, which phosphorylates MAPK. Through the downstream activation of many kinases, enzymes, and transcription factors, this phosphorylation cascade influences gene expression by transmitting signals to the nucleus from the cell surface receptor. This phosphorylation cascade affects gene expression by transmitting signals from the cell surface receptor to the nucleus, which in turn activates more kinases, enzymes, and transcription factors downstream. (d) TGF-β can bind to TGF-βR II directly or indirectly, which activates the TGF-β signaling pathway. TGF-βR II phosphorylates and recruits TGF-βR I, which phosphorylates R-SMAD. Subsequently, Co-SMAD forms a trimeric complex with the phosphorylated R-SMAD. After entering the nucleus, this complex acts as a transcription factor, influencing how genes are expressed. (This figure is adapted from Lee et al., 2024.)

II then phosphorylates and recruits TGF-βR I to activate it. The signaling cascade is then started by the phosphorylation and activation of receptor-regulated SMADs (R-SMADs, SMAD 1, SMAD 2, SMAD 3, SMAD 5, and SMAD 8). R-SMAD then interacts with the common-mediator SMAD (Co-SMAD, SMAD 4) to create a trimeric complex. As transcription factors, this complex moves into the nucleus and attaches itself to DNA binding sites to carry out its function in gene expression. By stopping cell division, TGF-β suppresses tumors early in colorectal cancer development. But mutations in SMAD4 and TGFBR2, which are frequently observed in advanced colorectal cancer, cause TGF-β to lose its tumor-suppressive properties, which encourages tumor growth and metastasis (Itatani et al., 2019).

3.6 VARIOUS MODELS USED TO RESEARCH CRC

3.6.1 In-Vitro Models

Therapeutic studies in cell-based models are usually the first steps in the drug discovery process. The two most used in vitro models are human and mouse cancer cell lines. Strong in vitro systems are essentially required for mechanistic studies that identify cancer treatment targets.

3.6.1.1 Human Cell Lines and Mouse Cell Lines

After the human HeLa cervical cancer line was developed in 1951, cell lines have been crucial in helping to understand signaling pathways in cancer. Numerous commercially available human colorectal cancer cell lines have been widely used in drug development (Shoemaker, 2006). The genetic landscape of human CRC is represented by the majority of the widely used cell lines, which have been genomically described. As an alternative, cell lines can be genetically altered by CRISPR-Cas9 gene editing, homologous recombination, or short hairpin RNA (shRNA) gene knockdown (Golovko et al., 2015). Numerous tests that assess antitumor effectiveness can be automated, which facilitates drug discovery by allowing several compounds to be tested simultaneously against several cell lines.

Additionally, CRC cell lines can also be obtained from mice. MC38, an adenocarcinoma cell line produced from a C57BL/6 mouse (Corbett et al., 1975), and CT26, obtained from a BALB/c mouse (Brattain et al., 1980), are two of the frequently utilized mouse cell lines. The carcinogens N-nitro-N-methylurethane and 1,2-dimethylhydrazine dihydrochloride were repeatedly administered rectally and subcutaneously, respectively, to produce these cell lines in the 1970s. They are easily available from repositories and are easy to culture. However, with many advantages, cancer cell lines also have several significant drawbacks. Naturally chosen to multiply in culture plates and other media, cell lines are probably quite different from the original tumor. The functional and genetic variability of human cancers is not replicated by cell lines, and this is a major contributing reason to resistance for targeted therapy (Gerlinger et al., 2012). Lastly, cell lines cannot be produced from corresponding normal tissue and are challenging to generate from specific patient tumors (Chen et al., 1987). For this reason, conventional cancer cell lines are not ideal for individualized clinical use. There are potentially serious disadvantages to using mouse CRC cell lines in therapeutic development. It is typically not possible to obtain mouse cell lines that mimic less frequent cancer mutations.

3.6.1.2 Organoids

An organoid is a three-dimensional, reduced-size, simplified representation of an organ, created in vitro that replicates the main biological, structural, and functional features of that organ. It is composed of one or a few tissue cells, embryonic stem cells, or induced pluripotent stem cells that may self-organize in three-dimensional culture due to their capacity for differentiation and self-renewal.

The Clevers lab recently published a three-dimensional culture model that involves cultivating murine intestinal crypts, which contain self-renewing stem cells and adjacent niche Paneth cells, using special media containing Wnt3, the Wnt activator R-spondin1, the BMP inhibitor Noggin, and other growth factors (Sato et al., 2009). The crypts develop into organoid structures known as "mini-intestines," which house both the differentiated intestinal cell types and stem cells. It is also possible to cultivate human intestine and colon tissue into three-dimensional organoids using surgical resections, colonoscopy biopsy materials, or even individual human $EphB^{2+}$ stem cells (Jung et al., 2011). In contrast to mouse crypts, human intestinal crypts are more difficult to grow and require supplementation with Wnt, R-spondin1, and Noggin in addition to p38 and Alk pathway inhibitors for long-term culture (Sato et al., 2011). Because human colon biopsies can produce organoids that can double in population at least 100 times in culture, there are plenty of them for experimental investigation. For in vitro drug testing, the Clevers lab has created a library of organoids derived from human colorectal cancers and a corresponding normal colon (Sato & Clevers, 2013). Agents that are effective in the tumor organoid but non-toxic to the normal organoid can be chosen for additional research. Organoid culturing of both healthy and malignant colorectal tissue is therefore a new way to convert tumor genetic information into tailored treatment. Additionally, it is possible to cultivate organoids from primary and metastatic sites to uncover pathways or from various tumor sections to represent tumor heterogeneity.

3.6.2 In-Vivo Transplant Models

Even though in vitro cancer models have numerous benefits, in vivo systems are necessary to evaluate how the host immune system, angiogenesis, and tumor microenvironment (TME) affect the tumor's response to treatment. The murine xenograft is the most often used in vivo model. To address the shortcomings of the xenograft, orthotopic transplant models and patient-derived xenografts have recently been created.

3.6.2.1 Cell Line Xenograft

Subcutaneous injections of human colorectal cancer cells into immunocompromised mice typically result in the development of a tumor at the injection site. Classically, nude (athymic) and severe combined immunodeficient (SCID) mice are frequently employed; these mice lack T cells (Croy et al., 2001) or both B and T lymphocytes (Blunt et al., 1996).

Xenograft models have serious drawbacks. When a big, homogeneous cell population is injected to create a tumor, it does not represent the cellular heterogeneity seen in the majority of human colorectal cases (CRCs) (Golovko et al., 2015). The crucial function that the host immune system plays in carcinogenesis is eliminated when immunodeficient mice are used. The species difference makes it impossible to investigate the interactions between cancer cells and stromal cells. Furthermore, xenograft cells are usually injected into the subcutaneous area, a milieu that differs greatly from the gut. The histological characteristics of human cancer are not replicated in tumors produced from transplanted CRC cell lines (Voskoglou-Nomikos et al., 2003).

3.6.2.2 Organoid Xenografts

There have been reports of xenograft transplantation of Apc-deficient mouse colon organoids with extra driver mutations. Despite adenoma organoids lacking in Apc by themselves, they do not successfully engraft, tumors with histology resembling invasive adenocarcinoma are produced by combinatorial mutations of Apc, Tp53, KrasG12D, and Smad4 (Li et al., 2014). Furthermore, only metastatic CRC organoids effectively spread to the liver, not designed organoids (Matano et al., 2015). In a separate study, the Clevers group discovered that effective tumor growth in human small intestine organoid xenografts requires CRISPR-mediated deletion of APC, TP53, and SMAD4, as well as activation of KRASG12D. All colon organoids that had mutations in APC, KRAS, and TP53 (with or without SMAD4 deletion) produced adenomas after transplantation. Only xenografts with all four mutations showed localized invasive carcinoma in histology; xenografts with fewer mutations developed cystic formations devoid of crucial characteristics of colon adenocarcinoma (Drost et al., 2015). These investigations show how CRC organoid xenografts may be used for research and individualized treatment. However, low engraftment rates for less stage tumors may restrict the clinical value of CRC xenograft engraft.

3.6.2.3 Patient-Derived Xenografts

The Patient-Derived Xenograft (PDX) paradigm involves implanting parts of the patient's tumor tissue that were taken during surgery into the immunodeficient mice (P0). The tumor is surgically excised and reinserted into another mice (P1) once it has grown. Until enough animals (i.e., P2, P3, etc.) are obtained for the experiment, this procedure is repeated. In contrast to conventional xenografts, tumor-stroma cross-talk is possible in PDXs because the tumor stroma grows alongside the tumor cells (Rosfjord et al., 2014). Compared to monoclonal xenografts, the tumor's cellular and molecular heterogeneity is more noticeable in PDXs. Important features of primary CRCs like the vascularity, architecture, and tumor histology are preserved by PDXs.

Compared to conventional xenografts, PDX models are more accurate at predicting clinical response to therapy (Siolas & Hannon, 2013). When there is no defined treatment plan for patients with unusual tumor forms or combinations of driver mutations, or when patients are resistant to existing treatments, PDXs may play a significant role in their care (Morelli et al., 2012). Though their promise is exciting, PDX models have significant drawbacks. As with traditional xenografts, PDX models require host animals with weakened immune systems. PDX has the potential to be used in customized treatment, however, its therapeutic relevance may be limited due to its reported tumor start rate of about 70% (Aparicio et al., 2015). Furthermore, it might take six months or longer to implant the tumor and screen for possible treatments. Lastly, neither government nor commercial health insurers usually cover PDX models due to their high cost for clinical use.

3.6.2.4 Orthotopic Transplant Models

A significant drawback of conventional and PDX is that the subcutaneous space's microenvironment is very different from the colon's. Consequently, models of orthotopic transplants have been developed in which CRC cells are inserted straight into the colon. Tumor cells can be directly implanted into the cecum's serosal surface during a laparoscopy. In addition to a primary tumor, this method may cause metastases to the liver, lung, and local lymphatics (Céspedes et al., 2007). After transplantation, optic colonoscopy can be used to monitor tumors created by orthotopic transplantation.

3.6.2.5 Carcinogen-Induced Mouse Models

Mice exposed to 1,2-dimethylhydrazine (DMH) or its metabolite, azoxymethane (AOM), can develop colon cancer. In the liver, AOM undergoes N-oxidation and hydroxylation before being eliminated in the bile (Reddy et al., 1974). Microbial flora in the colon subsequently converts it into a form that encourages the development of cancer. When AOM is administered, tumors frequently exhibit mutations in the genes for beta-catenin (Ctnnbl) and Kras (Kobaek-Larsen et al., 2000). However, this model rarely shows distant metastases or mucosal invasion. MC38, a frequently used mouse CRC cell line, was isolated from a tumor generated by DMH(Corbett et al., 1975). One drawback of AOM is that tumor formation may need a maximum 30-week delay time. Additionally, dextran sulfate sodium (DSS), a drug that causes colitis, can be used with AOM (Neufert et al., 2007). This combination reduces the latency period to 10 weeks and is used to simulate CRC linked to inflammatory bowel illness, which has a nearly 100% tumor incidence (Clapper et al., 2007). Although DMH and AOM are the most well-characterized carcinogens, additional carcinogens such as heterocyclic amines, aromatic amines, and alkylnitrosamide chemicals are also utilized, either alone or in combination with DSS, to imitate CRC (Nishikawa et al., 2005).

There are several benefits to carcinogen-induced models, including low multiplicity, easy and affordable tumor start, excellent repeatability, visual colonoscopy for monitoring distal colon cancers, and the capacity to create tumors in mice with varying genetic backgrounds. But rather than sporadic CRC, the AOM/DSS model would be more suited as a model of CRC linked to inflammatory bowel illness, which accounts for a small percentage of CRCs (Golovko et al., 2015).

3.6.2.6 Genetically Engineered Mouse Models

Mice that have undergone genetic engineering are effective models because they replicate the development of adenoma or carcinoma in the natural, immunocompetent colon milieu. Since the Apc^Min mouse was first discovered in 1990, the number of genetically modified animals for colon cancer modeling has grown significantly.

Hereditary nonpolyposis colorectal cancer (HNPCC), commonly referred to as Lynch syndrome, is brought on by mutations in DNA mismatch repair genes, mainly in MSH2 and MLH1 (Edelmann et al., 2000). These mutations are typified by microsatellite instability. Mice lacking Mlh1, Msh2, Msh3, Msh6, or Pms2 develop skin cancers, lymphomas, and small intestinal tumors. Due to their high levels of microsatellite instability, these models might be beneficial for treating human cancer subtypes like medullary, mucinous, undifferentiated, and signet-ring carcinomas (Boivin et al., 2003).

By conditionally targeting gene deletion or activation to the intestine, the Cre recombinase system helps prevent lethality and extraintestinal symptoms that germline mutations may cause. This can be accomplished by mating *Apc* floxed mice with mice that have an intestinal Cre. Villin-Cre mice, which are constitutive and tamoxifen-inducible, are frequently used to model intestinal carcinogenesis and restrict *Apc* and other gene loss to the intestinal epithelium (el Marjou et al., 2004).

Additional Cre mice lines that are used to simulate CRC with different polyp burdens in the colon and small intestine include Fabp-Cre (Robanus-Maandag et al., 2010), Ah-Cre (Sansom et al., 2004) (Cyp1A promoter), CDX2P-NLS-Cre (Hinoi et al., 2007), and CAC-Cre (Xue et al., 2010) (carbonic anhydrase I gene promoter, expressed only in the large intestine).

3.6.3 *In-silico* Models

In silico modeling refers to the use of computational simulations to study biological systems. In the context of CRC, these models replicate tumor behavior, interactions within the tumor microenvironment, and responses to various treatments. The integration of patient-specific data into these models allows for personalized medicine approaches, enhancing the precision of therapeutic interventions. The continuous evolution of computational techniques promises to further enhance the precision and efficacy of CRC management.

3.6.3.1 Molecular Docking and Dynamics Simulations

Molecular docking and dynamics simulations are computer methods that forecast how compounds, such as medications, will interact with their targets. These techniques simulate the binding affinity and stability of prospective therapeutic compounds with certain cancer-related proteins, which aids in the identification of such compounds in CRC research.

Resveratrol Derivatives as Potential CRC Inhibitors: The APC protein, a major contributor to the formation of CRC, was the target of molecular docking investigations of resveratrol derivatives. Strong binding affinities were found in several of the compounds, indicating that they may have use as therapeutic agents. Molecular dynamics simulations further validated the durability of these interactions, highlighting the potential of medicines based on resveratrol in the treatment of CRC(Akash et al., 2024).

3.6.3.2 Personalized Therapeutics through Computational Modeling

The goal of personalized medicine is to modify care according to each patient's unique traits. By mimicking tumor behaviors and treatment responses unique to each patient, in silico models make this possible.

In Silico Drosophila Patient Model (DPM): The molecular control of Drosophila midgut epithelial cells was represented by a Boolean network model that the researchers created. They developed an in silico drug combination prediction model (DPM) by using mutation data unique to CRC patients. This approach effectively found synergistic therapy pairings, including regorafenib and paclitaxel, providing a guide for individualized CRC treatment plans (Gondal et al., 2021).

3.6.3.3 Target-Based Small Molecule Drug Discovery

It is essential for successful medication development to identify and target certain molecular pathways implicated in the growth of CRC. Because they can simulate interactions at the molecular level, in silico techniques are essential to this process.

Review of Molecular Pathways and in Silico Studies: A thorough analysis emphasized the use of in silico techniques to pinpoint important biological pathways linked to CRC. In order to speed up the development of targeted medicines, the study highlighted the importance of computer modeling in identifying tiny compounds that can efficiently target these pathways (Moshawih et al., 2022).

3.6.3.4 Artificial Intelligence and Radiomics in CRC

The development of radiomics and artificial intelligence (AI) has transformed medical imaging analysis, opening the door to noninvasive CRC prognostics and diagnosis.

Radiomics-Based AI Models: The use of AI models that examine radiomic characteristics taken from imaging modalities such as CT and MRI images has advanced recently. By helping to differentiate between benign and malignant lesions, evaluate tumor stages, and forecast treatment outcomes, these models improve the accuracy of CRC care (Karami & Elahi, 2024).

3.6.3.5 Mathematical Modeling of Tumorigenesis

In CRC, mathematical models offer a framework for comprehending the intricate dynamics of tumor development and therapy response.

Modeling the MAPK Pathway: A research study offered a mathematical model that concentrated on the MAPK pathway, which is essential for the survival and proliferation of cells. The model demonstrated the value of mathematical techniques in CRC research by giving insights into tumor behavior and possible treatment approaches by simulating different mutations and therapeutic treatments within this pathway (Sommariva et al., 2023).

3.6.3.6 Integration of Genomic and Morphological Data

To better understand CRC subtypes and their behaviors, morphological analysis and genetic data are combined.

Deep Learning for Genomic and Morphological Correlation: Researchers investigated the connection between cellular morphology and genetic variations in CRC subtypes using deep learning techniques. This method helped with the categorization and treatment planning of CRC by making it easier to identify clear patterns that link genetic alterations with histopathological characteristics (Hezi et al., 2024).

A thorough grasp of tumor biology, drug discovery, and the creation of customized therapies are all made possible by in silico models, which are essential to the advancement of CRC research. It is anticipated that the ongoing development of computational methods will improve the accuracy and effectiveness of CRC treatment.

3.7 TUMOR HETEROGENEITY IN COLORECTAL CANCER

Genetic and Epigenetic Diversity: The significant genetic and epigenetic heterogeneity that defines CRC complicates diagnosis, prognosis, and treatment. Mutations in important oncogenes (including KRAS, PIK3CA, and BRAF) and tumor suppressor genes (such as APC, TP53, and SMAD4) dominate the genetic landscape of CRC. Histone modifications and DNA methylation are examples of epigenetic changes that introduce an additional level of intricacy. A thorough understanding of these changes and their part in CRC development is made possible by the combination of whole-genome sequencing (WGS), exome sequencing (WES), and DNA methylation profiling within systems biology frameworks (Malki et al., 2020).

Tumor Microenvironment (TME): Comprising immune cells, stromal cells, endothelial cells, and extracellular matrix components, the TME has a major impact on the development of CRC and treatment resistance. Transcriptomic, proteomic, and metabolomic information from the TME is integrated by systems biology to comprehend how these elements work together to promote tumor growth and immune evasion (Wu et al., 2023).

Inter-patient heterogeneity in CRC is driven by genetic, molecular, and phenotypic differences, influencing prognosis and treatment response. The majority (85%) of sporadic CRCs arise from adenomas and exhibit CIN, microsatellite stability (MSS), and KRAS mutations without BRAF mutations. The remaining 15% originate from serrated adenomas, characterized by (CIMP-high, MSI, and BRAF mutations. MSI-high (MSI-H) tumors, although less common, have a better prognosis in early-stage disease (Molinari et al., 2018).

Tumor location also plays a role in heterogeneity. Right-sided tumors are linked to a higher frequency of BRAF mutations, MSI, and genetic predispositions such as Lynch syndrome, while left-sided tumors exhibit CIN and EGFR pathway activation. Right-sided tumors also have a higher mutation burden and distinct inflammatory and angiogenic marker profiles compared to left-sided tumors. Additionally, microbial composition varies by tumor location, further contributing to tumor behavior and treatment response. CRC can be classified into four consensus molecular subtypes (CMS): CMS1 (MSI-immune), CMS2 (epithelial with WNT/MYC activation), CMS3 (metabolic, KRAS-mutated), and CMS4 (mesenchymal, associated with poor prognosis). The TME, including stromal, inflammatory, and microbial influences, contributes to further genomic instability and phenotypic diversity. Recent findings highlight the microbiome's role in shaping tumor characteristics, particularly in a newly identified metabiome-dependent CIN subtype (Molinari et al., 2018).

Spatial heterogeneity in CRC manifests at multiple levels: within one tumor (intratumoral heterogeneity (ITH)), within the same patient between several metastatic lesions (inter-metastatic heterogeneity), and within individual metastatic lesions. This heterogeneity arises from genetic evolution, the coexistence of differentiated cancer cells and cancer stem cells, functional differences among tumor cells, and variations in immune cell distribution influenced by the TME, including microbiota.

The "big bang" model of tumor evolution, proposed by Sottoriva et al., suggests that ITH in CRC results from early mutations rather than clonal selection, with diverse clones stabilizing and progressing toward metastasis. This implies that profiling the primary tumor may provide prognostic insights. However, single-needle biopsies often underestimate the mutational complexity, leading to inaccurate treatment decisions (Sottoriva et al., 2015). Spatial heterogeneity also contributes to treatment resistance, as chemotherapy-induced selective pressure can promote the survival and proliferation of resistant subclones. Studies have linked higher tumor heterogeneity with poorer outcomes. According to Sveen et al., individuals with CRC who had less heterogeneity fared considerably better in terms of overall survival (OS) and progression-free survival (PFS) than those who had more heterogeneity (Sveen et al., 2016). Currently, the lack of standardized methods to assess ITH limits its clinical application. To better comprehend tumor heterogeneity, cutting-edge methods like single-cell sequencing, computational deconvolution of tumor gene expression, and multiomic integration (genetic, epigenetic, transcriptomic, and proteomic data) are becoming available. The conventional methods for evaluating geographic heterogeneity are still whole-exome sequencing and multi-gene panels from many biopsy sites; however, despite their expense and technical difficulties, digital PCR and ultra-deep sequencing provide greater sensitivity. Although the usefulness of targeted next-generation sequencing (NGS) panels is still up for dispute, they have been suggested for clinical use. Currently, genetic testing primarily focuses on RAS and BRAF mutations, highlighting the need for improved strategies to address CRC heterogeneity in patient management (Molinari et al., 2018).

The effect of tumor heterogeneity on clinical molecular diagnostics in CRC is critical for accurate patient selection and treatment optimization, particularly for molecular-targeted therapies. Identifying all relevant mutant subpopulations is essential, as even low-frequency RAS mutations may influence the response to anti-EGFR therapy. International guidelines recommend performing RAS mutation analysis at diagnosis using the most recent and representative tumor tissue, but selecting an appropriate specimen can be challenging due to factors like poor tumor cellularity, ITH, and prior therapy. To minimize false negatives, pooling multiple tumor sites for genetic testing is recommended. However, in metastatic CRC (mCRC), many diagnoses rely solely on radiological assessments, meaning that only primary tumor samples are available for molecular testing, despite the goal of treating metastases. While primary and metastatic tumor tissues are often concordant in mutational status, some patients exhibit discordance, potentially leading to suboptimal treatment choices. Additionally, in patients who relapse after surgery and adjuvant therapy, chemotherapy-induced mutational changes may further complicate treatment decisions. To reevaluate molecular changes after chemotherapy resistance, fresh biopsies of original or metastatic tumors may be required for non-chemo-naive patients. Given the complexity of CRC progression, an individualized approach to diagnostics and treatment is crucial, considering prior surgical, clinical, and molecular history (Molinari et al., 2018).

3.8 MECHANISMS OF TREATMENT RESISTANCE

The three main treatments for CRC are chemotherapy, radiation therapy, and surgery. Traditional chemotherapy drugs, including oxaliplatin, irinotecan, 5-fluorouracil, and capecitabine, an oral version of 5-fluorouracil, are used to treat CRC.

Drug resistance is regarded as one of the main issues in chemotherapy. There are several different reasons that treatment resistance might develop, including changes in the TME growth dynamics, genetic alterations, and tumor heterogeneity (Daskın et al., 2021).

a. **Chemoresistance**: Drug resistance is a significant problem, yet chemotherapy is still the mainstay of CRC treatment. Changes in the DNA damage response, cell cycle checkpoints, and drug efflux mechanisms are frequently the cause of resistance to widely used chemotherapeutic drugs, including oxaliplatin and 5-fluorouracil (5-FU). Systems biology techniques that integrate genomic, transcriptomic, and proteomic data have revealed important chemoresistance drivers, including TP53 gene alterations and overexpression of ABCB1 (P-glycoprotein) (Blondy et al., 2020).
b. **Immunotherapy Resistance**: Immunotherapy has transformed the treatment of cancer, particularly with regard to immune checkpoint inhibitors (ICIs). However, immune checkpoint blockade treatments (e.g., PD-1/PD-L1 inhibitors) have a limited effect on microsatellite stable (MSS) CRC tumors. Utilizing systems biology, recent research has demonstrated that immune evasion mechanisms inside the TME, including TGF-β signaling, altered tumor-associated macrophages (TAMs), and overexpression of IDO1 (indoleamine 2,3-dioxygenase 1), contribute to resistance (Pancione et al., 2014).
c. **Targeted Therapy Resistance**: For certain CRC patients, targeted medicines that disrupt particular pathways—like the EGFR pathway—are employed. Acquired resistance to these drugs, however, is frequent. According to studies, resistance is largely caused by mutations in KRAS, NRAS, and PIK3CA, as well as the activation of other signaling pathways (such as MAPK/ERK and PI3K-AKT-mTOR). Resistance mechanisms are increasingly being identified and predicted using systems biology approaches like route modeling and network analysis (Voutsina ct al., 2013).

3.9 CONCLUSION

Targeted therapeutic interventions have been made possible by advances in systems biology and molecular diagnostics, which have illuminated the complex signaling pathways that drive tumor initiation and metastasis. The significance of genetic, epigenetic, and environmental variables in the development of CRC and the necessity of customized treatment strategies were highlighted in the discussion. The heterogeneity of CRC, spanning inter-patient, spatial, and molecular levels, presents a major challenge for effective treatment, as tumors exhibit diverse mutation profiles and varying responses to therapy. Emerging technologies, such as patient-derived organoids, xenograft models, and multiomics approaches, hold promise for refining CRC research and clinical management. However, despite these advancements, treatment resistance remains a critical obstacle, necessitating further research into adaptive resistance mechanisms and combination therapies. Overall, this chapter underscores the importance of integrating molecular insights with clinical applications to improve patient outcomes and move toward a more personalized approach in CRC treatment.

REFERENCES

Ait Ouakrim, D., Pizot, C., Boniol, M., Malvezzi, M., Negri, E., Bota, M.,…, & Autier, P. (2015). Trends in colorectal cancer mortality in Europe: Retrospective analysis of the WHO mortality database. *British Medical Journal*, 351, h4970. https://doi.org/10.1136/bmj.h4970

Akash, S., Islam, M. R., Bhuiyan, A. A., Islam, M. N., Bayıl, I., Saleem, R. M.,…, & Abdel-Daim, M. M. (2024). In silico evaluation of anti-colorectal cancer inhibitors by resveratrol derivatives targeting Armadillo repeats domain of APC: Molecular docking and molecular dynamics simulation. *Frontiers in Oncology*, 14, 1360745. https://doi.org/10.3389/fonc.2024.1360745

Aparicio, S., Hidalgo, M., & Kung, A. L. (2015). Examining the utility of patient-derived xenograft mouse models. *Nature Reviews Cancer*, 15, 311–316. https://doi.org/10.1038/nrc3944

Arnold, M., Sierra, M. S., Laversanne, M., Soerjomataram, I., Jemal, A., & Bray, F. (2017). Global patterns and trends in colorectal cancer incidence and mortality. *Gut*, 66(4), 683–691. https://doi.org/10.1136/gutjnl-2015-310912

Blondy, S., David, V., Verdier, M., Mathonnet, M., Perraud, A., & Christou, N. (2020). 5-Fluorouracil resistance mechanisms in colorectal cancer: From classical pathways to promising processes. *Cancer Science*, 111(9), 3142–3154. https://doi.org/10.1111/cas.14532

Blunt, T., Gell, D., Fox, M., Taccioli, G. E., Lehmann, A. R., Jackson, S. P., & Jeggo, P. A. (1996). Identification of a nonsense mutation in the carboxyl-terminal region of DNA-dependent protein kinase catalytic subunit in the scid mouse. *Proceedings of the National Academy of Sciences of the United States of America*, 93(19), 10285–10290. https://doi.org/10.1073/pnas.93.19.10285

Boivin, G. P., Washington, K., Yang, K., Ward, J. M., Pretlow, T. P., Russell, R.,…, & Coffey, R. J. (2003). Pathology of mouse models of intestinal cancer: Consensus report and recommendations. *Gastroenterology*, 124(3), 762–777. https://doi.org/10.1053/gast.2003.50094

Boughdady, I. S., Kinsella, A. R., Haboubi, N. Y., & Schofield, P. F. (1992). K-ras gene mutation in colorectal adenomas and carcinomas from familial adenomatous polyposis patients. *Surgical Oncology*, 1(4), 269–274. https://doi.org/10.1016/0960-7404(92)90087-2

Brattain, M. G., Strobel-Stevens, J., Fine, D., Webb, M., & Sarrif, A. M. (1980). Establishment of mouse colonic carcinoma cell lines with different metastatic properties. *Cancer Research*, 40(7), 2142–2146.

Bray, F., Ferlay, J., Soerjomataram, I., Siegel, R. L., Torre, L. A., & Jemal, A. (2018). Global cancer statistics 2018: GLOBOCAN estimates of incidence and mortality worldwide for 36 cancers in 185 countries. *CA: A Cancer Journal for Clinicians*, 68(6), 394–424. https://doi.org/10.3322/caac.21492

Calle, E. E., & Kaaks, R. (2004). Overweight, obesity and cancer: Epidemiological evidence and proposed mechanisms. *Nature Reviews Cancer*, 4(8), 579–591. https://doi.org/10.1038/nrc1408

Chen, T. R., Drabkowski, D., Hay, R. J., Macy, M., & Peterson, W. (1987). WiDr is a derivative of another colon adenocarcinoma cell line, HT-29. *Cancer Genetics and Cytogenetics*, 27(1), 125–134. https://doi.org/10.1016/0165-4608(87)90267-6

Clapper, M. L., Cooper, H. S., & Chang, W. C. (2007). Dextran sulfate sodium-induced colitis-associated neoplasia: A promising model for the development of chemopreventive interventions. *Acta Pharmacologica Sinica*, 28(9), 1450–1459. https://doi.org/10.1111/j.1745-7254.2007.00695.x

Corbett, T. H., Griswold, D. P., Jr., Roberts, B. J., Peckham, J. C., & Schabel, F. M., Jr. (1975). Tumor induction relationships in development of transplantable cancers of the colon in mice for chemotherapy assays, with a note on carcinogen structure. *Cancer Research*, 35(9), 2434–2439.

Croy, B. A., Linder, K. E., & Yager, J. A. (2001). Primer for non-immunologists on immune-deficient mice and their applications in research. *Medical Components*, 51(4), 300–313.

Céspedes, M. V., Espina, C., García-Cabezas, M. A., Trias, M., Boluda, A., Gómez del Pulgar, M. T.,..., & Mangues, R. (2007). Orthotopic microinjection of human colon cancer cells in nude mice induces tumor foci in all clinically relevant metastatic sites. *American Journal of Pathology*, 170(3), 1077–1085. https://doi.org/10.2353/ajpath.2007.060773

Danielsen, S. A., Eide, P. W., Nesbakken, A., Guren, T., Leithe, E., & Lothe, R. A. (2015). Portrait of the PI3K/AKT pathway in colorectal cancer. *Biochimica et Biophysica Acta*, 1855(1), 104–121. https://doi.org/10.1016/j.bbcan.2014.09.008

Daskın, E., Kurter, H., Yeşil, J., Çalıbaşı Koçal, G., Ellidokuz, H., & Başbınar, Y. (2021). Drug resistance mechanisms on colorectal cancer. *Journal of Basic and Clinical Health Sciences*, 5(1), 88–93. https://doi.org/10.30621/jbachs.869310

Davies, H., Bignell, G. R., Cox, C., Stephens, P., Edkins, S., Clegg, S.,..., & Futreal, P. A. (2002). Mutations of the BRAF gene in human cancer. *Nature*, 417(6892), 949–954. https://doi.org/10.1038/nature00766

Drost, J., van Jaarsveld, R. H., Ponsioen, B., Zimberlin, C., van Boxtel, R., Buijs, A.,..., & Clevers, H. (2015). Sequential cancer mutations in cultured human intestinal stem cells. *Nature*, 521(7550), 43–47. https://doi.org/10.1038/nature14415

Edelmann, W., Umar, A., Yang, K., Heyer, J., Kucherlapati, M., Lia, M.,..., & Kucherlapati, R. (2000). The DNA mismatch repair genes Msh3 and Msh6 cooperate in intestinal tumor suppression. *Cancer Research*, 60(4), 803–807.

el Marjou, F., Janssen, K. P., Chang, B. H., Li, M., Hindie, V., Chan, L.,..., & Robine, S. (2004). Tissue-specific and inducible Cre-mediated recombination in the gut epithelium. *Genesis*, 39(3), 186–193. https://doi.org/10.1002/gene.20042

Fearon, E. R., & Vogelstein, B. (1990). A genetic model for colorectal tumorigenesis. *Cell*, 61(5), 759–767. https://doi.org/10.1016/0092-8674(90)90186-i

Gerlinger, M., Rowan, A. J., Horswell, S., Math, M., Larkin, J., Endesfelder, D.,..., & Swanton, C. (2012). Intratumor heterogeneity and branched evolution revealed by multiregion sequencing. *New England Journal of Medicine*, 366(10), 883–892. https://doi.org/10.1056/NEJMoa1113205

Golovko, D., Kedrin, D., Yilmaz, Ö. H., & Roper, J. (2015). Colorectal cancer models for novel drug discovery. *Expert Opinion on Drug Discovery*, 10(11), 1217–1229. https://doi.org/10.1517/17460441.2015.1079618

Gondal, M. N., Butt, R. N., Shah, O. S., Sultan, M. U., Mustafa, G., Nasir, Z.,..., & Chaudhary, S. U. (2021). A personalized therapeutics approach using an in silico drosophila patient model reveals optimal chemo- and targeted therapy combinations for colorectal cancer [original research]. *Frontiers in Oncology*, 11, 692592.

Grothey, A., Van Cutsem, E., Sobrero, A., Siena, S., Falcone, A., Ychou, M.,..., & Laurent, D. (2013). Regorafenib monotherapy for previously treated metastatic colorectal cancer (CORRECT): An international, multicentre, randomised, placebo-controlled, phase 3 trial. *Lancet*, 381(9863), 303–312. https://doi.org/10.1016/s0140-6736(12)61900-x

Guerrero, R. M., Labajos, V. A., Ballena, S. L., Macha, C. A., Lezama, M. S., Roman, C. P.,..., & Torrejon, A. F. (2022). Targeting BRAF V600E in metastatic colorectal cancer: Where are we today? *Ecancermedicalscience*, 16, 1489. https://doi.org/10.3332/ecancer.2022.1489

He, K., & Gan, W. J. (2023). Wnt/β-catenin signaling pathway in the development and progression of colorectal cancer. *Cancer Management and Research*, 15, 435–448. https://doi.org/10.2147/cmar.s411168

Hezi, H., Shats, D., Gurevich, D., Maruvka, Y. E., & Freiman, M. (2024). Exploring the interplay between colorectal cancer subtypes genomic variants and cellular morphology: A deep-learning approach. *PLoS One*, 19(9), e0309380. https://doi.org/10.1371/journal.pone.0309380

Hinoi, T., Akyol, A., Theisen, B. K., Ferguson, D. O., Greenson, J. K., Williams, B. O.,…, & Fearon, E. R. (2007). Mouse model of colonic adenoma-carcinoma progression based on somatic Apc inactivation. *Cancer Research*, 67(20), 9721–9730. https://doi.org/10.1158/0008-5472.can-07-2735

Hinoue, T., Weisenberger, D. J., Pan, F., Campan, M., Kim, M., Young, J.,…, & Laird, P. W. (2009). Analysis of the association between CIMP and BRAF in colorectal cancer by DNA methylation profiling. *PLoS One*, 4(12), e8357. https://doi.org/10.1371/journal.pone.0008357

Itatani, Y., Kawada, K., & Sakai, Y. (2019). Transforming growth factor-β signaling pathway in colorectal cancer and its tumor microenvironment. *International Journal of Molecular Sciences*, 20(23), 5822. https://doi.org/10.3390/ijms20235822

Jung, P., Sato, T., Merlos-Suárez, A., Barriga, F. M., Iglesias, M., Rossell, D.,…, & Batlle, E. (2011). Isolation and in vitro expansion of human colonic stem cells. *Nature Medicine*, 17(10), 1225–1227. https://doi.org/10.1038/nm.2470**

Kandoth, C., McLellan, M. D., Vandin, F., Ye, K., Niu, B., Lu, C.,…, & Ding, L. (2013). Mutational landscape and significance across 12 major cancer types. *Nature*, 502(7471), 333–339. https://doi.org/10.1038/nature12634

Karami, P., & Elahi, R. (2024). *Radiomics-Based Artificial Intelligence (AI) Models in Colorectal Cancer (CRC) Diagnosis, Metastasis Detection, Prognosis, and Treatment Response*.

Kobaek-Larsen, M., Thorup, I., Diederichsen, A., Fenger, C., & Hoitinga, M. R. (2000). Review of colorectal cancer and its metastases in rodent models: Comparative aspects with those in humans. *Medical Components*, 50(1), 16–26.

Korn, A. R., Reedy, J., Brockton, N. T., Kahle, L. L., Mitrou, P., & Shams-White, M. M. (2022). The 2018 world cancer research fund/american institute for cancer research score and cancer risk: A longitudinal analysis in the NIH-AARP diet and health study. *Cancer Epidemiology, Biomarkers & Prevention*, 31(5), 1983–1992. https://doi.org/10.1158/1055-9965.epi-22-0044

Le, D. T., Uram, J. N., Wang, H., Bartlett, B. R., Kemberling, H., Eyring, A. D.,…, & Diaz, L. A., Jr. (2015). PD-1 blockade in tumors with mismatch-repair deficiency. *New England Journal of Medicine*, 372(26), 2509–2520. https://doi.org/10.1056/NEJMoa1500596

Lee, Y. J., Kim, W. R., Park, E. G., Lee, D. H., Kim, J. M., Shin, H. J.,…, & Kim, H. S. (2024). Exploring the key signaling pathways and ncRNAs in colorectal cancer. *International Journal of Molecular Sciences*, 25(8), 25084548. https://doi.org/10.3390/ijms25084548

Li, J., Li, Z. P., Ruan, W. J., & Wang, W. (2024). Colorectal cancer screening: The value of early detection and modern challenges. *World Journal of Gastroenterology*, 30(20), 2726–2730. https://doi.org/10.3748/wjg.v30.i20.2726

Li, Q., Geng, S., Luo, H., Wang, W., Mo, Y.-Q., Luo, Q.,…, & Xu, B. (2024). Signaling pathways involved in colorectal cancer: Pathogenesis and targeted therapy. *Signal Transduction and Targeted Therapy*, 9(1), 266. https://doi.org/10.1038/s41392-024-01953-7

Li, Q., Wu, H., Cao, M., Li, H., He, S., Yang, F.,…, & Chen, W. (2022). Colorectal cancer burden, trends and risk factors in China: A review and comparison with the United States. *Chinese Journal of Cancer Research*, 34(5), 483–495. https://doi.org/10.21147/j.issn.1000-9604.2022.05.08

Li, X., Nadauld, L., Ootani, A., Corney, D. C., Pai, R. K., Gevaert, O.,…, & Kuo, C. J. (2014). Oncogenic transformation of diverse gastrointestinal tissues in primary organoid culture. *Nature Medicine*, 20(7), 769–777. https://doi.org/10.1038/nm.3585

Lynch, H. T., & de la Chapelle, A. (2003). Hereditary colorectal cancer. *New England Journal of Medicine*, 348(10), 919–932. https://doi.org/10.1056/NEJMra012242

Malki, A., ElRuz, R. A., Gupta, I., Allouch, A., Vranic, S., & Al Moustafa, A. E. (2020). Molecular mechanisms of colon cancer progression and metastasis: Recent insights and advancements. *International Journal of Molecular Sciences*, 22(1), 22010130. https://doi.org/10.3390/ijms22010130

Matano, M., Date, S., Shimokawa, M., Takano, A., Fujii, M., Ohta, Y.,..., & Sato, T. (2015). Modeling colorectal cancer using CRISPR-Cas9–mediated engineering of human intestinal organoids. *Nature Medicine*, 21(3), 256–262. https://doi.org/10.1038/nm.3802

Molinari, C., Marisi, G., Passardi, A., Matteucci, L., De Maio, G., & Ulivi, P. (2018). Heterogeneity in colorectal cancer: A challenge for personalized medicine? *International Journal of Molecular Sciences*, 19(12), 19123733. https://doi.org/10.3390/ijms19123733

Morelli, M. P., Calvo, E., Ordoñez, E., Wick, M. J., Viqueira, B.-R., Lopez-Casas, P. P.,..., & Hidalgo, M. (2012). Prioritizing phase I treatment options through preclinical testing on personalized tumorgraft. *Journal of Clinical Oncology: Official Journal of the American Society of Clinical Oncology*, 30(4), e45–48. https://doi.org/10.1200/jco.2011.36.9678

Morin, P. J., Sparks, A. B., Korinek, V., Barker, N., Clevers, H., Vogelstein, B., & Kinzler, K. W. (1997). Activation of beta-catenin-Tcf signaling in colon cancer by mutations in beta-catenin or APC. *Science*, 275(5307), 1787–1790. https://doi.org/10.1126/science.275.5307.1787

Moshawih, S., Lim, A. F., Ardianto, C., Goh, K. W., Kifli, N., Goh, H. P.,..., & Ming, L. C. (2022). Target-based small molecule drug discovery for colorectal cancer: A review of molecular pathways and in silico studies. *Biomolecules*, 12(7), 878.

Neufert, C., Becker, C., & Neurath, M. F. (2007). An inducible mouse model of colon carcinogenesis for the analysis of sporadic and inflammation-driven tumor progression. *Nature Protocols*, 2(8), 1998–2004. https://doi.org/10.1038/nprot.2007.279

Nishikawa, A., Imazawa, T., Kuroiwa, Y., Kitamura, Y., Kanki, K., Ishii, Y.,..., & Hirose, M. (2005). Induction of colon tumors in C57BL/6J mice fed MeIQx, IQ, or PhIP followed by dextran sulfate sodium treatment. *Toxicological Sciences*, 84(2), 243–248. https://doi.org/10.1093/toxsci/kfi079

Ogino, S., Nosho, K., Kirkner, G. J., Kawasaki, T., Meyerhardt, J. A., Loda, M.,..., & Fuchs, C. S. (2009). CpG island methylator phenotype, microsatellite instability, BRAF mutation and clinical outcome in colon cancer. *Gut*, 58(1), 90–96. https://doi.org/10.1136/gut.2008.155473

Pancione, M., Giordano, G., Remo, A., Febbraro, A., Sabatino, L., Manfrin, E.,..., & Colantuoni, V. (2014). Immune escape mechanisms in colorectal cancer pathogenesis and liver metastasis. *Journal of Immunology Research, 2014*, 686879. https://doi.org/10.1155/2014/686879

Peker, Y. S., Can, M. F., Ozerhan, I. H., Yagci, G., Zeybek, N., Kavakli, K.,..., & Peker, Y. (2018). BRAF inhibitors for BRAF V600E mutant colorectal cancers: Literature survey and case report. *Case Reports in Surgery*, 2018, 8782328. https://doi.org/10.1155/2018/8782328

Porru, M., Pompili, L., Caruso, C., Biroccio, A., & Leonetti, C. (2018). Targeting KRAS in metastatic colorectal cancer: Current strategies and emerging opportunities. *Journal of Experimental & Clinical Cancer Research*, 37(1), 57. https://doi.org/10.1186/s13046-018-0719-1

Potocki, P., & Wysocki, P. (2018). BRAF — a new therapeutic target in colorectal cancer. *Oncology in Clinical Practice*, 14(2), 86–95. https://doi.org/{}

Reddy, B. S., Weisburger, J. H., Narisawa, T., & Wynder, E. L. (1974). Colon carcinogenesis in germ-free rats with 1,2-dimethylhydrazine and N-methyl-n′-nitro-N-nitrosoguanidine. *Cancer Research*, 34(9), 2368–2372.

Robanus-Maandag, E. C., Koelink, P. J., Breukel, C., Salvatori, D. C., Jagmohan-Changur, S. C., Bosch, C. A.,..., & Smits, R. (2010). A new conditional APC-mutant mouse model for colorectal cancer. *Carcinogenesis*, 31(5), 946–952. https://doi.org/10.1093/carcin/bgq046

Rosfjord, E., Lucas, J., Li, G., & Gerber, H. P. (2014). Advances in patient-derived tumor xenografts: From target identification to predicting clinical response rates in oncology. *Biochem Pharmacol*, 91(2), 135–143. https://doi.org/10.1016/j.bcp.2014.06.008

Sansom, O. J., Reed, K. R., Hayes, A. J., Ireland, H., Brinkmann, H., Newton, I. P.,..., & Winton, D. J. (2004). Loss of Apc in vivo immediately perturbs Wnt signaling, differentiation, and migration. *Genes & Development*, 18(12), 1385–1390. https://doi.org/10.1101/gad.287404

Sato, T., & Clevers, H. (2013). Growing self-organizing mini-guts from a single intestinal stem cell: Mechanism and applications. *Science*, 340(6137), 1190–1194. https://doi.org/10.1126/science.1234852

Sato, T., Stange, D. E., Ferrante, M., Vries, R. G., Van Es, J. H., Van den Brink, S.,…, & Clevers, H. (2011). Long-term expansion of epithelial organoids from human colon, adenoma, adenocarcinoma, and Barrett's epithelium. *Gastroenterology*, 141(5), 1762–1772. https://doi.org/10.1053/j.gastro.2011.07.050

Sato, T., Vries, R. G., Snippert, H. J., van de Wetering, M., Barker, N., Stange, D. E.,…, & Clevers, H. (2009). Single Lgr5 stem cells build crypt-villus structures in vitro without a mesenchymal niche. *Nature*, 459(7244), 262–265. https://doi.org/10.1038/nature07935

Shoemaker, R. H. (2006). The NCI60 human tumour cell line anticancer drug screen. *Nature Reviews Cancer*, 6, 813–823. https://doi.org/10.1038/nrc1951

Siegel, R. L., Giaquinto, A. N., & Jemal, A. (2024). Cancer statistics, 2024. *CA: A Cancer Journal for Clinicians*, 74(1), 12–49. https://doi.org/10.3322/caac.21820

Siegel, R. L., Miller, K. D., Goding Sauer, A., Fedewa, S. A., Butterly, L. F., Anderson, J. C.,…, & Jemal, A. (2020). Colorectal cancer statistics, 2020. *CA: A Cancer Journal for Clinicians*, 70(3), 145–164. https://doi.org/10.3322/caac.21601

Siegel, R. L., Wagle, N. S., Cercek, A., Smith, R. A., & Jemal, A. (2023). Colorectal cancer statistics, 2023. *CA: A Cancer Journal for Clinicians*, 73(3), 233–254. https://doi.org/10.3322/caac.21772

Siolas, D., & Hannon, G. J. (2013). Patient-derived tumor xenografts: Transforming clinical samples into mouse models. *Cancer Research*, 73(17), 5315–5319. https://doi.org/10.1158/0008-5472.can-13-1069

Sommariva, S., Berra, S., Biddau, G., Caviglia, G., Benvenuto, F., & Piana, M. (2023). In-silico modelling of the mitogen-activated protein kinase (MAPK) pathway in colorectal cancer: Mutations and targeted therapy [Original Research]. *Frontiers in Systems Biology*, 3, 1207898.

Song, M., & Chan, A. T. (2019). Environmental factors, gut microbiota, and colorectal cancer prevention. *Clinical Gastroenterology and Hepatology*, 17(2), 275–289. https://doi.org/10.1016/j.cgh.2018.07.012

Sottoriva, A., Kang, H., Ma, Z., Graham, T. A., Salomon, M. P., Zhao, J.,…, & Curtis, C. (2015). A big bang model of human colorectal tumor growth. *Nature Genetics*, 47(3), 209–216. https://doi.org/10.1038/ng.3214

Sveen, A., Løes, I. M., Alagaratnam, S., Nilsen, G., Høland, M., Lingjærde, O. C.,…, & Lothe, R. A. (2016). Intra-patient inter-metastatic genetic heterogeneity in colorectal cancer as a key determinant of survival after curative liver resection. *PLoS Genetics*, 12(7), e1006225. https://doi.org/10.1371/journal.pgen.1006225

Voskoglou-Nomikos, T., Pater, J. L., & Seymour, L. (2003). Clinical predictive value of the in vitro cell line, human xenograft, and mouse allograft preclinical cancer models. *Clinical Cancer Research*, 9(11), 4227–4239.

Voutsina, A., Tzardi, M., Kalikaki, A., Zafeiriou, Z., Papadimitraki, E., Papadakis, M.,…, & Georgoulias, V. (2013). Combined analysis of KRAS and PIK3CA mutations, MET and PTEN expression in primary tumors and corresponding metastases in colorectal cancer. *Modern Pathology*, 26(2), 302–313. https://doi.org/10.1038/modpathol.2012.150

Wang, R. T., Zhao, Y., Wang, A. L., Wang, Y. T., Yin, Z. P., & Chen, K. (2021). Efficacy and safety of regorafenib monotherapy among patients with previously treated metastatic colorectal cancer in a Chinese population: A real-world exploratory study. *International Journal of General Medicine*, 14, 5363–5373. https://doi.org/10.2147/ijgm.s325545

Wong, M. C., Ding, H., Wang, J., Chan, P. S., & Huang, J. (2019). Prevalence and risk factors of colorectal cancer in Asia. *Intestinal Research*, 17(3), 317–329. https://doi.org/10.5217/ir.2019.00021

Wu, X., Yan, H., Qiu, M., Qu, X., Wang, J., Xu, S.,…, & Liang, L. (2023). Comprehensive characterization of tumor microenvironment in colorectal cancer via molecular analysis. *Elife*, 12, 86032. https://doi.org/10.7554/eLife.86032

Xie, J., & Itzkowitz, S. H. (2008). Cancer in inflammatory bowel disease. *World Journal of Gastroenterology*, 14(3), 378–389. https://doi.org/10.3748/wjg.14.378

Xue, C., Li, G., Lu, J., & Li, L. (2021). Crosstalk between circRNAs and the PI3K/AKT signaling pathway in cancer progression. *Signal Transduction and Targeted Therapy*, 6(1), 400. https://doi.org/10.1038/s41392-021-00788-w

Xue, Y., Johnson, R., Desmet, M., Snyder, P. W., & Fleet, J. C. (2010). Generation of a transgenic mouse for colorectal cancer research with intestinal cre expression limited to the large intestine. *Molecular Cancer Research*, 8(8), 1095–1104. https://doi.org/10.1158/1541-7786.mcr-10-0195

Zhu, G., Pei, L., Xia, H., Tang, Q., & Bi, F. (2021). Role of oncogenic KRAS in the prognosis, diagnosis and treatment of colorectal cancer. *Molecular Cancer*, 20(1), 143. https://doi.org/10.1186/s12943-021-01441-4

4 Prostate Cancer Model

Garima Gupta, Roma Chandra, Muskan Chauhan, and Yatender Kumar

4.1 INTRODUCTION

Prostate cancer (PCa) is a type of cancer that occurs in the prostate, a small gland in the male reproductive system that produces seminal fluid, which nourishes and transports sperm. The prostate is located just below the bladder and in front of the rectum. PCa develops when cells in the prostate begin to grow uncontrollably. This uncontrolled growth can form a tumor, which may stay confined to the prostate (localized) or spread to other parts of the body (metastatic) (Rebello et al., 2021). PCa is one of the most common types of cancer in men, particularly affecting older men, although it can occur in younger men as well. PCa is highly treatable, especially when detected early, and many men live with it for years without it affecting their quality of life. Regular screening and early detection are crucial for improving outcomes (Sekhoacha et al., 2022).

4.2 STATISTICS ABOUT PROSTATE CANCER

PCa is the second most common cancer in men worldwide, after lung cancer, with 1,276,106 new cases and 358,989 deaths reported in 2018, accounting for 3.8% of all cancer-related deaths in men (Chu et al., 2025). The incidence and mortality rates are closely linked to age, with the average diagnosis age being 66 years. African-American men, in particular, have a higher incidence and more aggressive form of PCa compared to White men. In 2040, an estimated 2,293,818 new cases are expected, with a small increase in mortality rates (1.05%) (Wang et al., 2022) (Figure 4.1).

At early stages, PCa may be asymptomatic and typically follows an indolent course, requiring little to no treatment. The most common complaints include difficulty urinating, increased frequency, and nocturia, which can also be symptoms of benign prostatic hypertrophy (Miernik & Gratzke, 2020). More advanced stages may present with urinary retention and back pain, often due to metastasis to the bones, particularly the skeleton.

Elevated prostate-specific antigen (PSA) levels are commonly used to detect PCa, although PSA levels can also be elevated in men without cancer. A biopsy is the standard method to confirm the diagnosis. Diet and physical activity play a significant role in the development and progression of PCa, with dietary habits contributing to regional and ethnic variations in incidence rates (Oczkowski et al., 2021).

Research is focused not only on identifying genes involved in inherited PCa but also on understanding the mutations involved in the acquired form. A thorough analysis of PCa epidemiology, combined with risk factor evaluation, can help determine the

DOI: 10.1201/9781003583349-5

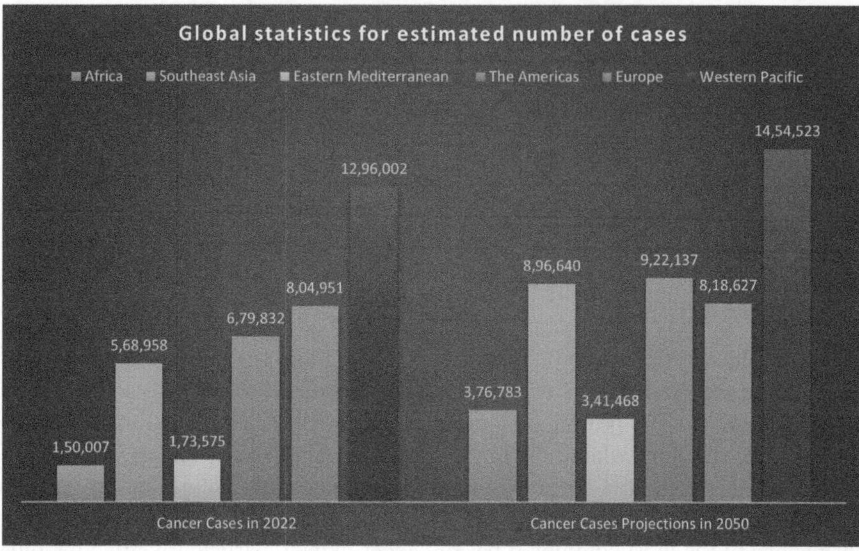

FIGURE 4.1 Global statistics for estimated number of new male cases from 2022 to 2050 between the age group of 0–85+ years (https://gco.iarc.who.int/tomorrow).

link between genetic mutations and environmental factors, aiding in the development of effective screening and prevention strategies (Bergengren et al., 2023).

4.3 EPIDEMIOLOGY

As of 2024, PCa remains one of the most common cancers worldwide, with an estimated 1.4 million new cases and approximately 375,000 deaths annually. The age-standardized incidence rate (ASR) is highest in regions like Oceania (79.1 per 100,000), North America (73.7 per 100,000), and Europe (62.1 per 100,000). Conversely, the incidence is much lower in Africa (26.6 per 100,000) and Asia (11.5 per 100,000), although both incidence and mortality have been steadily increasing in many parts of these regions due to aging populations and better diagnostic facilities (Siegel et al., 2024). The incidence of PCa rises significantly with age, with the highest rates seen in men over 65. African-American men have the highest risk globally, with rates significantly higher than those in other racial groups, partly due to genetic and environmental factors (Taitt, 2018). In contrast, incidence rates in countries like Bhutan remain very low. Although the mortality rate from PCa has declined in many Western countries due to improvements in early detection and treatment, this trend is not seen universally. In low-resource regions, PCa mortality is projected to rise, especially in Africa and Asia, due to limited access to screening and late-stage diagnosis.

PCa is increasingly prevalent in India, with rising incidence due to better awareness, improved diagnostics, and lifestyle changes. The Indian Council of Medical Research (ICMR) reports that PCa is the second most common cancer in Indian men, after oral cancers, with an increase in both diagnosed cases and mortality over recent years.

While its age-standardized incidence rate (ASR) of 4.3 per 100,000 is lower than in Western countries, the absolute number of cases is increasing due to population growth and aging (Rawla, 2019). PCa is now the second most common cancer in Indian men. It primarily affects men over 65 and is linked to Westernized lifestyles. Urban areas show higher rates due to better screening and diagnostics, while rural regions face challenges in access to treatment, impacting survival rates (Bhatia et al., 2022).

4.4 ETIOLOGY AND RISK FACTORS

The etiology of PCa remains the focus of extensive research, but it is still less understood compared to other common cancers. The primary risk factors for PCa are advanced age, ethnicity, genetic predisposition, and family history (Gann, 2002). Other factors positively linked to PCa include diet (higher intake of saturated animal fats, red meat, and lower consumption of fruits, vegetables, and coffee), obesity, physical inactivity, inflammation, hyperglycemia, infections, and environmental exposure to chemicals or ionizing radiation (Mandair et al., 2014) (Table 4.1).

TABLE 4.1
Etiology and the Risk Factors Associated with Prostate Cancer

Factor	Etiology	Remarks
Age	Prostate cancer (PCa) is the most commonly diagnosed cancer in older men. The rising number of diagnoses is partly due to increased life expectancy and the widespread use of prostate-specific antigen (PSA) screening. The risk of PCa increases significantly after the age of 50 in White men without a family history, and after 40 in Black men or those with a familial history of the disease. Scardino's research indicated that nearly 30% of men over 50 who died from other causes showed histological evidence of PCa at autopsy (Stangelberger et al., 2008).	This suggests that due to its slow progression, many elderly men with concurrent severe comorbidities are more likely to die from other health conditions rather than PCa itself.
Ethnicity:	The incidence of PCa varies significantly across different racial groups. In the U.S., the lowest incidence rates are observed among American Indian/Alaska Native (46.9) and Asian/Pacific Islander (52.4) populations, while White men have an incidence of 93.9. The highest rates are seen in African-American men (157.6) (Iyengar et al., 2020).	This disparity is thought to result from both socioeconomic and biological factors.
Family history and genetic factors:	Approximately 20% of PCa patients report a family history, which could be attributed to shared genetic factors, common exposure to environmental carcinogens, and similar lifestyle habits. African-American men are more likely to have certain chromosome 8q24 variants, which are associated with an increased risk of PCa. Studies also suggest that Black men exhibit a higher rate of variations in genes that suppress tumors, such as EphB2, or those regulating apoptosis, like BCL2 (Brandão et al., 2020).	Gene linkage studies have identified several susceptibility loci for PCa, like HPC1 (Chromosome 1q24-25), HPC2/ELAC2 (Chromosome 17p11), MSR1 (Chromosome 8p22), BRCA1 and BRCA2 mutations

(Continued)

TABLE 4.1 (*Continued*)
Etiology and the Risk Factors Associated with Prostate Cancer

Factor	Etiology	Remarks
Diet:	Diet influences PCa risk, as changes in lifestyle lead to a shift in incidence rates. African-American men had a 40-fold higher incidence of PCa compared to those living in Africa, while Chinese men in the USA had a 16-fold higher incidence compared to those in China. High consumption of saturated fat, including meat and dairy products, cooking meat at high temperatures, and high consumption of dairy products, particularly calcium supplements, increase the risk to PCa. Certain vegetables, especially cruciferous ones (e.g., broccoli, cauliflower, and cabbage), tomatoes, soy, and green tea have been linked to a reduced risk of PCa (Kaiser et al., 2019).	
Vitamin and mineral supplements	Men with low vitamin D levels may be at higher risk, especially in darker-skinned individuals who have a reduced ability to produce vitamin D from sunlight. Some studies indicate that vitamin D can inhibit PCa cell proliferation and promote apoptosis, although clinical trials using vitamin D analogs are still ongoing. Selenium, found in foods like meat and seafood, has been inversely associated with PCa risk (Song et al., 2018).	
Alcohol consumption and PCa:	Heavy alcohol consumption (more than 15 g of ethanol daily or more than three drinks a day) is thought to be a potential risk factor for PCa and other cancers (Macke & Petrosyan, 2022).	
Coffee consumption:	Heavy coffee drinkers had a significantly lower risk of developing advanced or lethal PCa compared to those who consumed less (Wilson et al., 2011).	
Obesity, insulin, and physical activity:	Obesity is associated with advanced and aggressive PCa, and a high body mass index (BMI) often correlates with worse outcomes. Obese men frequently show alterations in circulating metabolic and sex steroid hormones, which are involved in prostate development and oncogenesis. Obesity, particularly when coupled with physical inactivity, leads to insulin resistance and elevated blood insulin levels, which promote tumor growth and proliferation. Furthermore, adipose tissue can release inflammatory mediators, contributing to cancer progression. Additionally, obesity is linked to more advanced cancer at diagnosis due to "hemodilution" of prostate-specific antigen (PSA) levels, which affects detection. Studies have shown that veterans who exercise have a significantly lower risk of PCa, and PCa patients who engage in regular exercise tend to have lower PSA levels, a delay in starting androgen deprivation therapy (ADT), and a lower risk of high-grade disease (Wilson et al., 2021).	

(*Continued*)

TABLE 4.1 (*Continued*)
Etiology and the Risk Factors Associated with Prostate Cancer

Factor	Etiology	Remarks
Cigarette smoking:	Smoking may contribute to disease progression through hormonal or genetic mechanisms. Smokers typically have higher levels of circulating sex hormones, which could increase PCa risk, or they may have genetic variations affecting the metabolism of carcinogenic chemicals in cigarette smoke. However, studies examining smoking and PCa mortality indicate that smokers are twice as likely to die from the disease compared to nonsmokers, with a dose-response relationship between the number of cigarettes smoked and the risk of death from PCa (Al-Fayez and El-Metwally, 2023).	
Sex hormones:	The role of androgens in PCa development and progression is well established, with the "androgen hypothesis" dating back to 1941. Clinical treatments such as androgen deprivation therapy (ADT) are central to managing advanced PCa. Estrogen has been shown to induce genetic damage and may act as a carcinogen in PCa development (Michaud et al., 2015).	
Insulin and insulin-like growth factor	Insulin promotes cell growth and may contribute to cancer progression. High insulin levels decrease the production of insulin-like growth factor (IGF) binding proteins, leading to increased IGF-1 levels, which in turn promote cancerous growth (Zaorsky et al., 2016).	
Chronic inflammation and prostatitis	Inflammatory responses play a key role in the tumor microenvironment, contributing to extracellular matrix (ECM) remodeling and epithelial-mesenchymal transition (EMT). Inflammatory cells release growth factors and cytokines that promote angiogenesis, ECM remodeling, and EMT, further driving cancer progression. Patients with elevated PSA levels often exhibit intraprostatic inflammation, which can be detected through biopsy. Inflammatory effectors, such as pentraxin 3, have recently been identified as biomarkers for predicting PCa progression driven by prostatic inflammation. Chronic inflammation leads to proliferative inflammatory atrophy (PIA), which can develop into prostatic intraepithelial neoplasia (PIN), a known precursor to PCa. Men with prostatitis symptoms tend to undergo more frequent biopsies, leading to a higher likelihood of PCa diagnosis (Oseni et al., 2023).	

(*Continued*)

TABLE 4.1 (*Continued*)
Etiology and the Risk Factors Associated with Prostate Cancer

Factor	Etiology	Remarks
Sexually transmitted diseases	Several epidemiological studies have suggested that sexually transmitted diseases (STDs) and sexual behaviors may influence PCa risk. Some early studies proposed a connection between STDs and PCa, with mechanisms such as prostate inflammation or viral transformation of prostate cells being considered. Large population-based case-control studies have shown an increased risk of PCa in men with a history of gonorrhea or syphilis. HPV, detected in both PCa and benign prostatic tissues, has been found to transform prostate cells in laboratory studies. Trichomonas vaginalis, a common cause of vaginitis in women, can infect men and cause chronic prostatitis, potentially leading to PCa over time (Cheng et al., 2010)	

4.5 MEDICATIONS AND PROSTATE CANCER PREVENTION

Despite advances in understanding the causes of PCa, effective chemopreventive treatments remain elusive. Some drugs targeting testosterone and inflammatory pathways have been investigated for their potential to prevent PCa. Finasteride and dutasteride are the most studied chemopreventive agents for PCa. These drugs inhibit the 5-α reductase enzyme, which converts testosterone to dihydrotestosterone (DHT), the most potent androgen involved in prostate growth and cancer promotion. NSAIDs, including aspirin (ASA), have been shown to have potential in cancer prevention, particularly in reducing PCa risk. These drugs target cyclooxygenase (COX), with COX-2 playing a key role in prostate inflammation and proliferative inflammatory atrophy (PIA), a precursor to PCa. Statins are medications that inhibit lipid synthesis, particularly cholesterol, and recent studies have shown they can lower PSA levels, as well as reduce the risk of advanced or aggressive PCa (Rivero et al., 2018).

4.6 ENVIRONMENTAL CARCINOGENS

Prostate carcinogenesis can also be influenced by exposure to certain environmental factors, including insecticides, herbicides, and organic chemicals. Agent Orange (AO), a mixture of herbicides used as a defoliant during the Vietnam War (1962–1971), was contaminated with dioxins, particularly 2,3,7,8-tetrachlorodibenzo-p-dioxin (TCDD), which is considered a carcinogen. Studies show that veterans exposed to AO had double the risk of PCa, which was diagnosed at younger ages and often at a higher grade, suggesting that AO exposure may not only increase the risk of PCa but also contribute to more aggressive forms and metastatic disease. Chlordecone, an organochlorine insecticide used in the French West Indies, has estrogenic properties and a long biological half-life. It has been linked to an increased risk of PCa.

Bisphenol A (BPA) has been associated with an increased risk of PCa. Studies show that even low levels of BPA exposure may promote PCa by inducing centrosome abnormalities, a hallmark of cancerous cells. Vasectomy has been studied for its potential link to PCa. Growing evidence suggests that more frequent ejaculation may lower the risk of PCa as frequent ejaculation might alter prostate cell function and reduce the formation of prostatic intraluminal crystalloids, which are associated with higher PCa risk (Vigneswaran et al., 2021).

4.7 DIAGNOSTIC RADIOLOGIC PROCEDURE AND ULTRAVIOLET LIGHT EXPOSURE

Ionizing radiation from X-rays, CT scans, and nuclear imaging can penetrate the body's tissues to provide insights into internal organs. However, while these imaging techniques are valuable, they can also damage DNA. Although most of this damage is repaired by the body, some areas may remain altered, potentially leading to mutations that increase the risk of cancer years later. A study examining the relationship between low-dose ionizing radiation from diagnostic X-rays and PCa risk found that exposure to hip/pelvic X-rays significantly raised PCa risk, independent of other factors such as family history of cancer. However, the increased risk associated with medical radiation appears to be minimal unless men are exposed to high doses of radiation during cancer treatment in their youth. Since the rise in high-dose imaging began only after 1980 and the effects of radiation damage can take years to manifest, this may explain the weak association between ionizing radiation and PCa risk observed so far. Interestingly, exposure to solar UV radiation has been found to be inversely related to both the incidence and mortality of PCa, possibly due to its role in vitamin D synthesis, which may have protective effects (Myles et al., 2008; Moon et al., 2005).

4.8 PREVENTION

Effective prevention strategies for PCa could greatly benefit public health by reducing the high lifetime risk of PCa, minimizing the morbidity associated with cancer treatment—particularly in men with biologically indolent PCa who currently undergo curative therapies rather than active surveillance—and addressing the challenges of treating life-threatening metastatic PCa (Posdzich et al., 2023).

Epidemiological evidence points to lifestyle factors playing a crucial role in PCa development. Since prostate carcinogenesis typically spans decades, lifestyle changes could serve as a feasible and cost-effective way to delay its onset. While data on the impact of specific lifestyle factors on PCa risk have been inconsistent, many studies suggest that a diet rich in fruits, vegetables, and antioxidant micronutrients—while low in saturated fats and well-done red meats—can significantly reduce the risk of PCa, as well as other diseases prevalent in industrialized countries.

A deeper understanding of PCa etiology could open the door to new prevention opportunities. Several nutrients and pharmaceutical agents have been investigated as potential chemoprevention options. Vitamin E and selenium showed initial promise, but large trials like SELECT found that neither agent reduced PCa risk. Vitamin D analogs, NSAIDs, and toremifene (a selective estrogen receptor modulator) have also been explored in laboratory and observational studies. However, vitamin D has not yet been tested in primary prevention trials, and the NSAID rofecoxib (a COX-2 selective inhibitor) was withdrawn from the market during a trial due to safety concerns related to cardiovascular risks. Toremifene showed modest risk reduction in a phase II trial but no significant impact in a phase III trial. The prevention landscape also involves identifying molecular targets and personalized risk assessments, as well as finding sensitive and accurate surrogate molecular markers to serve as intermediate endpoints. Personalized, molecularly targeted approaches could improve outcomes and offer effective therapies tailored to individual patients with PCa.

4.9 *IN VITRO* MODELS FOR STUDYING PROSTATE CANCER

In vitro models are widely used to study PCa because they provide a controlled environment to explore the molecular mechanisms of cancer progression, test therapeutic agents, and investigate the tumor microenvironment. The detailed overview of the *in vitro* models for studying PCa is as follows:

4.9.1 Prostate Cancer Cell Lines

Cell lines are the most commonly used *in vitro* models for studying PCa. These are immortalized cells derived from human prostate tumors and are used to understand cancer biology, gene expression, and drug response. Some commonly used PCa cell lines are:

1. **LNCaP**: LNCaP cells are androgen-sensitive and were derived from a metastatic PCa lesion in a human patient. They are one of the most widely used PCa cell lines. LNCaP cells are typically used to study androgen receptor (AR) signaling, androgen deprivation therapy (ADT), and the development of castration-resistant prostate cancer (CRPC) (Tindall & Lonergan, 2011). These cells express a functional AR, and their growth is regulated by androgens, making them ideal for studies involving hormone dependency.
2. **PC3**: PC3 cells are derived from a metastatic PCa patient and are considered androgen-independent. They are frequently used in studies of metastatic PCa and CRPC. They are used for investigating the molecular basis of cancer progression and for testing drugs targeting pathways involved in metastasis (Zadra & Loda, 2017). PC3 cells do not express functional AR, making them useful for studying androgen-independent PCa.
3. **DU145**: DU145 cells are another androgen-independent PCa cell line, derived from a brain metastasis of a PCa patient. DU145 cells are often used for studying tumor progression, invasiveness, and the mechanisms driving metastatic disease. They are also used to test novel therapeutic strategies aimed at targeting androgen-independent PCa (Mori et al., 2006). These cells are less sensitive to androgen and are suitable for studying the androgen-independent progression of PCa, as well as therapeutic resistance.
4. **C4-2**: C4-2 cells were derived from LNCaP cells and represent a more advanced stage of PCa. These cells are androgen-sensitive in culture but

exhibit androgen-independent growth in vivo. C4-2 cells are valuable for studying the transition from androgen-sensitive to androgen-independent PCa. They are often used to investigate mechanisms of resistance to ADT and the development of CRPC (Gordon et al., 2019).

5. **22Rv1**: 22Rv1 cells are derived from a human PCa xenograft and are commonly used in studies of AR signaling and PCa progression. 22Rv1 cells are used to investigate mechanisms of resistance to ADT and for testing potential therapeutic strategies targeting AR signaling. They are often employed in studies focusing on AR mutations and their role in castration-resistant PCa (Bluemn et al., 2017).
6. **VCaP**: VCaP cells are derived from a human PCa bone metastasis and are androgen-sensitive. VCaP cells are often used to study the metastatic behavior of PCa, particularly its ability to metastasize to bone. They are also useful in research on AR signaling, bone metastasis, and androgen resistance mechanisms (Mori et al., 2006).
7. **ALVA-31**: ALVA-31 cells are androgen-independent PCa cells derived from a lymph node metastasis. These cells are primarily used for studying androgen-independent progression and the molecular mechanisms driving metastatic spread, particularly to lymph nodes. They are used in investigating therapeutic resistance in advanced PCa (Slaton et al., 2004)
8. **RWPE-1**: RWPE-1 is an immortalized, non-tumorigenic prostate epithelial cell line derived from normal prostate tissue. RWPE-1 cells are used as a control for comparing the genetic and phenotypic alterations that occur during the transformation from normal to cancerous prostate tissue. They are used in early-stage PCa research and investigations of tumor initiation (Hosseini-Beheshti et al., 2012).

Advantages and Limitations of Cell Lines: Cell lines can be easily cultured and genetically modified, making them highly versatile. They are amenable to high-throughput screening for drug discovery. Their limited applications are due to the lack of a complex tumor microenvironment, such as stromal and immune components. They are at loss of heterogeneity compared to primary tumors.

4.9.2 Prostate Cancer Organoids

Organoids are three-dimensional (3D) cell cultures that more accurately replicate the complexity and architecture of human tissues. Organoids derived from PCa patient tissues are highly relevant for understanding disease heterogeneity, progression, and treatment responses. These organoids closely mimic the architecture, heterogeneity, and cellular interactions found in prostate tumors. They can be generated by isolating PCa cells from patient samples or animal models, followed by culture in a matrix that supports 3D growth, such as Matrigel (Pauli et al., 2017).

Patient-Derived Prostate Cancer Organoids: PCa organoids, derived from patient biopsies, are increasingly used for drug screening and studying tumor biology in a more clinically relevant context. They maintain the original genetic and phenotypic characteristics of the primary tumor, enabling personalized medicine approaches.

CRISPR-Modified Organoids: Recent advances have allowed the generation of CRISPR-Cas9-modified PCa organoids, which enable the study of gene function in cancer progression. These models have been used to investigate tumor suppressors, oncogenes, and resistance mechanisms to therapies.

4.9.3 Applications of Prostate Cancer Organoids

- **Cancer Biology**: PCa organoids provide an *in vitro* model to investigate the genetic and epigenetic alterations driving tumorigenesis, tumor heterogeneity, and progression. They enable the study of specific mutations and pathways, such as those related to the AR signaling axis, which is critical in PCa biology (Gao et al., 2014).
- **Drug Screening and Personalized Medicine**: Organoids allow for the screening of therapeutic agents in a patient-specific manner. Drugs can be tested on organoids derived from individual patient samples, providing a personalized approach to treatment and potentially identifying effective therapies tailored to specific mutations or treatment resistance profiles.
- **Metastasis Modeling**: The use of organoids to study PCa metastasis, particularly to bone and lymph nodes, is an area of active research. PCa organoids can be cultured in the presence of extracellular matrix (ECM) components or cocultured with stromal cells to study metastatic processes (Boj et al., 2014).

4.9.4 Advantages and Limitations of Prostate Cancer Organoids

Unlike two-dimensional (2D) cultures, organoids maintain the genetic and phenotypic diversity of primary tumors, including stromal components and intratumoral heterogeneity. This provides a more accurate model for studying tumor progression and testing therapies (Boj et al., 2014). Organoids can be passaged and expanded over time, making them suitable for long-term studies. This reproducibility is advantageous for screening large numbers of compounds or testing different therapeutic regimens (Beshiri et al., 2018). Patient-derived organoids (PDOs) offer a powerful tool for tailoring treatments based on individual tumor profiles, as organoids retain the genetic characteristics and response to therapy of the original patient tumor (Gao et al., 2014).

While organoids mimic many aspects of the tumor, they do not fully replicate the complexity of the TME, such as interactions with the immune system or vasculature (Robinson et al., 2015). This limits their ability to model immune responses or the dynamics of tumor-associated inflammation. Most PCa organoid models are derived from immunodeficient sources (such as xenografts or PDXs), limiting their ability to study immune system interactions. However, coculture systems are being developed to address this gap (Gao et al., 2014). While organoids can be used to study metastasis *in vitro*, they may not fully recapitulate the intricate processes of metastasis that occur in vivo, such as the complex interactions between circulating tumor cells (CTCs), the ECM, and distant organs.

4.10 3D CELL CULTURE MODELS

3D cell culture models are rapidly gaining prominence in PCa research due to their ability to more closely mimic the in vivo tumor microenvironment compared to traditional 2D cell cultures. These 3D models allow for the study of tumor biology, drug responses, metastasis, and tumor-stroma interactions in a more physiologically relevant context.

4.10.1 Types of 3D Cell Culture Models for Prostate Cancer

Spheroid Cultures: One of the most common 3D cell culture models, spheroids are formed when PCa cells are cultured in non-adherent conditions, allowing them to aggregate into spherical structures. These spheroids replicate the 3D organization of tumors, with a central necrotic core surrounded by proliferating cells (Gao et al., 2014). Spheroid models are useful for studying drug resistance, cell-cell interactions, and the effects of hypoxia on tumor growth.

Organoids: PCa organoids are 3D structures derived from patient-derived tissues or cell lines that can self-organize into gland-like structures. They provide a more accurate model for studying the heterogeneity of PCa, tumor-stroma interactions, and metastasis (Beshiri et al., 2018). Organoids are increasingly used in precision medicine for drug testing and personalized therapies.

Fibrin-Based and Matrigel-Based Cultures: These models involve the use of scaffolds, such as Matrigel or fibrin, to support 3D cell culture. These extracellular matrices allow PCa cells to grow in a 3D configuration that mimics the in vivo microenvironment, facilitating studies on cellular behaviors such as invasion, migration, and angiogenesis (Matrigel) or tissue formation (fibrin) (Pauli et al., 2017, 2019).

4.10.2 Applications of 3D Cell Culture Models for Prostate Cancer

Drug Screening and Personalized Medicine: 3D PCa models allow for high-throughput screening of anticancer drugs and evaluation of their effects in a more clinically relevant setting. By testing drug responses on PCa spheroids or organoids derived from individual patients, these models offer a personalized approach to treatment (Kemper et al., 2020).

Studying Tumor-Stroma Interactions: The TME, including interactions between cancer cells and stromal cells, plays a crucial role in PCa progression and drug resistance. 3D cell cultures allow for the incorporation of stromal cells (e.g., fibroblasts, endothelial cells) into PCa models, providing a platform to study these interactions.

Metastasis Modeling: 3D models, particularly those based on coculture systems, are used to study PCa metastasis. By co-culturing PCa cells with endothelial or immune cells, researchers can investigate how the cancer cells invade the surrounding tissue and spread to distant organs, such as the bone, which is a common site for PCa metastasis (Boj et al., 2014).

4.10.3 Advantages and Limitations of 3D Cell Culture Models for Prostate Cancer

3D cultures more closely replicate the structural and functional aspects of in vivo tumors, including cell-cell and cell-matrix interactions, compared to 2D cultures. This enables better modeling of PCa growth, differentiation, and resistance mechanisms (Gao et al., 2014). For example, PCa spheroids replicate the hypoxic conditions found in tumors, which is crucial for studying cancer metabolism and drug resistance. 3D cell culture systems have been shown to more accurately predict drug efficacy compared to 2D models. This is because they better mimic the physical barriers and drug diffusion patterns encountered in actual tumors (Gao et al., 2014). Organoids, with their preserved tumor architecture, are particularly useful for screening drugs that affect tumor growth, apoptosis, or metastasis. Unlike 2D cell cultures, 3D models can incorporate various elements of the TME, such as stromal cells, immune cells, and ECM components, which are critical for studying cancer progression and response to therapies (Pauli et al., 2017, 2019).

While 3D cell culture models offer significant advantages, they are often more costly and technically challenging to establish and maintain compared to traditional 2D cultures. This can limit their widespread use, especially in high-throughput screening (Gao et al., 2014). Most 3D PCa models, including organoids and spheroids, are derived from human tumor tissues or cell lines, but they generally lack a fully functioning immune system and vascularization. This is a significant limitation for studying immune responses or testing immunotherapies (Bluemn, 2017).

4.11 COCULTURE SYSTEMS

Coculture systems in PCa research have become an essential tool for studying the complex interactions between tumor cells and various components of the TME, such as stromal cells, immune cells, endothelial cells, and fibroblasts. These systems allow researchers to better understand how the PCa cells interact with their surrounding environment, contributing to tumor progression, metastasis, drug resistance, and therapeutic response.

4.11.1 Types of Coculture Systems in Prostate Cancer

Cancer Cell + Stromal Cell Coculture: Stromal cells, including fibroblasts, myofibroblasts, and endothelial cells, are critical components of the PCa microenvironment. Co-culturing PCa cells with stromal cells allows the study of tumor-stroma interactions that influence tumor growth, invasion, and response to therapies. For example, PCa-associated fibroblasts (CAFs) secrete various growth factors and cytokines that promote tumor progression and drug resistance (Balkwill et al., 2012).

Cancer Cell + Immune Cell Coculture: Coculturing PCa cells with immune cells, such as T-cells, macrophages, or dendritic cells, is important for studying the tumor-immune interaction. These models help researchers understand immune evasion mechanisms, immunotherapy responses, and the development of resistance to immune checkpoint inhibitors (e.g., PD-1/PD-L1 blockade). Tumor-associated macrophages (TAMs) and regulatory T-cells (Tregs) are known to play roles in suppressing immune responses in PCa (Balkwill et al., 2012).

Cancer Cell + Endothelial Cell Coculture: Endothelial cells line blood vessels and play a crucial role in angiogenesis, which supports tumor growth and metastasis. PCa cells can be cocultured with endothelial cells to investigate the role of angiogenesis in tumor progression. These models are useful for studying the effects of anti-angiogenic drugs and understanding the mechanisms underlying vascularization in PCa (Balkwill et al., 2012).

4.11.2 Applications of Coculture Systems in Prostate Cancer

Studying Tumor Progression and Metastasis: Coculture systems can mimic the interactions between PCa cells and surrounding tissues, allowing researchers to investigate how these interactions contribute to cancer progression and metastasis. For example, the interaction between PCa cells and CAFs has been shown to promote epithelial-mesenchymal transition (EMT), a key process in metastasis. Cocultures with endothelial cells can be used to study angiogenesis and the formation of new blood vessels that support metastatic growth (Kamphaus et al., 2000).

Drug Resistance Mechanisms: Tumor–stroma interactions are important for developing resistance to chemotherapy and targeted therapies. Coculture models allow researchers to investigate how stromal cells, such as fibroblasts or TAMs, may protect cancer cells from drug-induced apoptosis. For instance, CAFs have been shown to secrete factors that enhance the survival of PCa cells under treatment with chemotherapy agents.

Immunotherapy Testing: Coculture systems involving PCa cells and immune cells are used to evaluate the efficacy of immune checkpoint inhibitors, such as anti-PD-1/PD-L1 therapies (Zhao et al., 2021). These models enable the investigation of how the TME, including immune cells like Tregs and TAMs, modulates the immune response to treatment. By co-culturing cancer cells with immune cells, researchers can study the dynamics of immune evasion and develop strategies to overcome resistance to immunotherapy.

4.11.3 Advantages and Limitations of Coculture Systems in Prostate Cancer Research

Coculture models incorporate multiple cell types, allowing for a more accurate representation of the PCa microenvironment than monoculture systems. These interactions are crucial for studying the complex behavior of cancer cells, including proliferation, invasion, and resistance to therapy (Zhao et al., 2019). Coculture systems are particularly useful for investigating how cancer cells interact with stromal components (fibroblasts, immune cells, endothelial cells) in the TME. This helps to uncover the cellular and molecular mechanisms that drive tumor progression, metastasis, and therapeutic resistance (Balkwill et al., 2012). Coculture systems are useful for evaluating the effects of therapeutic agents, including chemotherapy, targeted therapies, and immunotherapies, in the presence of stromal cells or immune cells. This allows researchers to test how therapies can overcome the protective effects of the tumor microenvironment and restore therapeutic sensitivity (Han et al., 2020).

While coculture systems provide a more realistic model of the tumor microenvironment than monoculture systems, they still cannot fully replicate all aspects of the in vivo environment. For instance, they often lack the full diversity of immune cells, ECM components, or complex tissue architecture found in tumors. Studying the interactions between different cell types in coculture systems can be challenging. The effects of specific cell-cell interactions on tumor progression, drug resistance, or metastasis are often difficult to isolate, requiring sophisticated techniques such as live-cell imaging or gene expression profiling (Balkwill et al., 2012). Coculture systems can be labor-intensive and costly to establish and maintain, especially when multiple cell types or primary patient-derived cells are involved. This limits their use in high-throughput screening or in large-scale studies (Zhao et al., 2019).

4.12 GENE EDITING MODELS

Gene editing models, particularly using technologies like CRISPR/Cas9, have revolutionized cancer research, including the study of PCa. These models allow researchers to modify the genome of PCa cell lines, animal models, or organoids to investigate the roles of specific genes in cancer initiation, progression, metastasis, and drug resistance. Gene editing also facilitates the study of genetic mutations commonly associated with prostate cancer, providing a better understanding of tumor biology and identifying potential therapeutic targets.

4.12.1 Types of Gene Editing Models in Prostate Cancer

CRISPR/Cas9 in PCa **Cell Lines**: This technology allows precise modifications in the genome, enabling the study of the effects of specific genetic alterations, such as mutations in the *PTEN*, *TP53*, or *BRCA2* genes, which are commonly associated with PCa (Ye et al., 2017). CRISPR-mediated gene editing is used to explore cancer cell proliferation, migration, apoptosis, and drug resistance mechanisms.

CRISPR/Cas9 in Prostate Cancer Mouse Models: These models are used to investigate the role of specific genes in PCa development and metastasis. For example, CRISPR/Cas9 has been employed to knock out the *PTEN* gene in mouse models, leading to the development of PCa that mimics the human disease, offering a valuable tool for studying tumorigenesis, metastasis, and therapeutic responses (Cunningham & You, 2015).

Gene Editing in Prostate Cancer Organoids: CRISPR/Cas9 can be used to introduce or correct mutations, allowing researchers to study how these genetic changes affect tumor growth and response to therapies. This is particularly useful for personalized medicine, where organoids derived from individual patients are edited to reflect their unique genetic landscape (Beshiri et al., 2018).

4.12.2 Applications of Gene Editing Models in Prostate Cancer

Studying Cancer Pathogenesis and Genetic Alterations: Gene editing allows for the investigation of specific genetic mutations that drive PCa. Common mutations in PCa include loss of *PTEN*, mutations in the *AR*, and alterations in tumor suppressor genes such as *TP53* (Taylor et al., 2010). By editing these genes in PCa models, researchers can directly examine how each mutation contributes to tumor initiation and progression.

Metastasis Modeling: Gene editing can be used to study the role of specific genes in the metastatic spread of PCa. For instance, knocking out or overexpressing genes involved in epithelial-mesenchymal transition, such as *TWIST1*, can help reveal how PCa cells acquire the ability to invade surrounding tissues and spread to distant organs (Xu et al., 2006).

Drug Resistance Mechanisms: Gene editing in PCa models enables the study of how genetic alterations contribute to resistance to chemotherapy or ADT. For example, mutations in *AR* or *TP53* are frequently implicated in resistance to ADT, and gene editing can be used to introduce these mutations into PCa cell lines to study the resulting resistance mechanisms (Hull et al., 2002).

4.12.3 Advantages and Limitations of Gene Editing Models in Prostate Cancer

Gene editing technologies like CRISPR/Cas9 allow for precise modifications in the PCa genome, enabling the investigation of the specific role of individual genes in cancer progression (Ye et al., 2017). Gene editing can be used to generate PCa models that accurately replicate the complex genetic alterations seen in human PCa. This is particularly useful for studying mutations that occur in key genes like *PTEN*, *TP53*, and *BRCA2*, which are involved in tumor suppression and therapy resistance (Cunningham & You, 2015). By using patient-derived organoids or cell lines, gene editing can be applied to reflect the unique genetic profiles of individual PCa patients. This allows for the development of personalized treatment strategies and the evaluation of how specific genetic alterations influence drug responses (Beshiri et al., 2018).

One of the major challenges with CRISPR/Cas9 gene editing is the potential for off-target effects, where unintended genes are modified. These off-target mutations can complicate the interpretation of experimental results and may introduce errors that mask the true biological effects of the targeted gene alterations (Ye et al., 2017). The development of knock-in or knock-out models often requires extensive optimization of transfection protocols or careful management of gene-editing events (Ye et al., 2017). PCa is a genetically heterogeneous disease, tumor heterogeneity can impact treatment responses and metastasis, which may not be fully modeled by single-genetic edits (Taylor et al., 2010).

4.12.4 In Vivo Models for Studying Prostate Cancer

In vivo models are essential for studying PCa as they provide a more accurate representation of human cancer biology, including interactions with the TME, immune system, and metastatic processes. These models enable researchers to investigate the development, progression, metastasis, and therapeutic responses of PCa in a living organism.

The main **in vivo models** used in PCa research, including their characteristics, advantages, limitations, and relevant citations are as follows:

4.13 MOUSE MODELS

Mouse models are the most commonly used in vivo systems for studying PCa, as they allow for the manipulation of genes, the study of tumor progression, and the testing of therapeutic strategies.

a. **Syngeneic Mouse Models**: Syngeneic mouse models involve transplanting PCa cells from one mouse strain into another of the same genetic background, allowing for the study of tumor progression and immune interactions without the confounding factors of an immune response against the host (Wculek et al., 2019). These models are typically used in immunocompetent mice, meaning the immune system is functional and can mount a response against the tumor, mimicking human cancer immunology more closely than in xenograft models.

4.13.1 Types of Syngeneic Mouse Models

Transgenic Adenocarcinoma of the Mouse Prostate (TRAMP): The TRAMP mouse model is genetically engineered to express the SV40 T antigen under the control of the prostate-specific PSA promoter, leading to the development of PCa from hyperplasia to metastasis. TRAMP mice are commonly used for studying tumor progression, immune evasion, and therapeutic interventions.

PTEN−/− Mouse Model: These mice have a heterozygous knockout of **PTEN**, a key tumor suppressor gene frequently mutated in human PCa. PTEN loss leads to the development of PCa and mimics the molecular alterations seen in human disease, making it a valuable model for studying cancer progression and metastasis.

4.13.2 Applications of Syngeneic Mouse Models in Prostate Cancer

Studying Immune Interactions in Prostate Cancer: Syngeneic mouse models enable the study of how tumors interact with immune cells such as T cells, macrophages, and dendritic cells. They are widely used to evaluate immune checkpoint inhibitors, such as anti-PD-1 or anti-CTLA-4 therapies, that aim to re-activate the immune system against the tumor (Takimoto et al., 2019).

Evaluating Immunotherapy: These models have been used to test novel immune-based therapies for PCa, such as tumor vaccines, immune checkpoint inhibitors, and adoptive T-cell therapies. For instance, *TRAMP-C2* cells implanted into syngeneic mice have been used to study immune response-based therapies, including those that activate or suppress the immune system's ability to target prostate tumors.

Testing Combination Therapies: Syngeneic models are used to assess the effectiveness of combination treatments, such as combining immune checkpoint inhibitors with conventional therapies like chemotherapy, radiation, or ADT. The *TRAMP-C2* model, for example, has been used to explore the synergy between immune therapies and ADT, demonstrating how AR receptor signaling affects immune cell recruitment to the tumor (Zhao et al., 2021).

4.13.3 Advantages and Limitations of Syngeneic Mouse Models in Prostate Cancer

One of the major advantages of syngeneic models is that they involve an immunocompetent mouse, meaning that the immune system is functional. This allows researchers to study the effects of immune responses on tumor progression and evaluate the effectiveness of immunotherapies in an intact immune environment, which is critical for understanding how these therapies might perform in human patients (Takimoto et al., 2019). By studying PCa in the context of an intact immune system and native microenvironment, researchers can better understand how the immune system affects tumor progression, resistance, and metastasis. Syngeneic models help elucidate how PCa tumors develop immune evasion strategies, such as immune checkpoint upregulation, and test the efficacy of novel immune-modulating treatments (Yue et al., 2019).

While syngeneic models provide valuable insights into immune-tumor interactions, they may not fully capture the genetic and phenotypic heterogeneity found in human PCa (Bhardwaj et al., 2017). Although syngeneic models are immunocompetent, the mouse immune system differs from the human immune system in several key ways. This can limit the ability to fully translate findings from mouse models to human therapies (Takimoto et al., 2019). While orthotopic syngeneic models can replicate many aspects of human PCa, they still cannot fully reproduce all components of the human TME, particularly the endothelial and stromal cells that interact with cancer cells. These differences may limit the model's ability to predict human responses to treatments (Yue et al., 2019).

4.14 XENOGRAFT MODELS

Xenograft models are widely used in PCa research to study tumor growth, metastasis, drug efficacy, and molecular mechanisms in a setting that more closely resembles human PCa. These models involve the implantation of human PCa cells or tissues into immunodeficient mice, which lack a functional immune system, thereby preventing rejection of the human tumor cells.

4.14.1 Types of Xenograft Models in Prostate Cancer

Subcutaneous Xenografts: In subcutaneous xenograft models, PCa cells are injected into the subcutaneous tissue of immunodeficient mice. This model is commonly used for rapid tumor growth and drug testing but does not fully recapitulate the tumor's natural microenvironment or metastatic potential (Lawrence et al., 2013).

Orthotopic Xenografts: In orthotopic xenograft models, PCa cells are implanted directly into the prostate tissue of immunodeficient mice. This method is preferred for studying tumor growth, metastasis, and interactions with the local microenvironment because it allows the cancer cells to grow in the tissue of origin, simulating human PCa more accurately than subcutaneous xenografts.

Patient-Derived Xenografts (PDXs): Patient-derived xenografts (PDXs) are generated by implanting tumor tissue from human PCa patients into immunodeficient mice. PDX models preserve the histopathological and molecular characteristics of the original patient tumor, making them particularly valuable for studying individual tumor heterogeneity, drug resistance, and personalized therapies (Nguyen et al., 2017).

4.14.2 Applications of Xenograft Models in Prostate Cancer

Studying Tumor Growth and Metastasis: These models are widely used to explore the molecular mechanisms underlying metastasis and to test potential inhibitors of metastatic progression (Zhu et al., 2014).

Testing Chemotherapeutic Agents and Targeted Therapies: Xenograft models are extensively used in preclinical testing of new therapeutic agents, including chemotherapy, immunotherapy, and targeted therapies. The subcutaneous xenograft model is often used to evaluate the antitumor activity of small molecule inhibitors, such as those targeting AR signaling (e.g., enzalutamide), PI3K/Akt pathways, and other molecular targets (Nguyen et al., 2017).

Patient-Derived Xenografts for Personalized Medicine: PDX models provide a more clinically relevant platform for evaluating the responses of individual patient tumors to various treatments. This approach allows researchers to study tumor-specific characteristics, such as gene mutations and expression patterns, and to predict how a particular patient might respond to specific therapies (Nguyen et al., 2017).

4.14.3 Advantages and Limitations of Xenograft Models in Prostate Cancer

Unlike traditional cell line models, xenograft models allow for the study of tumor behavior in the context of a whole organism with an intact TME. This is particularly valuable for studying tumor growth, angiogenesis, and metastasis in PCa (Zhu et al., 2014). Xenografts enable the use of human PCa cells in a living mouse model, which allows for more accurate translation of findings to human clinical settings (Nguyen et al., 2017). Xenograft models are widely used in drug discovery and preclinical testing. These models provide a platform to test the efficacy of potential therapies before clinical trials, enabling the evaluation of therapeutic agents under conditions that better mimic human cancer (Lawrence et al., 2013).

One of the major limitations of xenograft models is the lack of an intact immune system in the host mice, particularly in immunodeficient models (e.g., NOD/SCID, athymic nude). This prevents the study of tumor-immune system interactions and the evaluation of immune therapies, such as immune checkpoint inhibitors, in a normal immune context (Nguyen et al., 2017). Factors such as the immune landscape, vasculature, and stromal components may not behave exactly as they do in humans, potentially limiting the translatability of some findings (Zhu et al., 2014). Some xenograft models, particularly cell line-based models, may not fully reflect the genetic and phenotypic heterogeneity observed in human PCas. Even PDX models, while more representative of human tumors, can still suffer from issues such as clonal selection during tumor propagation in mice (Nguyen et al., 2017).

4.15 TRANSGENIC MOUSE MODELS

Transgenic mouse models are engineered to express or overexpress genes that drive PCa development. These models enable researchers to study the molecular mechanisms of PCa initiation, progression, metastasis, and response to therapies in vivo.

4.15.1 Types of Transgenic Mouse Models

The most common methods for generating transgenic PCa models include **oncogene expression** and **tumor suppressor gene loss** (e.g., prostate-specific overexpression of Myc, PTEN deletion, or mutant Ras expression) (Foster et al., 1997).

Transgenic Adenocarcinoma of the Mouse Prostate (TRAMP) Model: One of the most widely used transgenic models is the TRAMP model, which expresses the **Simian virus 40 (SV40)** large T-antigen under the control of the **PSA promoter**. This model develops PCa that progresses from low-grade prostatic intraepithelial neoplasia (PIN) to invasive carcinoma and metastasis, making it a valuable tool for studying PCa initiation, progression, and therapy.

PTEN-Deficient Models: Another popular transgenic model involves the deletion of the **PTEN tumor suppressor gene** specifically in the prostate, leading to PCa development. PTEN loss is commonly observed in human PCa, and the PTEN-deficient model mimics human disease in terms of ARsignaling, cell survival, and PI3K-Akt pathway activation (Yoshimoto et al., 2008).

Myc Transgenic Models: Overexpression of the **Myc oncogene** in the prostate of mice leads to the development of PCa. The Myc transgenic models are commonly used to study cell cycle regulation, apoptosis, and the oncogenic effects of Myc in PCa progression (Mehta et al., 2011).

4.15.2 Applications of Transgenic Mouse Models in Prostate Cancer

Studying Prostate Cancer Progression: These models mimic the stages of human PCa, enabling the investigation of genetic alterations and molecular pathways involved in each stage (Foster et al., 1997).

Investigating Tumor Microenvironment: Transgenic mouse models are also valuable for studying the interactions between tumor cells and the surrounding microenvironment. For example, the TRAMP model provides insight into how PCa cells interact with stromal cells, endothelial cells, and immune cells in the tumor microenvironment (Kim et al., 2015).

Evaluating Therapeutic Responses: These models allow for the preclinical evaluation of new therapeutic strategies. For instance, in the **PTEN-deficient model**, researchers can test the effectiveness of targeted therapies that inhibit the PI3K-Akt pathway, which is often activated in human PCa due to PTEN loss (Yoshimoto et al., 2008).

4.15.3 Advantages and Limitations of Transgenic Mouse Models in Prostate Cancer

One of the major advantages of transgenic mouse models is their ability to replicate genetic mutations commonly found in human PCa, such as PTEN loss and Myc overexpression. This allows for understanding the molecular mechanisms of PCa development and progression (Yoshimoto et al., 2008). Transgenic models often capture the heterogeneity seen in human tumors. For example, the TRAMP model mimics the heterogeneity of human PCa by showing a spectrum of disease, from low-grade to high-grade cancer (Foster et al., 1997). Unlike xenograft models, where tumors are implanted into mice, transgenic mouse models allow for spontaneous tumor development. This is advantageous for studying tumor initiation and the early stages of cancer, as well as for evaluating long-term effects of genetic mutations or therapies (Kim et al., 2015).

One of the major limitations of transgenic mouse models is the long period required for tumor development. Tumors may take several months to develop, making them less suitable for high-throughput screening or for quickly evaluating the efficacy of drugs (Foster et al., 1997). While transgenic models can mimic certain aspects of human PCa, they may not fully replicate the complexity of the human TME, immune response, or interactions with the stroma (Kim et al., 2015). Many transgenic models rely on the overexpression of specific oncogenes (e.g., Myc) or the deletion of tumor suppressors (e.g., PTEN). While these genetic alterations are common in human cancer, they do not always reflect the full spectrum of mutations or the genetic variability seen in clinical PCa, potentially limiting their generalizability (Mehta et al., 2011).

4.16 BONE METASTASIS MODELS

Description: Bone metastasis is a hallmark of advanced PCa. In vivo models are used to study the mechanisms by which PCa spreads to bone and to test therapies that target bone metastases.

4.16.1 Types of Bone Metastasis Models

PCa frequently metastasizes to the bones, particularly to the vertebrae, pelvis, and long bones. The **bone metastasis models** in PCa are designed to mimic this process, allowing researchers to study the interaction between PCa cells and the bone microenvironment. There are several types of models used to study bone metastasis, including **in vivo mouse models, cell line-based models**, and **PDXs**.

In Vivo Bone Metastasis Models: These models involve the injection of PCa cells into immunocompromised mice, often directly into the bone or via the tail vein to simulate the hematogenous spread of PCa to the bones.

The most common sites for metastasis are the **femur**, **tibia**, **lumbar vertebrae**, and **pelvis**.

Orthotopic Bone Models: PCa cells can be implanted orthotopically into the prostate of mice, allowing the tumor to grow in the prostate and metastasize to the bone naturally. This model mimics the human disease more accurately by enabling spontaneous metastasis from the primary site (human prostate) to the bone (e.g., using *LNCaP* or *PC-3* cells) (Wang et al., 2014).

Tail Vein Injection Models: Another approach involves injecting PCa cells into the tail vein of mice, which can subsequently circulate to the bone. This model is particularly useful for studying the early stages of metastasis and understanding the metastatic cascade, including the interaction between CTCs and bone tissue (Thalmann et al., 1994).

4.16.2 Applications of Bone Metastasis Models in Prostate Cancer

Studying Tumor-Bone Microenvironment Interactions: Bone metastasis models are essential for investigating how PCa cells interact with the bone microenvironment, including osteoblasts, osteoclasts, and bone matrix components. This interaction is critical for the growth and progression of metastatic lesions in the bone. For example, in models where PCa cells are injected into the bone, researchers can study how tumor cells stimulate bone resorption or mineralization, which contributes to the osteolytic (bone-degrading) or osteoblastic (bone-forming) lesions seen in patients (Thalmann et al., 1994).

Evaluating Therapeutic Strategies: Bone metastasis models provide a platform for evaluating novel therapeutic approaches targeting both PCa cells and the bone microenvironment. For instance, treatments targeting **receptor activator of nuclear factor kappa-B ligand** signaling, which is involved in osteoclast activation, have shown promise in preventing bone metastasis and bone-related complications in PCa (Wang et al., 2014).

Studying Molecular Pathways: These models are valuable for studying the molecular pathways that drive PCa bone metastasis, such as the **Wnt/β-catenin**, **TGF-β**, and **EGF** signaling pathways. Bone metastasis models allow researchers to explore how these pathways contribute to tumor growth and the regulation of the bone microenvironment, and to identify potential biomarkers and therapeutic targets (Thalmann et al., 1994).

4.16.3 Advantages and Limitations of Bone Metastasis Models in Prostate Cancer

Bone metastasis is a major cause of morbidity in PCa patients, making these models highly relevant for studying the metastatic spread of PCa in a clinically meaningful context. By studying bone metastasis in vivo, researchers can better understand the pathophysiology of the disease and identify potential targets for therapeutic intervention (Thalmann et al., 1994). Bone metastasis models provide a platform for investigating the role of osteoblasts, osteoclasts, and other bone-related factors that contribute to tumor growth and bone destruction (Wang et al., 2014). Models that involve the injection of PCa cells into the tail vein or orthotopic implantation allow for the study of the metastatic cascade, including the role of CTCs and how they interact with endothelial cells in the bone vasculature. This is valuable for understanding how PCa cells invade and colonize bone tissue (Thalmann et al., 1994).

While these models are valuable for studying bone metastasis, they do not fully replicate the complexity of human PCa metastasis. For example, the immune system is often absent or compromised in many of the commonly used models (e.g., immunodeficient mice), which can affect the ability to study the immune response to bone metastasis or evaluate immunotherapies (Wang et al., 2014). Most bone metastasis models use a single cell line (e.g., *PC-3*, *LNCaP*), which may not fully reflect the genetic and phenotypic heterogeneity observed in human PCa (Thalmann et al., 1994). While these models allow for the study of tumor-bone interactions, the bone microenvironment in mice differs from that in humans (Wang et al., 2014).

4.17 METASTASIS MODELS

Models specifically designed to study metastasis allow researchers to track the spread of PCa cells from primary tumors to secondary sites like the lymph nodes, liver, and lungs. These models help in investigating the metastatic cascade, interactions with the microenvironment, and evaluating potential therapeutic strategies to prevent or treat metastasis.

4.17.1 Types of Metastasis Models

Metastasis models in PCa generally involve introducing PCa cells into the bloodstream or specific tissues of mice to mimic the metastatic process seen in human patients. These models help in understanding tumor cell dissemination, organ-specific colonization, and metastasis-related signaling pathways.

Orthotopic Injection Models: In **orthotopic injection models**, PCa cells are implanted directly into the prostate of mice, which allows for the spontaneous metastasis of cancer cells to distant organs, particularly the bone. This method accurately replicates the primary site of the tumor and provides a more natural progression of metastasis to organs like the **bone**, **lymph nodes**, and **lungs** (Stephenson et al., 1992). These models are particularly useful for studying the early stages of metastasis and the role of the tumor microenvironment in promoting metastasis.

Tail Vein Injection Models: In **tail vein injection models**, PCa cells are injected into the tail vein of mice, allowing them to circulate in the bloodstream and potentially metastasize to distant organs like the **lungs**, **liver**, and **bone** (Stephenson et al., 1992). This model enables the monitoring of tumor cell arrest, extravasation, and colonization in target organs.

Patient-Derived Xenograft (PDX) Models: In **PDX models**, tissue from human PCa patients is implanted into immunodeficient mice, which allows the cancer to grow and metastasize in a more clinically relevant environment. These models can closely replicate the heterogeneity and complexity of human PCa metastasis, making them valuable for studying metastasis in various organs, including bone and lymph nodes (Linxweiler et al., 2020).

Genetically Engineered Mouse Models (GEMMs): **Genetically engineered mouse models (GEMMs)**, such as those with prostate-specific expression of oncogenes (e.g., Myc or Ras) or loss of tumor suppressors (e.g., PTEN), allow for the development of spontaneous metastasis. These models provide insights into how specific genetic alterations contribute to PCa metastasis and help identify potential therapeutic targets (Kim et al., 2015).

4.17.2 Applications of Metastasis Models in Prostate Cancer

Studying Metastatic Cascade: Metastasis models are vital for understanding the **metastatic cascade**, which involves the steps of primary tumor cell invasion, intravasation into the bloodstream, extravasation into distant tissues, and colonization (Stephenson et al., 1992). These models provide insights into how PCa cells breach vascular barriers, survive in the bloodstream, and establish metastatic lesions in distant organs.

Understanding Bone Metastasis: PCa is known for preferentially metastasizing to the bone. Bone metastasis models, such as **orthotopic injection models** and **tail vein injection models**, are widely used to study the interaction between PCa cells and the bone microenvironment, including the roles of osteoclasts and osteoblasts in tumor growth and bone remodeling (Kim et al., 2015).

Evaluating Therapeutic Interventions: Metastasis models are frequently used to evaluate the efficacy of potential therapies designed to inhibit metastasis. For example, **receptor activator of nuclear factor kappa-B ligand inhibitors** have been tested in PCa metastasis models to prevent bone metastasis by inhibiting osteoclast-mediated bone resorption (Linxweiler et al., 2020).

4.17.3 Advantages and Limitations of Metastasis Models in Prostate Cancer

Metastasis models are highly relevant to human PCa, as they allow researchers to study how cancer cells disseminate to different organs, particularly the **bone**, which is a common site of metastasis in PCa patients. These models help identify factors that promote metastasis and establish new therapeutic targets (Linxweiler et al., 2020). These models also allow researchers to investigate how PCa cells interact with the TME, including endothelial cells, immune cells, fibroblasts, and bone stromal cells. Understanding these interactions is crucial for developing strategies to block metastasis or target specific microenvironmental factors that facilitate the spread of cancer (Kim et al., 2015). Metastasis models provide a platform for testing the efficacy of new drugs and therapies aimed at reducing metastatic spread or treating metastatic lesions. These include **targeted therapies** (e.g., inhibitors of the AR or PI3K pathway), **immunotherapies**, and **bone-targeted therapies** (e.g., bisphosphonates or denosumab) (Stephenson et al., 1992).

Many metastasis models, particularly **xenograft** and **PDX models**, are performed in immunodeficient mice, which do not have a functional human immune system.

This limits the ability to study immune responses to metastatic tumors or evaluate the efficacy of immunotherapies (Stephenson et al., 1992). Differences between human and mouse TMEs can limit the ability to fully replicate human PCa metastasis (Kim et al., 2015). Although **PDX models** closely replicate human tumor heterogeneity, they may not fully capture all genetic variations found in clinical PCa, limiting their utility in some studies (Linxweiler et al., 2020).

4.18 HUMANIZED MODELS

4.18.1 Humanized Mice

4.18.1.1 Development and Applications of Humanized Mouse Models

Humanized models are experimental systems that bridge the gap between preclinical and clinical research by integrating human genes, cells, or tissues into nonhuman creatures to replicate human biological processes. Humanized mice, a prominent example, are genetically modified or engrafted with human tissues, genes, or immune system components, allowing researchers to examine biological and pathological processes unique to humans. Currently, these models are being used by numerous researchers to assess the effectiveness of immunotherapies and investigate the relationships between human tumors and the immune system (Yin et al., 2020).

Humanized mouse models' creation includes engrafting human immune components into immunodeficient mice, allowing the generation of a functional human immune system within a murine host. There are three main approaches to achieve this. To achieve this, there are three primary methods. In the first method, represented by Hu-PBL-SCID, immunodeficient mice are injected with human peripheral blood lymphocytes (PBLs), creating a temporary immune system that is dominated by human T cells. Although this approach is rapid and simple, it is constrained by a risk of graft-versus-host disease and a short model lifespan (King et al., 2009). The second method, the Hu-SRC-SCID model, involves injecting immunodeficient mice with human hematopoietic stem cells (HSCs), which are obtained from bone marrow, cord blood, or peripheral blood. By differentiating inside the mouse, these cells produce a range of human immune cells and offer a more reliable model (Bernard et al., 2008). The third and most comprehensive method, known as the BLT method, combines the injection of HSC with the transplanting of human fetal liver and thymus tissues beneath the mouse's renal capsule. As a result, a completely functional immune system with appropriate T-cell maturation and thymic education is developed. Strong platforms for researching disease mechanisms, human immunity, and therapeutic interventions are offered by these methodical approaches (Shultz et al., 2012) (Figure 4.2 and Table 4.2).

In biomedical research, humanized mouse models are vital, particularly when studying human diseases, assessing novel treatments, and developing customized medicine. These models enable the in vivo study of biological processes unique to humans that are otherwise difficult to simulate in traditional mouse models due to species differences. To study human immune responses and disease mechanisms, humanized mice typically occur by transplanting mice that lack immunity with

TABLE 4.2
Mouse Strains for Immune Deficiency and Engraftment Studies

Mouse Strain	Immune Phenotype	Source of Reconstitution	Engraftment and Immune Response
SCID	Absence of T and B cells	Fetal hematopoietic tissue	Minimal engraftment levels
NOD/SCID	T and B cell deficiency	PBMCs, HSCs	Improved engraftment, residual innate immunity
Elevated NK Cell Activity	High NK cell activity, reduced T-cell activity	PBMCs, Spleen	Low T-cell engraftment, residual NK cells, and immunity
$Rag1^{null}$ or $Rag2^{null}$	Absence of T- and B-cells	HSCs	Absence of T and B cells
NOD/SCID/ IL-$2r\gamma^{null}$	T and B cell absence, innate immune defects	PBMCs, HSCs	High engraftment and restoration of immune cells

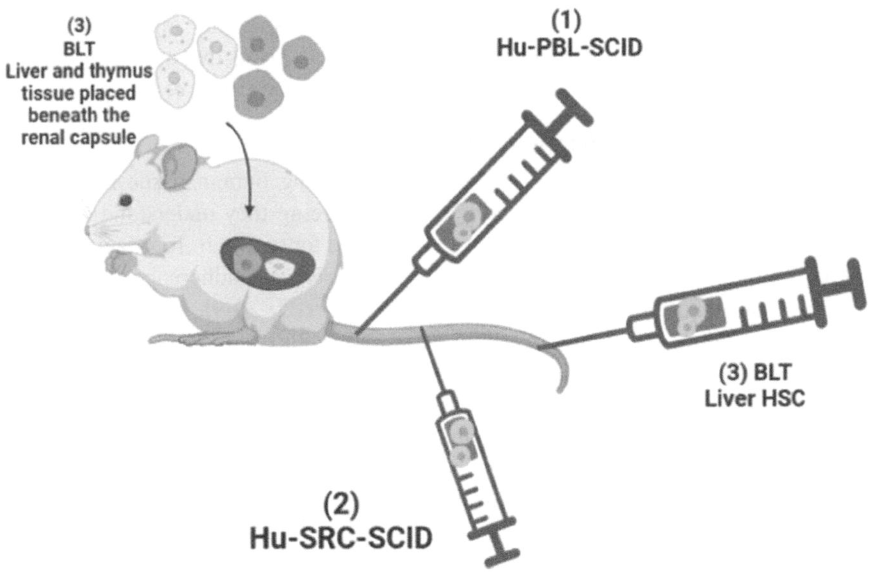

FIGURE 4.2 Developmental methods for humanized mice models.

human tissues, such as fetal tissues, adult immune cells, or hematopoietic stem cells. Their uses include infectious disease research, where humanized mice provide a more realistic depiction of human pathogen-host interactions, and cancer research, where they are used to test novel immunotherapies like checkpoint inhibitors. These models additionally help in the development of vaccines and the examination of drugs toxicity and efficacy in an immunological environment similar to that of humans, making the shift from preclinical to clinical trials easier (Karnik et al., 2023).

4.18.1.2 Human Tissue Explants

Human tissue explants are samples of human tissue that have been removed from the body and are being cultivated in a lab for scientific study. More physiologically relevant than traditional cell culture, these explants are usually utilized to research the behavior of tissues or organs. In a system that more closely resembles the in vivo environment, explants enable researchers to examine intricate biological processes like cancer formation, tissue regeneration, immunological responses, and medication interactions by maintaining the tissue's architecture and cellular makeup (Grivel & Margolis, 2009).

4.18.1.3 Use of Human Prostate Tissue in Research

The transplants of human prostate tissue are useful in medical studies, especially when investigating tissue regeneration, hormone control, and PCa. Since these explants preserve the prostate's cellular composition and structural integrity, scientists can watch as immune, stromal, and cancer cells interact in real time (Amornsupak et al., 2014). Explants of prostate tissue are used to investigate the molecular reasons behind the course of disease, evaluate the effects of possible therapeutic drugs, and compare the effects of various treatments in a model that is more useful for humans. This method offers important insights into the biology of cancer, especially regarding interactions between the TME and treatment resistance mechanisms that are challenging to reproduce in conventional cell lines or animal models (Levesque & Nelson, 2017).

4.18.1.4 Ethical Considerations and Limitations

There are important ethical questions raised by using human tissue in research. Donors must give their informed consent, guaranteeing they understand how their tissue will be utilized, preserved, and perhaps shared for research. To safeguard donor rights and privacy, ethical principles also mandate rigorous adherence to regulatory agencies and institutional review boards (IRBs) (Grizzl, 2019). The need for open procedures and precise rules governing tissue ownership and use is underscored by worries about the commercialization of human tissue and the possibility of exploitation. Human tissue explants are also limited by sample variability, the difficulty of sustaining tissue viability over time, and the difficulty of accurately simulating the entire complexity of in vivo settings. All these factors can have an impact on the dependability and repeatability of results, hence rigorous experimental techniques and meticulous standardization are required (Aronson & Rehm, 2015).

4.18.1.5 Translational Applications

The process of turning fundamental scientific findings into useful medical interventions is known as translational applications (Guin et al., 2021). The advancement of healthcare depends on this connection between test results and clinical practice. To create novel treatments, diagnostic instruments, and preventative measures, translational research frequently takes a multidisciplinary approach, integrating laboratory research, clinical trials, and technological advancements (Kristensen et al., 2015). It seeks to guarantee that advances in genomics, molecular biology, and other scientific fields are immediately implemented to improve patient care. Prior to clinical trials, human tissue explants, animal models, and sophisticated screening technologies are essential for testing and refining new medicines (Coleman, 2011). This procedure guarantees that potential treatments are carefully examined and improved, enhancing their efficacy and safety for use in humans.

4.18.1.6 Drug Development and Screening

Translational research turns fundamental scientific discoveries into possible therapeutic treatments through drug development and screening, a process of finding, evaluating, and launching a new pharmaceutical product. After determining the biological targets linked to a disease, a vast number of compounds are screened to discover those with promising biological activity. Following their identification, possible medication candidates are optimized to increase their safety and effectiveness (Mori et al., 2006). Preclinical testing, which includes *in vitro* and in vivo research, is the next step in the process to evaluate safety and biological effects. The medication next goes through clinical studies to assess its effectiveness, safety, and ideal dosage in people. Before being on sale, successful medications must be approved by the regulatory body. To help researchers find the most promising candidates and guarantee their efficacy and safety, drug screening is essential to this procedure (Imran et al., 2023) (Figure 4.3).

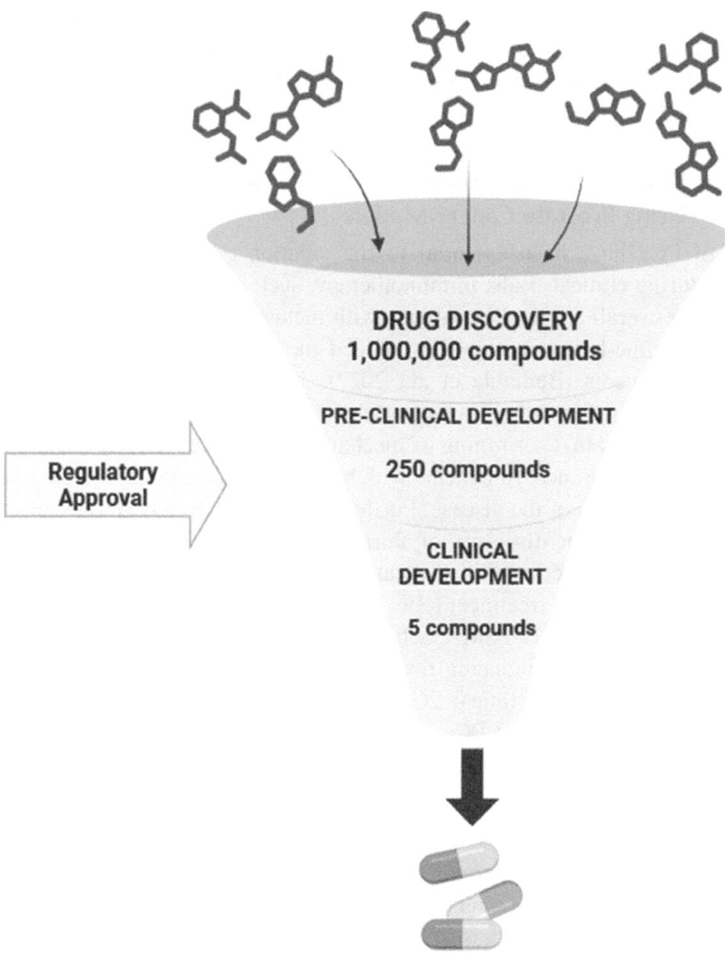

FIGURE 4.3 Drug discovery pipeline.

4.18.1.7 Role of Models in Preclinical Drug Testing

Prior to clinical trials, preclinical drug testing mainly uses a variety of experimental models to assess the pharmacokinetics, safety, and effectiveness of new therapy candidates. These models, which include in vivo animal experiments and *in vitro* cell-based assays, aim to mimic human physiological settings to forecast drug action. Due to their genetic similarity to humans and simplicity of manipulation, rodent models—such as transgenic mice—are frequently employed to investigate disease processes and treatment responses (Negi et al., 2023). Furthermore, because of their greater anatomical and physiological similarity to humans, large animal models, such as pigs, are frequently used for cardiovascular and metabolic research. Pharmacokinetics (PK) and pharmacodynamics (PD) studies, in which researchers evaluate drug absorption, distribution, metabolism, and excretion (ADME) parameters to guarantee optimal dosage techniques, are also supported by preclinical models (Mittal et al., 2023). Humanized mice models and organ-on-a-chip are examples of emerging technologies that improve translational accuracy by integrating immunological or metabolic pathways unique to humans. The limitations of these models, such as species-specific variations that could result in conflicting findings in human trials, persist even though they greatly enhance drug testing. To close the gap between preclinical and clinical research stages, these models must be continuously improved and validated (Schuck et al., 2015).

4.18.1.8 Case Studies of Successful Drug Development Using Prostate Cancer Models

Models of PCa have been essential to the creation of innovative treatments. For example, during clinical trials, immunotherapy, such as Sipuleucel-T, was successful in increasing overall survival in patients with metastatic CRPC (mCRPC). Targeting PSA, this vaccine-based strategy has opened the door for additional immunotherapeutic advancements (Belkahla et al., 2022). The second-generation AR inhibitor enzalutamide is another example of a successful treatment. PDXs and other preclinical models were crucial in determining its mechanism and effectiveness, which ultimately resulted in FDA clearance. In patients with mCRPC, enzalutamide increased survival and slowed the course of the disease. Furthermore, research in hormone-sensitive and resistant mice led to the discovery of abiraterone acetate, a CYP17 inhibitor. By preventing androgen production, this medication has had a major impact on clinical results and is still a mainstay of treatment (Rice et al., 2019). The testing of radiopharmaceuticals such as Lutetium-177-PSMA, which have demonstrated encouraging results in extending survival while reducing toxicity, has also been made easier by advanced imaging techniques in preclinical PCa animals (Jia et al., 2023). These illustrations highlight the significance that PCa models play in connecting preclinical research to successful clinical applications, hence expanding therapy options.

4.18.1.9 Biomarker Discovery

Biomarker identification is a rapidly expanding field of scientific study that aims to pinpoint the distinctive characteristics of a certain illness or condition (Schwartz & Caprioli, 2008). It is the process of finding biological molecules—such as proteins, genes, metabolites, or microRNAs—that serve as indicators for normal biological processes, pathological states, or therapeutic responses. Biomarkers such as HER2 for breast cancer and PSA for PCa have transformed clinical diagnostics and treatment monitoring, as the majority of individuals undergoing PSA screening have localized illness, which has led to early diagnosis (Citrin et al., 2013). However, the quest for new biomarkers has been fueled by its drawbacks, including low specificity. New molecular markers that are showing promise in enhancing diagnostic precision and detecting aggressive disease characteristics include PCA3, CTCs, and TMPRSS2-ERG gene fusions (Salagierski & Schalken, 2010). Other than that, preclinical models, including PDXs) and GEMMs, have been crucial in locating and confirming these biomarkers (Janitri et al., 2024). For example, PDX models recapitulate tumor heterogeneity, providing insights into biomarker expression in metastatic PCa (Shi et al., 2019). Moreover, advanced technologies, such as liquid biopsies and single-cell sequencing, are being integrated with these models to uncover biomarkers in noninvasive samples, like blood and urine. These advancements are expected to address the challenges of PCa heterogeneity and improve patient outcomes.

4.18.1.10 How Models Contribute to the Identification of Biomarkers for Early Detection and Prognosis

Identifying and validating biomarkers that enhance PCa prognosis and early detection is made easier by preclinical models. GEMMs and PDXs enable researchers to discover molecular markers such as PCA3 and TMPRSS2-ERG fusion genes by reproducing the genetic and phenotypic variability of human PCa (Choi et al., 2018). When compared to PSA alone, these markers have improved the specificity of PCa diagnosis. Furthermore, high-throughput screening of biomarkers linked to aggressive disease types, such as extracellular vesicles and CTCs, is made possible by organoid cultures obtained from PCa tissues (Zhou et al., 2024). The noninvasive detection of prognostic markers, such as AR-V7 in metastatic PCa, that predict resistance to AR-targeting therapy has been made possible by recent developments in liquid biopsy technologies combined with preclinical models. In organoids and PDX models, single-cell sequencing has also revealed new genetic alterations that act as early markers of prostate carcinogenesis. In the end, these models bridge the gap between laboratory results and clinical practice by facilitating the validation of biomarkers and offering a platform for testing their clinical application.

4.18.1.11 Precision Medicine

Building individualized treatment plans according to a patient's genetic, molecular, and environmental characteristics is the main goal of precision medicine. Precision medicine uses genetic and biomarker data to stratify patients and inform treatment choices in PCa. Prostate tumor genomic analysis, for example, finds actionable mutations like BRCA1/2 that show a tumor's vulnerability to PARP inhibitors like Olaparib. Because they can duplicate patient-specific tumor characteristics and allow for tailored drug testing, advanced preclinical models—such as organoids and PDXs—are essential to precision medicine. Studying patient immunological responses to new immunotherapies, including checkpoint inhibitors that target the PD-1/PD-L1 pathways, is made easier with the use of humanized mouse models.

Advanced preclinical models are essential for improving precision medicine, such as organoids and PDXs. Personalized medication regimens can be tested using these models, which reproduce tumor characteristics unique to each patient (Durinikova et al., 2021). This is not possible with regular cell lines. Additionally, humanized mice models are necessary to assess immune-based treatments, such as immune checkpoint inhibitors that target the PD-1/PD-L1 pathways, to better understand patient-specific immune responses and increase the efficacy of immunotherapies for PCa (Guo et al., 2021). Preclinical testing and genetic profiling combined provide precision medicine with a promising strategy to enhance treatment results while reducing side effects.

4.18.1.12 Integration of Models in Personalized Treatment Strategies

PCa treatment advancement depends on the incorporation of preclinical models into individualized treatment plans. GEMMs, organoids, and PDXs are useful methods for modeling human PCa and determining the response of individual tumors to therapy. To match treatments with the unique genetic and molecular characteristics of each patient's tumor, these models assist in identifying molecular markers and therapeutic targets. PDX models, for example, can show how well new treatments that are adapted to the genetic changes or mutations seen in a patient's tumor—like BRCA1/2 mutations or the presence of AR-V7, a sign of resistance to several AR-targeting treatments—work (Konda et al., 2023). Furthermore, immune-based therapies, such as immune checkpoint inhibitors that target the PD-1/PD-L1 pathway, may be evaluated in a system that more closely resembles human immune responses thanks to humanized mouse models. This is especially crucial for PCa, as immunotherapy is showing growing promise as a treatment. Combining these models with multi-omics data (genomic, transcriptomic, and proteomic) allows for more precise treatment response predictions, providing a tailored approach to PCa treatment. By incorporating these models into clinical decision-making, the trial-and-error method typically used in cancer treatment is reduced and most effective strategies for specific patients are determined.

4.18.1.13 Challenges and Limitations

PCa remains a major public health concern, and despite progress in both its diagnosis and treatment, several obstacles persist that limit the ability to provide optimal care. The main challenges and limitations include:

Early Detection and Diagnosis: The use of PSA testing for early PCa detection is contentious. PSA testing can result in false positives (elevated PSA levels due to non-cancerous issues) and false negatives (missed cancers in individuals with normal PSA levels), leading to unnecessary biopsies or delayed diagnoses. The standard diagnostic method, transrectal ultrasound-guided biopsy, also has its drawbacks, such as missing hidden or small tumors, potentially resulting in underdiagnosis or incorrect grading.

Aggressiveness Variability and Risk Stratification: PCa displays considerable heterogeneity, with some cases growing slowly while others are highly aggressive. A major challenge lies in determining which cancers need immediate treatment and which can be safely monitored. While Gleason grading and PSA levels are valuable, they do not always provide accurate predictions about whether a tumor will progress rapidly or remain indolent.

Treatment Challenges and Side Effects: Treatments like radical prostatectomy, radiation therapy, and ADT can result in long-term side effects, such as erectile dysfunction, urinary incontinence, and fatigue, significantly affecting a patient's quality of life. Moreover, the development of CRPC, where cancer cells stop responding to hormonal therapies, poses a major challenge and leads to continued tumor progression.

Advanced Staging and Metastasis: PCa frequently spreads to the bones, causing complications like pain and fractures. Despite advances in imaging technologies, detecting bone metastasis at an early stage remains difficult. Current imaging techniques, such as CT, MRI, and bone scans, still have limited sensitivity for detecting early metastasis. Emerging methods like PSMA PET/CT imaging are improving detection but are not yet widely available.

Health Disparities: African American men are disproportionately impacted by PCa, experiencing both higher incidence rates and more aggressive disease forms. Socioeconomic factors also contribute to disparities in access to screening, diagnosis, and treatment. Additionally, unequal access to healthcare continues to result in later-stage diagnoses and worse outcomes, particularly in underserved communities.

Psychological and Emotional Impact: The emotional and psychological effects of PCa treatment can be substantial, especially with the impact of erectile dysfunction, which can significantly affect the patients' quality of life.

Research and Development: While there have been advancements, challenges such as drug resistance and the complex biology of PCa continue to hinder the development of new therapies. More targeted drugs, particularly those focusing on specific genetic mutations, are needed. Furthermore, participation in clinical trials remains low among certain demographic groups (e.g., minorities and older adults), limiting the applicability of research findings across different populations.

4.19 FUTURE DIRECTIONS

4.19.1 Emerging Technologies and Methodologies in Modeling

Research on PCa is being revolutionized by emerging technologies, which offer sophisticated models for examining drug efficacy, tumor dynamics, and therapy responses. The organoid model is a significant advancement that creates 3D cultures from PCa cells taken from patients. Because of their genetic and architectural similarities to real malignancies, these organoids provide a more accurate platform for evaluating medication responses and individualized treatments. Additionally, organoids enable researchers to examine how genetic abnormalities, including changes in BRCA1/2, affect tumor growth and medication resistance, which is essential for precision medicine in PCa (Corsini & Knoblich, 2022). Creating genetically modified models of PCa using CRISPR-Cas9 technology is another innovation allowing researchers to manipulate genes linked to cancer, such as PTEN, and develop models that more closely resemble human PCa subtypes, including metastatic ones, thanks to this technique (Tabibian et al., 2024). By examining the molecular processes behind tumor growth and metastasis, these altered models can reveal new potential treatment targets. PCa immunotherapy research is also being advanced by humanized mouse models that include human immune cells or tissues. Because they can replicate human immune responses in a manner that traditional models cannot, these animals are essential for testing immune checkpoint inhibitors, such as PD-1/PD-L1 blockers (Olson et al., 2018). They aid in assessing immunotherapies' effectiveness and any adverse effects in a setting similar to that of humans.

4.19.2 Opportunities for Interdisciplinary Collaboration

Research on PCa must advance through interdisciplinary collaboration since the disease's complexity requires expertise from a variety of disciplines. For instance, by working together to study genomic data, bioinformaticians, molecular biologists, and clinical oncologists can find new biomarkers that can guide treatment plans. Clinical oncologists steer these discoveries toward useful applications, while bioinformatics tools assist in processing enormous volumes of data from methods like next-generation sequencing, offering deeper insights into genetic mutations and their role in cancer progression (Boehm et al., 2021). Furthermore, immunologists and pharmacologists work together to advance immunotherapies, especially when testing immune checkpoint inhibitors for PCa using humanized mice models. Working together with engineers may encourage the development of innovative tools like high-throughput drug screening microfluidic devices, which enable more precise testing of possible treatments. This multidisciplinary approach can be further supported by imaging technologies that track tumor responses and progression (Li et al., 2020). Additionally, epidemiologists and clinical researchers provide crucial practical insights that guarantee laboratory results are successfully transferred into clinical settings, improving patient care through individualized treatment plans (Gandaglia et al., 2021). Research on PCa is thorough thanks to these varied partnerships, which focus on everything from molecular causes to treatment options.

4.19.3 Advances in Model Development

New developments in PCa models have made it possible to investigate tumor biology and treatment responses with greater accuracy and efficacy. Because they preserve important aspects of the original human tumor, PDXs have gained popularity and are perfect for examining the intricacy of PCa. For example, PDX models have been used to assess how well new treatments, such as AR inhibitors, work and have been crucial in figuring out how drug resistance works (Abdolahi et al., 2022). Furthermore, there is great potential for simulating the PCa microenvironment using the organoid model, which generates 3D tumor cultures. In addition to maintaining genetic diversity, organoids facilitate drug screening, providing a strong foundation for customized therapy (Robinson et al., 2015). Another significant advancement is the use of CRISPR/Cas9 gene editing technology. By precisely altering genes linked to PCa, like TP53 and PTEN, this technique improves models for examining the development, spread, and metastasis of tumors. These modified models aid in determining novel treatment targets and assessing how genetic alterations affect the behavior of cancer cells. There is also a lot of promise for testing immunotherapies using humanized mice models that incorporate human immune systems. In an environment that closely mimics human immune responses, they offer an in vivo platform for assessing the effectiveness of immune checkpoint inhibitors, such as PD-1 and CTLA-4 blockers (Cogels et al., 2021).

4.19.4 Emerging Technologies (e.g., CRISPR, Single-Cell Sequencing)

Research on PCa is being revolutionized by emerging technologies like single-cell sequencing and CRISPR gene editing, which open up new possibilities for model building and treatment discovery. More accurate models of PCa can be produced thanks to the precise genetic changes made possible by CRISPR/Cas9 technology. Important genes that are frequently altered in PCa, such as PTEN, AR, and TP53, have been edited using this technique. Our knowledge of tumor progression and drug resistance mechanisms has increased because of the ability to modify these genes, which has also helped us create targeted therapies (Wang et al., 2020). Conversely, single-cell sequencing has become a game-changing technique for examining tumor heterogeneity at a never-before-seen level of detail. By profiling individual cancer cells, it enables researchers to see how PCa changes and responds to treatments on a cellular level. This method is especially helpful for finding uncommon but clinically important tumor subpopulations that bulk sequencing frequently overlooks. Researchers can find new biomarkers and treatment targets by looking at the molecular characteristics of these cells (Song et al., 2022). Additionally, a more thorough understanding of the TME is made possible by combining single-cell sequencing with other technologies, such as spatial transcriptomics, which directs the creation of more individualized and efficient treatments (Xiao et al., 2024) (Figure 4.4).

4.19.5 Integration of AI and Machine Learning in Model Development

PCa model development that incorporates artificial intelligence (AI) and machine learning (ML) has greatly enhanced drug discovery and tailored treatment approaches, as well as our understanding of the illness. Complex datasets, including genomic, proteomic, and clinical data, can be processed more efficiently with the help of AI and ML. These datasets are essential for building predictive models for the course of PCa. By spotting patterns and connections in massive amounts of data that conventional statistical techniques can overlook, AI algorithms might provide fresh perspectives on tumor biology and potential treatment targets (Mori et al., 2006). To design computational models that mimic tumor activity, forecast medication responses, and evaluate patient outcomes, ML techniques such as support vector machines (SVM), random forests, and neural networks are employed. For instance, using patient data like PSA levels, clinical features, and genetic profiles, ML algorithms have been used to predict the recurrence of PCa. This helps with treatment planning decision-making (Elmarakeby et al., 2021). To further improve predictions and customize treatments for specific patients, AI models are also being paired with preclinical models (Singh et al., 2023).

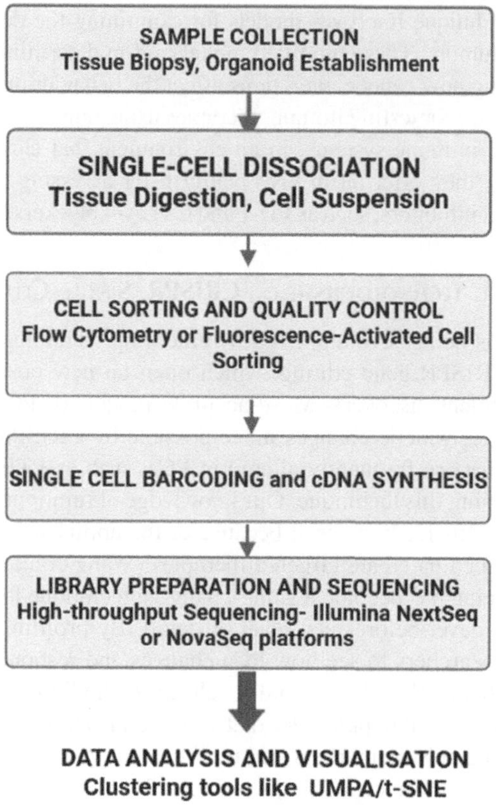

FIGURE 4.4 Drug development and screening pipeline.

4.19.6 Personalized and Precision Models

PCa therapies are customized for each patient according to their genetic and molecular traits using personalized and precision medicine. Drug response testing and more precise therapy selection are made possible by models that mimic patient-specific tumor biology, such as organoids and PDXs (Baydoun et al., 2023). Targeted treatments like PARP inhibitors are made possible by tumor profiling's ability to detect actionable mutations like BRCA1/2 (Palacios-Eito et al., 2019). Furthermore, by evaluating sizable datasets to forecast patient-specific treatment outcomes, AI and ML improve precision. This integration lowers needless side effects and increases therapy efficacy.

4.19.7 Next-Generation Models for Personalized Medicine

PCa personalized medicine is being revolutionized by next-generation models, that provide more precise and customized treatment plans. Patient-specific tumor situations can be simulated thanks to advancements like tumor-on-chip technologies, PDXs, and 3D organoid cultures. By modeling the genetic, epigenetic, and anatomical features of various cancers, these models can improve medication response testing and the identification of biomarkers. Additionally, to improve the knowledge of tumor heterogeneity and find new treatment targets, CRISPR-based gene editing and single-cell RNA sequencing are being incorporated into these models (Tabibian et al., 2024). By analyzing comprehensive patient data and creating more precise, individualized treatment plans, combining AI with next-generation models also aids in forecasting patient-specific outcomes. These cutting-edge models are essential for determining the best treatments, minimizing adverse effects, and eventually enhancing patient outcomes.

4.19.8 Potential for Using Models in Clinical Decision-Making

Models contribute significantly to improving clinical decision-making for PCa by offering more precise and individualized treatment outcome forecasts. Clinicians can now test drug responses in a manner that replicates the patient's tumor environment thanks to the integration of patient-derived models like organoids, tumor-on-chip systems, and PDXs. This gives them important information about the best treatments (Zhao et al., 2021). Additionally, comprehensive patient data, such as genetic, histological, and clinical data, can be analyzed using sophisticated AI and ML approaches to forecast a patient's response to particular treatments. By enabling physicians to customize medicines to each patient's unique illness characteristics, these models are assisting in the transition away from trial-and-error methods and toward better clinical outcomes while reducing needless side effects (Robinson et al., 2015). For instance, physicians can assess the efficacy of targeted treatments, such as PARP inhibitors, in patients with BRCA1/2 mutations by employing models such as genomic-based models and CRISPR-edited organoids (Philp, 2024). AI algorithms can also help with risk stratification and early diagnosis by forecasting illness development and discovering prognostic biomarkers (Mohsin et al., 2023). It is anticipated that this model-driven decision-making will transform the treatment of PCa by providing more precise, customized treatment regimens.

4.19.9 Interdisciplinary Approaches

Interdisciplinary approaches have become crucial in advancing PCa research, as the complexity of cancer biology demands expertise from multiple fields. More efficient diagnostic methods, therapeutic approaches, and customized models are being developed by integrating insights from molecular biology, genetics, bioinformatics, and clinical oncology. Strong patient-derived models, like organoids and PDXs, that imitate the unique TME and enable improved drug testing and biomarker discovery require cooperation between biologists, data scientists, engineers, and clinicians (Laliwala et al., 2024). Furthermore, when combined with clinical and genomic data, developments in AI and ML have the potential to improve prognosis prediction, risk assessment, and treatment strategy decision-making (Robinson et al., 2015). The investigation of novel immunotherapies is fueled by the crucial roles that immunology and genetics play in comprehending tumor heterogeneity, immune evasion, and resistance mechanisms in PCa. Patients with PCa can receive more accurate and efficient treatments thanks to this interdisciplinary teamwork, which speeds up the translation of research from bench to bedside (Boehm et al., 2021). These cooperative initiatives, supported by modern methods like single-cell sequencing, CRISPR gene editing, and sophisticated computational modeling, offer fresh approaches to resolving issues with PCa diagnosis and treatment, ultimately improving patient care (Durinikova et al., 2021).

4.19.10 Collaboration between Clinicians, Biologists, and Computational Scientists

Collaboration between biologists, computer scientists, and doctors is essential for doing effective PCa research. While biologists investigate the molecular causes of PCa, such as genetic mutations and TMEs, clinicians provide useful insights about patient responses, treatment effectiveness, and disease progression. In order to improve therapy options, computational scientists help by creating sophisticated models and analyzing clinical, genomic, and imaging data using data-driven techniques like ML. This synergy benefits patient-derived models, including organoids and PDXs, which mimic patient-specific malignancies for more precise therapy testing (He et al., 2023). Advances in personalized medicine are also fueled by the application of AI algorithms to forecast reactions to immunotherapies and targeted medicines.

4.19.11 Importance of Interdisciplinary Research

The development of PCa models depends heavily on interdisciplinary cooperation. While biologists concentrate on comprehending the disease's basic processes, clinicians offer insights into how patients respond to treatment and how the disease progresses. To create prediction models, computational scientists use AI and ML to evaluate massive datasets (Mori et al., 2006, 2023). When combined, these initiatives contribute to the development of more precise preclinical models that better represent tumor heterogeneity and can be utilized for efficient drug testing, such as organoids and PDXs. Furthermore, incorporating information from different fields speeds up the discovery of new biomarkers and improves early detection, which improves clinical judgment and individualized treatment regimens (Ye & Mori et al., 2006, 2022). In the end, multidisciplinary methods make it possible to convert lab research into clinical settings, which improves treatment results (Hull et al., 2002) (Figure 4.5).

4.20 RECENT STUDIES

PCa models were initially created to assess the impact of screening and treatment changes on PCa mortality. These models have been expanded to reconcile varying estimates of lead times and overdiagnosis associated with PSA screening, compare the natural histories and risks of cancer progression, examine the impact of contamination in the prostate section of the Prostate, Lung, Colorectal, and Ovarian (PLCO) cancer screening trial, and project expected disease trends if PSA screening were discontinued. Ongoing research is focused on understanding the differential natural history of PCa in African-American and White men and reconciling the effects of screening on mortality in the PLCO cancer screening trial and the European Randomized Study of Screening for Prostate Cancer (ERSPC).

4.21 CONCLUSION

Both preclinical and clinical PCa models have greatly enhanced our understanding of the disease. These models have allowed researchers to investigate the disease's underlying biology, explore its diversity, and identify potential biomarkers for early detection, risk assessment, and treatment response. Advances in genetic profiling, imaging technologies, and tumor modeling have provided more accurate representations of PCa's behavior and progression, especially in cases of CRPC and metastatic disease. Furthermore, these models have facilitated the testing of new therapies, offering insights into more personalized treatment approaches. Despite these advancements, however, translating the knowledge gained from these models into practical clinical applications remains a challenge. Differences between preclinical models and human patients, along with the complexity of the PCa microenvironment, have made it difficult to consistently replicate outcomes, limiting our ability to predict patient responses to treatments accurately. To improve both research and clinical results, a multidisciplinary approach combining genetics, immunology, advanced imaging, and personalized medicine is crucial. More targeted therapies tailored to an individual's genetic profile have the potential to transform treatment strategies by reducing side effects while increasing effectiveness. Moreover, progress in immunotherapy and gene editing techniques, like CRISPR, offers promise for overcoming some of the key challenges, especially in advanced stages of the disease. Early detection continues to be essential for reducing PCa mortality. Ongoing research to identify reliable biomarkers for early detection, along with more accurate

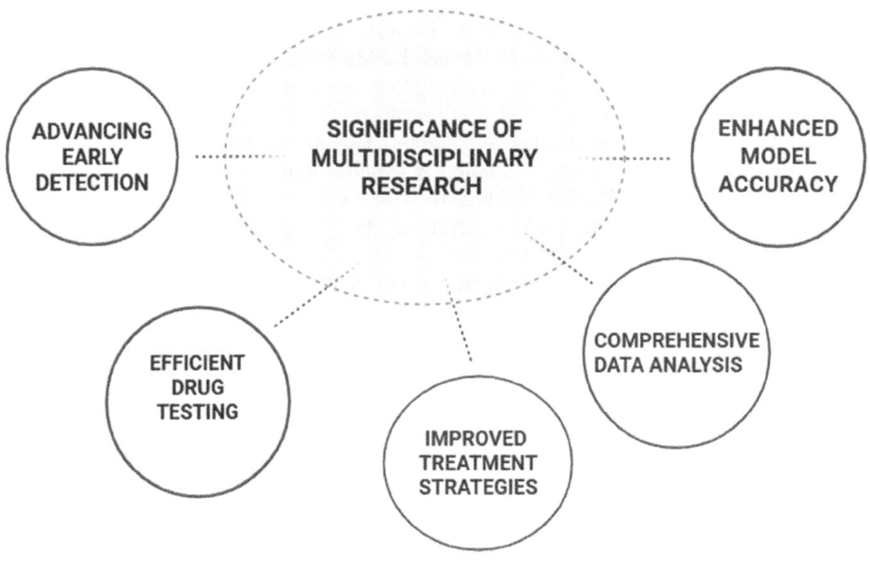

FIGURE 4.5 Significance of multidisciplinary research.

imaging tools, will improve our ability to detect the disease at curable stages. In conclusion, while significant progress has been made, considerable challenges still exist in enhancing PCa diagnosis, treatment, and patient outcomes. With ongoing innovation, collaborative research, and a focus on personalized care, we can continue to make strides in combating this widespread and complex disease.

REFERENCES

Abdolahi, S., Ghazvinian, Z., Muhammadnejad, S., Saleh, M., Aghdaei, H. A., & Baghaei, K. (2022). Patient-derived xenograft (PDX) models, applications and challenges in cancer research. *Journal of Translational Medicine*, 20(1), 206. https://doi.org/10.1186/s12967-022-03405-8

Al-Fayez, S., & El-Metwally, A. (2023). Cigarette smoking and prostate cancer: A systematic review and meta-analysis of prospective cohort studies. *Tobacco Induced Diseases*, 21, 1–12. https://doi.org/10.18332/tid/157231

Amornsupak, K., Insawang, T., Thuwajit, P., Charoenrat, O. P., Eccles, S. A., & Thuwajit, C. (2014). Cancer-associated fibroblasts induce high mobility group box 1 and contribute to resistance to doxorubicin in breast cancer cells. *BMC Cancer*, 14(1), 955. https://doi.org/10.1186/1471-2407-14-955

Aronson, S. J., & Rehm, H. L. (2015). Building the foundation for genomics in precision medicine. *Nature*, 526(7573), 336–342. https://doi.org/10.1038/nature15816

Balkwill, F. R., Capasso, M., & Hagemann, T. (2012). The tumor microenvironment at a glance. *Journal of Cell Science*, 125(23), 5591–5596. https://doi.org/10.1242/jcs.116392

Baydoun, A., Jia, A. Y., Zaorsky, N. G., Kashani, R., Rao, S., Shoag, J. E., Vince, R. A., Bittencourt, L. K., Zuhour, R., Price, A. T., Arsenault, T. H., & Spratt, D. E. (2023). Artificial intelligence applications in prostate cancer. *Prostate Cancer and Prostatic Diseases*, 27(1), 37–45. https://doi.org/10.1038/s41391-023-00684-0

Belkahla, S., Nahvi, I., Biswas, S., Nahvi, I., & Amor, N. B. (2022). Advances and development of prostate cancer, treatment, and strategies: A systemic review. *Frontiers in Cell and Developmental Biology*, 10, 991330. https://doi.org/10.3389/fcell.2022.991330

Bergengren, O., Pekala, K. R., Matsoukas, K., Fainberg, J., Mungovan, S. F., Bratt, O., Bray, F., Brawley, O., Luckenbaugh, A. N., Mucci, L., Morgan, T. M., & Carlsson, S. V. (2023). 2022 Update on prostate cancer epidemiology and risk factors—A systematic review. *European Urology*, 84(2), 191–206. https://doi.org/10.1016/j.eururo.2023.04.021

Bernard, D., Peakman, M., & Hayday, A. C. (2008). Establishing humanized mice using stem cells: Maximizing the potential. *Clinical & Experimental Immunology*, 152(3), 406–414. https://doi.org/10.1111/j.1365-2249.2008.03659.x

Beshiri, M. L., Tice, C. M., Tran, C., Nguyen, H. M., Sowalsky, A. G., Agarwal, S., Jansson, K. H., Yang, Q., McGowen, K. M., Yin, J., Alilin, A. N., Karzai, F. H., Dahut, W. L., Corey, E., & Kelly, K. (2018). A PDX/Organoid biobank of advanced prostate cancers captures genomic and phenotypic heterogeneity for disease modeling and therapeutic screening. *Clinical Cancer Research*, 24(17), 4332–4345. https://doi.org/10.1158/1078-0432.ccr-18-0409

Bhatia, S., Landier, W., Paskett, E. D., Peters, K. B., Merrill, J. K., Phillips, J., & Osarogiagbon, R. U. (2022). Rural–urban disparities in cancer outcomes: Opportunities for future research. *JNCI Journal of the National Cancer Institute*, 114(7), 940–952. https://doi.org/10.1093/jnci/djac030

Bluemn, E. G., Coleman, I. M., Lucas, J. M., Coleman, R. T., Hernandez-Lopez, S., Tharakan, R., Bianchi-Frias, D., Dumpit, R. F., Kaipainen, A., Corella, A. N., Yang, Y. C., Nyquist, M. D., Mostaghel, E., Hsieh, A. C., Zhang, X., Corey, E.,

Brown, L. G., Nguyen, H. M., Pienta, K., … Nelson, P. S. (2017). Androgen receptor pathway-independent prostate cancer is sustained through FGF signaling. *Cancer Cell*, 32(4), 474–489e6. https://doi.org/10.1016/j.ccell.2017.09.003

Boehm, K. M., Khosravi, P., Vanguri, R., Gao, J., & Shah, S. P. (2021). Harnessing multimodal data integration to advance precision oncology. *Nature Reviews Cancer*, 22(2), 114–126. https://doi.org/10.1038/s41568-021-00408-3

Boj, S. F., Hwang, C., Baker, L. A., Chio, I. I. C., Engle, D. D., Corbo, V., Jager, M., Ponz-Sarvise, M., Tiriac, H., Spector, M. S., Gracanin, A., Oni, T., Yu, K. H., Van Boxtel, R., Huch, M., Rivera, K. D., Wilson, J. P., Feigin, M. E., Öhlund, D., … Tuveson, D. A. (2014). Organoid models of human and mouse ductal pancreatic cancer. *Cell*, 160(1–2), 324–338. https://doi.org/10.1016/j.cell.2014.12.021

Brandão, A., Paulo, P., & Teixeira, M. R. (2020). Hereditary predisposition to prostate cancer: From genetics to clinical implications. *International Journal of Molecular Sciences*, 21(14), 5036. https://doi.org/10.3390/ijms21145036

Cheng, I., Witte, J. S., Jacobsen, S. J., Haque, R., Quinn, V. P., Quesenberry, C. P., Caan, B. J., & Van Den Eeden, S. K. (2010). Prostatitis, sexually transmitted diseases, and prostate cancer: The California men's health study. *PLoS One*, 5(1), e8736. https://doi.org/10.1371/journal.pone.0008736

Choi, Y., Lee, S., Kim, K., Kim, S., Chung, Y., & Lee, C. (2018). Studying cancer immunotherapy using patient-derived xenografts (PDXs) in humanized mice. *Experimental & Molecular Medicine*, 50(8), 1–9. https://doi.org/10.1038/s12276-018-0115-0]

Chu, F., Chen, L., Guan, Q., Chen, Z., Ji, Q., Ma, Y., Ji, J., Sun, M., Huang, T., Song, H., Zhou, H., Lin, X., & Zheng, Y. (2025). Global burden of prostate cancer: Age-period-cohort analysis from 1990 to 2021 and projections until 2040. *World Journal of Surgical Oncology*, 23(1), 98. https://doi.org/10.1186/s12957-025-03733-1

Citrin, D., Hudak, K., & Camphausen, K. A. (2013). Biomarkers to guide therapy or surveillance for prostate cancer. *Biomarkers in Medicine*, 7(6), 827–829. https://doi.org/10.2217/bmm.13.123

Cogels, M. M., Rouas, R., Ghanem, G. E., Martinive, P., Awada, A., Van Gestel, D., & Krayem, M. (2021). Humanized mice as a valuable pre-clinical model for cancer immunotherapy research. *Frontiers in Oncology*, 11, 784947. https://doi.org/10.3389/fonc.2021.784947

Coleman, R. A. (2011). Human tissue in the evaluation of safety and efficacy of new medicines: A viable alternative to animal models? *ISRN Pharmaceutics*, 2011, 1–8. https://doi.org/10.5402/2011/806789

Corsini, N. S., & Knoblich, J. A. (2022). Human organoids: New strategies and methods for analyzing human development and disease. *Cell*, 185(15), 2756–2769. https://doi.org/10.1016/j.cell.2022.06.051

Cunningham, D., & You, Z. (2015). In vitro and in vivo model systems used in prostate cancer research. *Journal of Biological Methods*, 2(1), 1. https://doi.org/10.14440/jbm.2015.63

Durinikova, E., Buzo, K., & Arena, S. (2021). Preclinical models as patients' avatars for precision medicine in colorectal cancer: Past and future challenges. *Journal of Experimental & Clinical Cancer Research*, 40(1), 185. https://doi.org/10.1186/s13046-021-01981-z

Elmarakeby, H. A., Hwang, J., Arafeh, R., Crowdis, J., Gang, S., Liu, D., AlDubayan, S. H., Salari, K., Kregel, S., Richter, C., Arnoff, T. E., Park, J., Hahn, W. C., & Van Allen, E. M. (2021). Biologically informed deep neural network for prostate cancer discovery. *Nature*, 598(7880), 348–352. https://doi.org/10.1038/s41586-021-03922-4

Foster, B. A., Gingrich, J. R., Kwon, E. D., Madias, C., & Greenberg, N. M. (1997). Characterization of prostatic epithelial cell lines derived from transgenic adenocarcinoma of the mouse prostate (TRAMP) model. *Cancer Research*, 57(16), 3325–3330. PMID: 9269988.

Gandaglia, G., Leni, R., Bray, F., Fleshner, N., Freedland, S. J., Kibel, A., Stattin, P., Van Poppel, H., & La Vecchia, C. (2021). Epidemiology and prevention of prostate cancer. *European Urology Oncology*, 4(6), 877–892. https://doi.org/10.1016/j.euo.2021.09.006

Gann, P. H. (2002). Risk factors for prostate cancer. *Nature Reviews Urology*, 4(Suppl 5), S3–S10. PMID: 16986064; PMCID: PMC1476014.

Gao, D., Vela, I., Sboner, A., Iaquinta, P. J., Karthaus, W. R., Gopalan, A., Dowling, C., Wanjala, J. N., Undvall, E. A., Arora, V. K., Wongvipat, J., Kossai, M., Ramazanoglu, S., Barboza, L. P., Di, W., Cao, Z., Zhang, Q. F., Sirota, I., Ran, L., … Chen, Y. (2014). Organoid cultures derived from patients with advanced prostate cancer. *Cell*, 159(1), 176–187. https://doi.org/10.1016/j.cell.2014.08.016

Gordon, J. A., Noble, J. W., Midha, A., Derakhshan, F., Wang, G., Adomat, H. H., Guns, E. S. T., Lin, Y., Ren, S., Collins, C. C., Nelson, P. S., Morrissey, C., Wasan, K. M., & Cox, M. E. (2019). Upregulation of scavenger receptor B1 is required for steroidogenic and non-steroidogenic cholesterol metabolism in prostate cancer. *Cancer Research*, 79(13), 3320–3331. https://doi.org/10.1158/0008-5472.can-18-2529

Grivel, J., & Margolis, L. (2009). Use of human tissue explants to study human infectious agents. *Nature Protocols*, 4(2), 256–269. https://doi.org/10.1038/nprot.2008.245

Guin, D., Thakran, S., Singh, P., Ramachandran, S., Hasija, Y., & Kukreti, R. (2021). Translational biotechnology: A transition from basic biology to evidence-based research. In Hasija, Y. (Eds.), *Translational Biotechnology* (pp. 3–24), Elsevier eBooks, Amsterdam, Netherlands. https://doi.org/10.1016/b978-0-12-821972-0.00006-x

Guo, W., Zhang, C., Qiao, T., Zhao, J., & Shi, C. (2021). Strategies for the construction of mouse models with humanized immune system and evaluation of tumor immune checkpoint inhibitor therapy. *Frontiers in Oncology*, 11, 673199. https://doi.org/10.3389/fonc.2021.673199

He, J., Zhang, C., Ozkan, A., Feng, T., Duan, P., Wang, S., Yang, X., Xie, J., & Liu, X. (2023). Patient-derived tumor models and their distinctive applications in personalized drug therapy. *Mechanobiology in Medicine*, 1(2), 100014. https://doi.org/10.1016/j.mbm.2023.100014

Hosseini-Beheshti, E., Pham, S., Adomat, H., Li, N., & Guns, E. S. T. (2012). Exosomes as biomarker enriched microvesicles: Characterization of exosomal proteins derived from a panel of prostate cell lines with distinct AR phenotypes. *Molecular & Cellular Proteomics*, 11(10), 863–885. https://doi.org/10.1074/mcp.m111.014845

Hull, G. W., Rabbani, F., Abbas, F., Wheeler, T. M., Kattan, M. W., & Scardino, P. T. (2002). Cancer control with radical prostatectomy alone in 1,000 consecutive patients. *The Journal of Urology*, 167(2 Part 1), 528–534. https://doi.org/10.1016/s0022-5347(01)69079-7

Imran, H., Tang, Y., Wang, S., Yan, X., Liu, C., Guo, L., Wang, E., & Xu, C. (2023). Optimized DOX drug deliveries via chitosan-mediated nanoparticles and stimuli responses in cancer chemotherapy: A review. *Molecules*, 29(1), 31. https://doi.org/10.3390/molecules29010031

Iyengar, S., Hall, I. J., & Sabatino, S. A. (2020). Racial/Ethnic disparities in prostate cancer incidence, distant stage diagnosis, and mortality by U.S. census region and age group, 2012–2015. *Cancer Epidemiology Biomarkers & Prevention*, 29(7), 1357–1364. https://doi.org/10.1158/1055-9965.epi-19-1344

Janitri, V., ArulJothi, K. N., Mythili, V. M. R., Singh, S. K., Prasher, P., Gupta, G., Dua, K., Hanumanthappa, R., Kaliappan, K., & Anand, K. (2024). The roles of patient-derived xenograft models and artificial intelligence toward precision medicine. *Medicine Communications*, 5(10), e745. https://doi.org/10.1002/mco2.745

Jia, A. Y., Kiess, A. P., Li, Q., & Antonarakis, E. S. (2023). Radiotheranostics in advanced prostate cancer: Current and future directions. *Prostate Cancer and Prostatic Diseases*, 27(1), 11–21. https://doi.org/10.1038/s41391-023-00670-6

Kaiser, A., Haskins, C., Siddiqui, M. M., Hussain, A., & D'Adamo, C. (2019). The evolving role of diet in prostate cancer risk and progression. *Current Opinion in Oncology*, 31(3), 222–229. https://doi.org/10.1097/cco.0000000000000519

Kamphaus, G. D., Colorado, P. C., Panka, D. J., Hopfer, H., Ramchandran, R., Torre, A., Maeshima, Y., Mier, J. W., Sukhatme, V. P., & Kalluri, R. (2000). Canstatin, a novel matrix-derived inhibitor of angiogenesis and tumor growth. *Journal of Biological Chemistry*, 275(2), 1209–1215. https://doi.org/10.1074/jbc.275.2.1209

Karnik, I., Her, Z., Neo, S. H., Liu, W. N., & Chen, Q. (2023). Emerging preclinical applications of humanized mouse models in the discovery and validation of novel immunotherapeutics and their mechanisms of action for improved cancer treatment. *Pharmaceutics*, 15(6), 1600. https://doi.org/10.3390/pharmaceutics15061600

King, M. A., Covassin, L., Brehm, M. A., Racki, W., Pearson, T., Leif, J., Laning, J., Fodor, W., Foreman, O., Burzenski, L., Chase, T. H., Gott, B., Rossini, A. A., Bortell, R., Shultz, L. D., & Greiner, D. L. (2009). Human peripheral blood leucocyte non-obese diabetic-severe combined immunodeficiency interleukin-2 receptor gamma chain gene mouse model of xenogeneic graft-versus-host-like disease and the role of host major histocompatibility complex. *Clinical & Experimental Immunology*, 157(1), 104–118. https://doi.org/10.1111/j.1365-2249.2009.03933.x

Konda, P., Garinet, S., Van Allen, E. M., & Viswanathan, S. R. (2023). Genome-guided discovery of cancer therapeutic targets. *Cell Reports*, 42(8), 112978. https://doi.org/10.1016/j.celrep.2023.112978

Kristensen, N., Nymann, C., & Konradsen, H. (2015). Implementing research results in clinical practice- the experiences of healthcare professionals. *BMC Health Services Research*, 16(1), 48. https://doi.org/10.1186/s12913-016-1292-y

Laliwala, A., Pant, A., Svechkarev, D., Sadykov, M. R., & Mohs, A. M. (2024). Advancements of paper-based sensors for antibiotic-resistant bacterial species identification. *Deleted Journal*, 1(1), 9. https://doi.org/10.1038/s44328-024-00016-9

Lawrence, M. G., Taylor, R. A., Toivanen, R., Pedersen, J., Norden, S., Pook, D. W., Frydenberg, M., Papargiris, M. M., Niranjan, B., Richards, M. G., Wang, H., Collins, A. T., Maitland, N. J., & Risbridger, G. P. (2013). A preclinical xenograft model of prostate cancer using human tumors. *Nature Protocols*, 8(5), 836–848. https://doi.org/10.1038/nprot.2013.043

Levesque, C., & Nelson, P. S. (2017). Cellular constituents of the prostate stroma: Key contributors to prostate cancer progression and therapy resistance. *Cold Spring Harbor Perspectives in Medicine*, 8(8), a030510. https://doi.org/10.1101/cshperspect.a030510

Li, J., Xu, C., Lee, H. J., Ren, S., Zi, X., Zhang, Z., Wang, H., Yu, Y., Yang, C., Gao, X., Hou, J., Wang, L., Yang, B., Yang, Q., Ye, H., Zhou, T., Lu, X., Wang, Y., Qu, M., … Sun, Y. (2020). A genomic and epigenomic atlas of prostate cancer in Asian populations. *Nature*, 580(7801), 93–99. https://doi.org/10.1038/s41586-020-2135-x

Linxweiler, J., Hajili, T., Körbel, C., Berchem, C., Zeuschner, P., Müller, A., Stöckle, M., Menger, M. D., Junker, K., & Saar, M. (2020). Cancer-associated fibroblasts stimulate primary tumor growth and metastatic spread in an orthotopic prostate cancer xenograft model. *Scientific Reports*, 10(1), 12575. https://doi.org/10.1038/s41598-020-69424-x

Macke, A. J., & Petrosyan, A. (2022). Alcohol and prostate cancer: Time to draw conclusions. *Biomolecules*, 12(3), 375. https://doi.org/10.3390/biom12030375

Mandair, D., Rossi, R. E., Pericleous, M., Whyand, T., & Caplin, M. E. (2014). Prostate cancer and the influence of dietary factors and supplements: A systematic review. *Nutrition & Metabolism*, 11(1), 30. https://doi.org/10.1186/1743-7075-11-30

Mehta, H. H., Gao, Q., Galet, C., Paharkova, V., Wan, J., Said, J., Sohn, J. J., Lawson, G., Cohen, P., Cobb, L. J., & Lee, K. (2011). IGFBP-3 is a metastasis suppression gene in prostate cancer. *Cancer Research*, 71(15), 5154–5163. https://doi.org/10.1158/0008-5472.can-10-4513

Michaud, J. E., Billups, K. L., & Partin, A. W. (2015). Testosterone and prostate cancer: An evidence-based review of pathogenesis and oncologic risk. *Therapeutic Advances in Urology*, 7(6), 378–387. https://doi.org/10.1177/1756287215597633

Miernik, A., & Gratzke, C. (2020). Current treatment for benign prostatic hyperplasia. *Deutsches* Ärzteblatt *International*, 117(49), 843–854. https://doi.org/10.3238/arztebl.2020.0843

Mittal, A., Ghai, R., Srivastava, A., Ghosh, D.P., Nagarajan, K. (2023). Pharmacokinetics and pharmacodynamics: Fundamentals and role(s) in drug discovery and development. In Singh, P. P. (Eds.), *Recent Advances in Pharmaceutical Innovation and Research*. Springer, Singapore. https://doi.org/10.1007/978-981-99-2302-1_15

Mohsin, S. N., Gapizov, A., Ekhator, C., Ain, N. U., Ahmad, S., Khan, M., Barker, C., Hussain, M., Malineni, J., Ramadhan, A., & Nagaraj, R. H. (2023). The role of artificial intelligence in prediction, risk stratification, and personalized treatment planning for congenital heart diseases. *Cureus*, 15(8), e44374. https://doi.org/10.7759/cureus.44374

Moon, S. J., Fryer, A. A., & Strange, R. C. (2005). Ultraviolet radiation: Effects on risks of prostate cancer and other internal cancers. *Mutation Research/Fundamental and Molecular Mechanisms of Mutagenesis*, 571(1–2), 207–219. https://doi.org/10.1016/j.mrfmmm.2004.09.015

Mori, K., Goff, B. L., Charrier, C., Battaglia, S., Heymann, D., & Rédini, F. (2006). DU145 human prostate cancer cells express functional receptor activator of NFκB: New insights in the prostate cancer bone metastasis process. *Bone*, 40(4), 981–990. https://doi.org/10.1016/j.bone.2006.11.006

Myles, P., Evans, S., Lophatananon, A., Dimitropoulou, P., Easton, D., Key, T., Pocock, R., Dearnaley, D., Guy, M., Edwards, S., O'Brien, L., Gehr-Swain, B., Hall, A., Wilkinson, R., Eeles, R., & Muir, K. (2008). Diagnostic radiation procedures and risk of prostate cancer. *British Journal of Cancer*, 98(11), 1852–1856. https://doi.org/10.1038/sj.bjc.6604370

Negi, S., Kumar, S., & Singh, A. (2023). Preclinical in vivo drug development studies: Limitations, model organisms, and techniques. In Rajput, V. S., & Runthala, A. (Eds.), *Drugs and a Methodological Compendium* (pp. 149–171). Springer, Singapore. https://doi.org/10.1007/978-981-19-7952-1_6

Nguyen, H. M., Vessella, R. L., Morrissey, C., Brown, L. G., Coleman, I. M., Higano, C. S., Mostaghel, E. A., Zhang, X., True, L. D., Lam, H., Roudier, M., Lange, P. H., Nelson, P. S., & Corey, E. (2017). LuCaP prostate cancer patient-derived xenografts reflect the molecular heterogeneity of advanced disease and serve as models for evaluating cancer therapeutics. *The Prostate*, 77(6), 654–671. https://doi.org/10.1002/pros.23313

Oczkowski, M., Dziendzikowska, K., Pasternak-Winiarska, A., Włodarek, D., & Gromadzka-Ostrowska, J. (2021). Dietary factors and prostate cancer development, progression, and reduction. *Nutrients*, 13(2), 496. https://doi.org/10.3390/nu13020496

Olson, B., Li, Y., Lin, Y., Liu, E. T., & Patnaik, A. (2018). Mouse models for cancer immunotherapy research. *Cancer Discovery*, 8(11), 1358–1365. https://doi.org/10.1158/2159-8290.cd-18-0044

Oseni, S. O., Naar, C., Pavlović, M., Asghar, W., Hartmann, J. X., Fields, G. B., Esiobu, N., & Kumi-Diaka, J. (2023). The molecular basis and clinical consequences of chronic inflammation in prostatic diseases: Prostatitis, benign prostatic hyperplasia, and prostate cancer. *Cancers*, 15(12), 3110. https://doi.org/10.3390/cancers15123110

Palacios-Eito, A., Béjar-Luque, A., Rodríguez-Liñán, M., & García-Cabezas, S. (2019). Oligometastases in prostate cancer: Ablative treatment. *World Journal of Clinical Oncology*, 10(2), 38–51. https://doi.org/10.5306/wjco.v10.i2.38

Pauli, C., Hopkins, B. D., Prandi, D., Shaw, R., Fedrizzi, T., Sboner, A., Sailer, V., Augello, M., Puca, L., Rosati, R., McNary, T. J., Churakova, Y., Cheung, C., Triscott, J., Pisapia, D., Rao, R., Mosquera, J. M., Robinson, B., Faltas, B. M., … Rubin, M. A. (2017). Personalized in vitro and in vivo cancer models to guide precision medicine. *Cancer Discovery*, 7(5), 462–477. https://doi.org/10.1158/2159-8290.cd-16-1154

Philp, L. K. (2024). Patient-derived xenograft models for translational prostate cancer research and drug development. *Methods in Molecular Biology*, 2806, 153–185. https://doi.org/10.1007/978-1-0716-3858-3_12

Posdzich, P., Darr, C., Hilser, T., Wahl, M., Herrmann, K., Hadaschik, B., & Grünwald, V. (2023). Metastatic prostate cancer—A review of current treatment options and promising new approaches. *Cancers*, 15(2), 461. https://doi.org/10.3390/cancers15020461

Rawla, P. (2019). Epidemiology of prostate cancer. *World Journal of Oncology*, 10(2), 63–89. https://doi.org/10.14740/wjon1191

Rebello, R. J., Oing, C., Knudsen, K. E., Loeb, S., Johnson, D. C., Reiter, R. E., Gillessen, S., Van Der Kwast, T., & Bristow, R. G. (2021). Prostate cancer. *Nature Reviews Disease Primers*, 7(1), 9. https://doi.org/10.1038/s41572-020-00243-0

Rice, M. A., Malhotra, S. V., & Stoyanova, T. (2019). Second-generation antiandrogens: From discovery to standard of care in castration resistant prostate cancer. *Frontiers in Oncology*, 9, 801 https://doi.org/10.3389/fonc.2019.00801

Rivero, J. R., Thompson, I. M., Liss, M. A., & Kaushik, D. (2018). Chemoprevention in prostate cancer: Current perspective and future directions. *Cold Spring Harbor Perspectives in Medicine*, 8(10), a030494. https://doi.org/10.1101/cshperspect.a030494

Robinson, D., Van Allen, E. M., Wu, Y., Schultz, N., Lonigro, R. J., Mosquera, J., Montgomery, B., Taplin, M., Pritchard, C. C., Attard, G., Beltran, H., Abida, W., Bradley, R. K., Vinson, J., Cao, X., Vats, P., Kunju, L. P., Hussain, M., Feng, F. Y., … Chinnaiyan, A. M. (2015). Integrative clinical genomics of advanced prostate cancer. *Cell*, 161(5), 1215–1228. https://doi.org/10.1016/j.cell.2015.05.001

Salagierski, M., & Schalken, J. A. (2010). PCA3 and TMPRSS2-ERG: Promising biomarkers in prostate cancer diagnosis. *Cancers*, 2(3), 1432–1440. https://doi.org/10.3390/cancers2031432

Schuck, E., Bohnert, T., Chakravarty, A., Damian-Iordache, V., Gibson, C., Hsu, C., Heimbach, T., Krishnatry, A. S., Liederer, B. M., Lin, J., Maurer, T., Mettetal, J. T., Mudra, D. R., Nijsen, M. J., Raybon, J., Schroeder, P., Schuck, V., Suryawanshi, S., Su, Y., … Wong, H. (2015). Preclinical pharmacokinetic/pharmacodynamic modeling and simulation in the pharmaceutical industry: An IQ consortium survey examining the current landscape. *The AAPS Journal*, 17(2), 462–473. https://doi.org/10.1208/s12248-014-9716-2

Schwartz, S. A., & Caprioli, R. M. (2008). Molecular imaging by mass spectrometry. *Nature Reviews Cancer*, 10, 533–554. https://doi.org/10.1016/b978-044451980-1.50025-2

Sekhoacha, M., Riet, K., Motloung, P., Gumenku, L., Adegoke, A., & Mashele, S. (2022). Prostate cancer review: Genetics, diagnosis, treatment options, and alternative approaches. *Molecules*, 27(17), 5730. https://doi.org/10.3390/molecules27175730

Shi, C., Chen, X., & Tan, D. (2019). Development of patient-derived xenograft models of prostate cancer for maintaining tumor heterogeneity. *Translational Andrology and Urology*, 8(5), 519–528. https://doi.org/10.21037/tau.2019.08.31

Shultz, L. D., Brehm, M. A., Garcia-Martinez, J. V., & Greiner, D. L. (2012). Humanized mice for immune system investigation: Progress, promise and challenges. *Nature Reviews. Immunology*, 12(11), 786–798. https://doi.org/10.1038/nri3311

Siegel, R. L., Giaquinto, A. N., & Jemal, A. (2024). Cancer statistics, 2024. *CA: A Cancer Journal for Clinicians*, 74(1), 12–49. https://doi.org/10.3322/caac.21820

Singh, S., Kumar, R., Payra, S., & Singh, S. K. (2023). Artificial intelligence and machine learning in pharmacological research: Bridging the gap between data and drug discovery. *Cureus*, 15(8), e44359. https://doi.org/10.7759/cureus.44359

Slaton, J. W., Unger, G. M., Sloper, D. T., Davis, A. T., & Ahmed, K. (2004). Induction of apoptosis by antisense CK2 in human prostate cancer xenograft model. *Molecular Cancer Research*, 2(12), 712–721. https://doi.org/10.1158/1541-7786.712.2.12

Song, H., Weinstein, H. N. W., Allegakoen, P., Wadsworth, M. H., Xie, J., Yang, H., Castro, E. A., Lu, K. L., Stohr, B. A., Feng, F. Y., Carroll, P. R., Wang, B., Cooperberg, M. R., Shalek, A. K., & Huang, F. W. (2022). Single-cell analysis of human primary prostate cancer reveals the heterogeneity of tumor-associated epithelial cell states. *Nature Communications*, 13(1), 4. https://doi.org/10.1038/s41467-021-27322-4

Song, Z., Yao, Q., Zhuo, Z., Ma, Z., & Chen, G. (2018). Circulating vitamin D level and mortality in prostate cancer patients: A dose–response meta-analysis. *Endocrine Connections*, 7(12), R294–R303. https://doi.org/10.1530/ec-18-0283

Stangelberger, A., Waldert, M., & Djavan, B. (2008). Prostate cancer in elderly men. *Nature Reviews Urology*, 10(2), 111–119. PMID: 18660852; PMCID: PMC2483315.

Stephenson, R. A., Dinney, C. P. N., Gohji, K., Ordonez, N. G., Killion, J. J., & Fidler, I. J. (1992). Metastatic model for human prostate cancer using orthotopic implantation in nude mice. *JNCI Journal of the National Cancer Institute*, 84(12), 951–957. https://doi.org/10.1093/jnci/84.12.951

Tabibian, M., Moghaddam, F. S., Motevaseli, E., & Ghafouri-Fard, S. (2024). Targeting mRNA-coding genes in prostate cancer using CRISPR/Cas9 technology with a special focus on androgen receptor signaling. *Cell Communication and Signaling*, 22(1), 504. https://doi.org/10.1186/s12964-024-01833-1

Taitt, H. E. (2018). Global trends and prostate cancer: A review of incidence, detection, and mortality as influenced by race, ethnicity, and geographic location. *American Journal of Men S Health*, 12(6), 1807–1823. https://doi.org/10.1177/1557988318798279

Takimoto, C., Chao, M., Gibbs, C., McCamish, M., Liu, J., Chen, J., Majeti, R., & Weissman, I. (2019). The Macrophage 'Do not eat me' signal, CD47, is a clinically validated cancer immunotherapy target. *Annals of Oncology*, 30(3), 486–489. https://doi.org/10.1093/annonc/mdz006

Taylor, B. S., Schultz, N., Hieronymus, H., Gopalan, A., Xiao, Y., Carver, B. S., Arora, V. K., Kaushik, P., Cerami, E., Reva, B., Antipin, Y., Mitsiades, N., Landers, T., Dolgalev, I., Major, J. E., Wilson, M., Socci, N. D., Lash, A. E., Heguy, A., … Gerald, W. L. (2010). Integrative genomic profiling of human prostate cancer. *Cancer Cell*, 18(1), 11–22. https://doi.org/10.1016/j.ccr.2010.05.026

Thalmann, G. N., Anezinis, P. E., Chang, S. M., Zhau, H. E., Kim, E. E., Hopwood, V. L., Pathak, S., von Eschenbach, A. C., & Chung, L. W. (1994). Androgen-independent cancer progression and bone metastasis in the LNCaP model of human prostate cancer. *Cancer Research*, 54(10), 2577–2581. Erratum in: *Cancer Research*, 54(14), 3953. PMID: 8168083.

Tindall, D., & Lonergan, P. (2011). Androgen receptor signaling in prostate cancer development and progression. *Journal of Carcinogenesis*, 10(1), 20. https://doi.org/10.4103/1477-3163.83937

Vigneswaran, H. T., Jagai, J. S., Greenwald, D. T., Patel, A. P., Kumar, M., Dobbs, R. W., Moreira, D. M., & Abern, M. R. (2021). Association between environmental quality and prostate cancer stage at diagnosis. *Prostate Cancer and Prostatic Diseases*, 24(4), 1129–1136. https://doi.org/10.1038/s41391-021-00370-z

Wang, D., Zhang, F., & Gao, G. (2020). CRISPR-based therapeutic genome editing: Strategies and in vivo delivery by AAV vectors. *Cell*, 181(1), 136–150. https://doi.org/10.1016/j.cell.2020.03.023

Wang, L., Lu, B., He, M., Wang, Y., Wang, Z., & Du, L. (2022). Prostate cancer incidence and mortality: Global status and temporal trends in 89 countries from 2000 to 2019. *Frontiers in Public Health*, 10, 811044. https://doi.org/10.3389/fpubh.2022.811044

Wang, N., Docherty, F. E., Brown, H. K., Reeves, K. J., Fowles, A. C., Ottewell, P. D., Dear, T. N., Holen, I., Croucher, P. I., & Eaton, C. L. (2014). Prostate cancer cells preferentially home to osteoblast-rich areas in the early stages of bone metastasis: Evidence from in vivo models. *Journal of Bone and Mineral Research*, 29(12), 2688–2696. https://doi.org/10.1002/jbmr.2300

Wculek, S. K., Amores-Iniesta, J., Conde-Garrosa, R., Khouili, S. C., Melero, I., & Sancho, D. (2019). Effective cancer immunotherapy by natural mouse conventional type-1 dendritic cells bearing dead tumor antigen. *Journal for ImmunoTherapy of Cancer*, 7(1), 5. https://doi.org/10.1186/s40425-019-0565-5

Wilson, K. M., Kasperzyk, J. L., Rider, J. R., Kenfield, S., Van Dam, R. M., Stampfer, M. J., Giovannucci, E., & Mucci, L. A. (2011). Coffee consumption and prostate cancer risk and progression in the health professionals follow-up study. *JNCI Journal of the National Cancer Institute*, 103(11), 876–884. https://doi.org/10.1093/jnci/djr151

Wilson, R. L., Taaffe, D. R., Newton, R. U., Hart, N. H., Lyons-Wall, P., & Galvão, D. A. (2021). Obesity and prostate cancer: A narrative review. *Critical Reviews in Oncology/Hematology*, 169, 103543. https://doi.org/10.1016/j.critrevonc.2021.103543

Xiao, J., Yu, X., Meng, F., Zhang, Y., Zhou, W., Ren, Y., Li, J., Sun, Y., Sun, H., Chen, G., He, K., & Lu, L. (2024). Integrating spatial and single-cell transcriptomics reveals tumor heterogeneity and intercellular networks in colorectal cancer. *Cell Death and Disease*, 15(5), 6. https://doi.org/10.1038/s41419-024-06598-6

Xu, J., Wang, R., Xie, Z. H., Odero-Marah, V., Pathak, S., Multani, A., Chung, L. W., & Zhau, H. E. (2006). Prostate cancer metastasis: Role of the host microenvironment in promoting epithelial to mesenchymal transition and increased bone and adrenal gland metastasis. *The Prostate*, 66(15), 1664–1673. https://doi.org/10.1002/pros.20488

Ye, R., Pi, M., Cox, J. V., Nishimoto, S. K., & Quarles, L. D. (2017). CRISPR/Cas9 targeting of GPRC6A suppresses prostate cancer tumorigenesis in a human xenograft model. *Journal of Experimental & Clinical Cancer Research*, 36(1), 90. https://doi.org/10.1186/s13046-017-0561-x

Yin, L., Wang, X. J., Chen, D. X., Liu, X. N., & Wang, X. J. (2020). Humanized mouse model: A review on preclinical applications for cancer immunotherapy. *American Journal of Cancer Research*, 10(12), 4568–4584.

Yoshimoto, M., Joshua, A. M., Cunha, I. W., Coudry, R. A., Fonseca, F. P., Ludkovski, O., Zielenska, M., Soares, F. A., & Squire, J. A. (2008). Absence of TMPRSS2:ERG fusions and PTEN losses in prostate cancer is associated with a favorable outcome. *Modern Pathology*, 21(12), 1451–1460. https://doi.org/10.1038/modpathol.2008.96

Zadra, G., & Loda, M. (2017). Metabolic vulnerabilities of prostate cancer: Diagnostic and therapeutic opportunities. *Cold Spring Harbor Perspectives in Medicine*, 8(10), a030569. https://doi.org/10.1101/cshperspect.a030569

Zaorsky, N. G., Shaikh, T., Ruth, K., Sharda, P., Hayes, S. B., Sobczak, M. L., Hallman, M. A., Smaldone, M. C., Chen, D. Y., & Horwitz, E. M. (2016). Prostate cancer patients with unmanaged diabetes or receiving insulin experience inferior outcomes and toxicities after treatment with radiation therapy. *Clinical Genitourinary Cancer*, 15(2), 326–335.e3. https://doi.org/10.1016/j.clgc.2016.08.020

Zhao, H., Wu, L., Yan, G., Chen, Y., Zhou, M., Wu, Y., & Li, Y. (2021). Inflammation and tumor progression: Signaling pathways and targeted intervention. *Signal Transduction and Targeted Therapy*, 6(1), 5. https://doi.org/10.1038/s41392-021-00658-5

Zhou, Y., Tao, L., Qiu, J., Xu, J., Yang, X., Zhang, Y., Tian, X., Guan, X., Cen, X., & Zhao, Y. (2024). Tumor biomarkers for diagnosis, prognosis and targeted therapy. *Signal Transduction and Targeted Therapy*, 9(1), 2. https://doi.org/10.1038/s41392-024-01823-2

5 Models of Ovarian Cancer Metastasis

Sourav Shukla, Astha Jadav, Shuvomoy Banerjee, Reena Agrawal Rajput, and Juni Banerjee

ABBREVIATIONS

AI	Artificial Intelligence
ATM	Ataxia-Telangiectasia Mutated
BRCA	BReast CAncer gene
BRIP1	BRCA1 Interacting Protein C-Terminal Helicase 1
BRPF1	Bromodomain and PHD Finger Containing 1
CAF	Cancer-Associated Fibroblasts
CAR	Chimeric Antigen Receptor)
CTLA-4	Cytotoxic T-Lymphocyte-Associated Protein 4
ECM	Extracellular Matrix
FDA	Food and Drug Administration
GANs	Generative Adversarial Networks
GEMMs	Genetically Engineered Mouse Models
GLOBOCAN	Global Cancer Observatory
HGSOC	High-Grade Serous Ovarian Carcinoma
HR	Homologous Recombination
HRD	Homologous Recombination Deficit
MMP	Matrix Metalloproteinase
NLP	Natural Language Processing
OC	Ovarian Cancer
PALB2	Partner and Localizer of BRCA2
PARP	Poly ADP-ribose Polymerase
PCMT1	Protein-L-isoaspartate (D-aspartate) O-methyltransferase 1
PD-L1	Programmed Death-Ligand 1
PDX	Patient-Derived Xenografts
TAM	Tumor-Associated Macrophages
TME	Tumor Microenvironment
TP53	Tumor Protein p53
TSG	Tumor Suppressor Genes

5.1 INTRODUCTION: OVARIAN CANCER AS THE SILENT INVADER

Ovarian Cancer (OC) is among the top three most severe gynecologic cancers worldwide, with the highest death rate among these malignancies (Jin 2025). As

DOI: 10.1201/9781003583349-6

one of the most common malignant tumors of the female reproductive system, OC presents a major challenge to healthcare systems worldwide. It is considered "the silent killer" because of its almost invisible development in the early stages. This process also results in a late diagnosis for 75% of its total cases.

The disease mainly begins in the ovaries and fallopian tubes. Unlike other solid tumors, which mainly spread through circulation, OC often spreads straight within the peritoneal cavity. The symptoms of early onset of this complex disease are vague and easily confused with benign conditions such as stomach edema, abdominal discomfort, trouble in eating, and pee urgency. OC differs from other cancers due to its specific histological and molecular characteristics. The disease is usually diagnosed at a median age of 63 years, mainly affecting postmenopausal women. Certain genetic predispositions, such as BReast CAncer gene 1 (BRCA1) and BReast CAncer gene 2 (BRCA2) mutations, can significantly raise a person's early age OC risk significantly. Seventy percent of the cases, which cause most of the OC-related deaths, happen due to high-grade serous ovarian carcinoma (HGSOC). HGSOC is one of the most known subtypes of OC. All subtypes pose major challenges in the diagnosis and therapy of OC.

Advanced OC, or the cause of death in OC, is due to a major feature of all cancers known as metastasis. Metastasis is the spread of cancer cells from the primary site or main tumor region to distant areas or secondary sites. Metastasis is often very difficult to discover. This difficulty is further complicated by many phases of early and late metastasis, like detachment from the main tumor, survival in the peritoneal fluid, adhesion to the peritoneal surface, invasion of the underlying tissue, and development of a new tumor. These events often create complications in the diagnosis and treatment plans.

There are many screening instruments for OC, among which transvaginal ultrasounds and serum CA-125 tests are the major ones. However, they lack the sensitivity and specificity needed for massive population applications. The variability in OC subtypes complicates early detection and uniform diagnosis. The majority of these cases are found at an advanced stage where prognosis is bad. Hence, to improve OC outcomes in the population, we need efficient screening strategies.

Scientists are creating in vivo models, two-dimensional (2D), and three-dimensional (3D) cell cultures to learn about OC metastasis. 3D models, also known as spheroid suspension cultures, form multicellular aggregates in ascites fluid, which help researchers study the penetration and efficacy of drugs with great efficiency. Like tissue engineering scaffolds and microfluidic systems, 3D-culture models replicate tumor-stromal interactions and medication responses. While genetically modified mouse models (GMMMs) and patient-derived xenografts (PDX) provide valuable in vivo systems for examining OC growth and metastases, organ-on-chip models mimic the dynamic character of the peritoneal cavity. Emerging to guide therapeutic approaches and forecast metastatic behavior are computational modeling techniques as well. Investigating these models would help to comprehend the tumor microenvironment (TME) influences on ovarian cancer metastasis, elucidate specific molecular pathways causing metastasis, and pinpoint the therapeutic targets.

This chapter attempts to give a thorough review of the several models applied to investigate ovarian cancer metastasis, assess the merits and drawbacks of present model systems, and address the translational possibilities of these models in the generation of novel treatments. Moreover, important research gaps, like the requirement of accurate preclinical models to recapitulate the heterogeneity of human OC, the difficulty of modeling the particular peritoneal metastases pattern of OC, integration of immune components in OC models to study immune evasion and possible immunotherapies, and development of patient-specific models for precision medicine approaches, have been highlighted.

5.2 EPIDEMIOLOGY AND STATISTICS OF THE "SILENT KILLER"

OC poses a great threat to women's health due to its often-asymptomatic course until advanced stages, which has given it its sinister name "the silent killer" and its high ranking among the deadliest gynecological cancers at a global level. With a significant worldwide burden and an ongoing case increase, OC presents a complicated epidemiological picture reflecting the interaction of genetic, environmental, and socioeconomic elements over different populations. Based on the recent global cancer statistics from GLOBOCAN 2020, OC ranks as the eighth most extensive cancer affecting women globally. The year 2020 witnessed around 313,959 new cases worldwide (Sung et al. 2021). More shockingly, in the same year, OC accounted for 4.7% of all cancer-related fatalities in women, or 207,252 deaths. These statistical data highlight the disproportionate death rate linked with OC, which reflects its generally late-stage diagnosis and difficulties in its treatment (Bray et al. 2018). The burden of OC is not homogeneous worldwide; rather, there are eye-catching differences seen depending on the location and nation (Table 5.1).

TABLE 5.1
Key Statistics and Projections for Ovarian Cancer

S. No.	Category	Description	Statistics/Projections	References
1	Five-year survival rate in Asian countries	Global 2025 data	59.6%	Jin (2025)
2	Global Incidence of OC	New cases in 2022	324,603	Hong et al. (2025)
3	Global Mortality due to OC	Deaths in 2022	206,956	Mazidimoradi et al. (2022)
4	Projected Incidence of OC	New cases by 2050	503,448 (55% increase)	Li et al. (2025)
5	Projected Mortality due to OC	Deaths by 2050	350,956 (almost 70% increase)	World Ovarian Cancer Coalition
6	Five-Year Survival Rate in developed nations	Global 2025 data	36%–46%	World Ovarian Cancer Coalition
7	Age-Standardized Incidence Rate	Global average (2022)	6.7 per 100,000	Ovarian cancer statistics
8	Countries with Highest Incidence	Top countries in 2022	China, India, U.S.	Ovarian cancer statistics
9	Risk Factors	Aging and genetics	High risk after 50 years, BRCA1/2 mutations	Ovarian Cancer Risk Factors

Usually, higher incidence rates were reported in high-income nations, especially those in North America and Europe. For example, in Western Europe, the age-standardized incidence rate is 7.5 per 100,000; in North America, it is 8.1 per 100,000. Lower rates are seen in many areas of Africa and Asia, where some nations document incidence rates as low as 3 per 100,000 women (Ferlay et al. 2021). Importantly, these lower rates in underdeveloped areas may represent underdiagnosis resulting from restricted healthcare access and poor cancer registries than they reflect actual lower incidence. Looking ahead, the GLOBOCAN 2040 forecasts show a worrying picture of the future OC burden. Rising to 424,966 cases by 2040, the worldwide prevalence of OC is expected to have jumped a shocking 36.6% from the year 2020. Low- and middle-income nations, fast population growth, genetic factors, and obesity are some of the direct or indirect influencing factors for growing OC burden. Likewise, the global deaths due to OC are presumed to increase from 207,201 in 2020 to 313,617 deaths by 2040; that is an increase of 51.3% within 20 years (Sung et al. 2021). These statistical estimations indicate the urgent need for developing better prevention strategies, detection technologies, and therapies for OC. Further, data in research shows that a few groups in the population are more susceptible to getting OC, including individuals more than 50 years of age. Usually diagnosed at 63, 70% of cases are those of women 55 years and above, making age another major risk factor (Torre et al. 2018). Racial and ethnic discrepancies abound as well; e.g., white Anglo-American women have the highest incidence rates recorded in the United States.

A significant portion of the risk attributed to OC stems from genetics; hereditary mutations constitute approximately 20%–25% of cases. The most usual genetic risk factors are the mutations in the BRCA1 and BRCA2 genes. Women's BRCA1 mutation places them at a significantly higher risk of developing OC (35%–70%) compared to those with the BRCA2 mutation (10%–30%) (Kurian et al. 2017). These statistics are extremely different from the rest of the people who have a lifespan risk of 1.3%. The other class of genetic syndromes with a higher risk of OC includes Lynch syndrome (hereditary nonpolyposis colorectal cancer) due to DNA mismatch repair genes and mutations in RAD51C, RAD51D, and BRIP1.

Unlike breast cancer and cervical cancer, which are prominent in females, OC does not have quality screening methods. Moreover, the methods and techniques used to detect OC differ around the globe, which explains the variation in results. Serum CA-125 testing and transvaginal ultrasound screening have not shown any remarkable decline in death rates. The optimal interval and age to start these diagnoses are still under research. Many countries have launched genetic testing programs to identify more prone individuals for targeted surveillance and risk-reducing interventions, such as the removal of fallopian tubes and ovaries (salpingo-oophorectomy) (Karlan et al. 2014). The proper access to healthcare, timely diagnosis, modern treatment, and active participation of people in clinical trials differ widely in parts of the world. Nonetheless, disparities in socioeconomic level, geography, and race or ethnicity persist even in these contexts, and the world topography of ovarian cancer epidemiology stresses the complex interplay of biological, environmental, and social factors influencing the trajectory of the disease. These data suggest that there is an urgent need for preclinical and clinical research to develop early detection, access to genetic testing, and personalized, targeted, and effective therapies.

5.3 UNRAVELING THE METASTATIC MAZE OF OC

Knowing the metastasis process in detail is important when it comes to understanding the course of OC and accurately formulating the appropriate therapeutic approaches. Genetic and molecular alterations in the TME are major drivers of the metastasis process. Research on the TME will help us in completing the puzzle of this disease. The findings may also open up new dimensions of treatment options. In this section, we will explore the genetic and molecular factors responsible for OC progression and mutations, as well as the cellular pathways underlying its aggressiveness.

5.3.1 OC Progression and Its Major Drivers

The important factors for cancer evolution are gene mutations, homologous recombination deficit (HRD), metastasis, and immune evasion mechanisms. The crosstalk of genetic and molecular alterations gives an aggressive nature to OC. 15%–20% of all the cases of OC are due to genetic mutations. The best-known genetic factors contributing to OC risk and progression are BRCA mutations. By their very nature, TSGs like BRCA1 and BRCA2 are DNA repair genes that signal through the homologous recombination (HR) pathway. If mutated, these genes impair double-stranded DNA break repair, leading to genomic instability and hence increased cancer risk. BRCA-1 carriers would have a 35%–70% lifetime risk of ovarian cancer, while the lifetime risk of OC associated with the BRCA-2 mutation is similar at somewhere between 10% and 30%. Besides their contribution to cancer susceptibility, BRCA mutations also affect the behavior of already existing tumors. Ovarian tumors that are BRCA altered sometimes show obvious and characteristic clinical as well as molecular features. Interestingly, while aggressive, BRCA-mutated tumors are, because of their impaired DNA repair machinery, often more sensitive to platinum-based chemotherapy and PARP inhibitors (Fong et al. 2009). This "BRCAness" property has turned into a key factor informing treatment decisions and engendered the development of novel targeted therapies designed to exploit that weakness. Another important genetic alteration in ovarian carcinogenesis is Tumor Protein p53 (TP53) mutations. Mostly called the "guardian of the genome" given its role in cell cycle control, DNA repair, and cell death, the TP53 gene codes for the p53 protein. Mutations in its TP53 gene affect about 95% of cases of HGSOC (Ahmed et al. 2010). Through its mutations, p53 sometimes has had a loss of ability to do its job as a cancer suppressor, unless in those rare cases the function is gained in favor of tumor growth and metastasis. Therefore, the incidence of TP53 mutations causes resistance against therapies and helps in the progression of OC, as mutant p53 could revoke the normal cellular stress response so that cancer cells live and grow even under environments that are fatal to normal cells.

HRD is a key biological characteristic for tumors beyond BRCA mutation. Mutations resulting from HRD are linked to the involvement of genes like RAD51, PALB2, and ATM in homologous recombination. This molecular defect causes a decline in the repair capacity of double-stranded breaks in DNA within the cell and is therefore a contributor to genomic instability that would raise the sensitivity of the cell lines to some DNA-damaging agents (Lord and Ashworth 2016). Thus, HRD in

OC has become a major biobanker for guiding treatment decisions and personalized medicine when anticipating response to both the PARP inhibitors and chemotherapy.

Another important feature of OC progression is immune evasion from the host's immune system. These processes include downregulation of tumor-associated antigens, expression of immune checkpoint proteins, such as PD-L1, and the recruitment of immunosuppressive cells to the TME, leading to escape in recognition and elimination by immune cells. The immune landscape of ovarian tumors is driven by both BRCA- and TP53-mutants. Higher mutation of BRCA-mutated malignancies results in increased immunogenicity, while p53 mutations have been linked to immune evasion strategies, thus stressing the intricate relationship between genetics and immunology in OC (Figure 5.1).

Metastasis is one of the major factors adding to the aggressive progression of OC. OC metastasis has a distinct pattern compared to others. It majorly spreads through the peritoneal cavity, leading to extensive abdominal metastasis (Lengyel 2010). Metastasis is mainly driven by interactions between the tumor stroma (a major component of the TME), extracellular matrix (ECM), and matrix metalloproteinases (MMPs, a family of proteolytic enzymes) (Shield et al. 2017). Collagen remodeling in the tumor stroma plays an important role in tumor progression by influencing cell adhesion, migration, and invasion. Spiked levels of MMPs can increase the

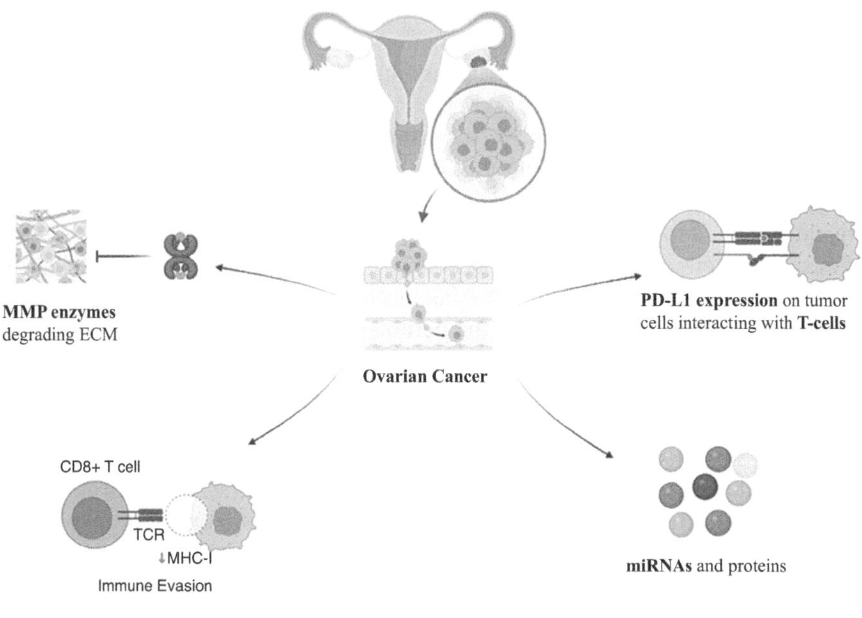

FIGURE 5.1 Schematic of key molecular pathways in ovarian cancer metastasis: This illustration outlines several critical mechanisms that drive ovarian cancer metastasis, including the degradation of the ECM by MMP enzymes, immune evasion through downregulation of MHC-I on tumor cells interacting with CD8+ T cells, PD-L1 expression on tumor cells interacting with T cells, and the roles of miRNAs and proteins in promoting cancer progression. The diagram visually summarizes the complex processes that contribute to the spread of ovarian cancer.

cancer cell invasion and metastasis by degrading ECM components and consequently reducing physical barriers to tumor dissemination (Davidson et al. 2014). In OC, metastatic spread with ECM degradation favors tumor cell progression. Investigating and targeting ECM remodeling and MMP activity has a future for developing novel biomarkers of OC progression and metastasis (Kenny and Lengyel 2009).

Integrating our knowledge of these major drivers of OC progression will be crucial for untangling the metastatic maze of ovarian cancer and creating more individualized and successful therapy paradigms for the patients.

5.3.2 The Metastatic Blueprint: How OC Spreads

The crosstalk among several molecular mediators, peritoneal milieu, and ovarian cancer cells gives rise to a unique metastatic blueprint for OC. To develop an actionable and effective approach toward this cancer, we need to understand and target its metastasis.

OC cells spread through the peritoneal cavity; there they can either survive alone or in small clusters. OC cells can survive an anchorage-independent growth in the peritoneal cavity. Previously, we already discussed the importance of ECM and MMPs in OC metastasis. Furthermore, genetic and epigenetic changes help the OC cells survive in the peritoneal cavity and avoid anoikis, a type of programmed cell death brought on by disturbance of cell-matrix interaction (Kim et al. 2012).

Of note, multicellular spheroids promote peritoneal cell distribution. These 3D cancer cell balls protect the cells from anoikis and other peritoneal fluid stresses. Spheroid formation requires cell aggregation, active cellular processes, and chemical changes. Spheroid development and maintenance also rely on the cell adhesion proteins, like E-cadherin. As a result, the microscopic spheroid structure generates oxygen, nourishment, and metabolic waste gradients, mimicking solid tumors and creating cell heterogeneity. These multicellular spheroids, also known as reservoirs of metastatic cells, float in the peritoneal fluid before binding to the mesothelial lining to colonize other areas. This tethering mechanism forms a complicated interaction between ECM components on the peritoneal surface and adhesion molecules, including integrins and β1-integrins on the spheroid surface (Shield et al. 2017).

Recently, researchers have gained considerable interest concerning the involvement of exosomes in OC metastasis. Exosomes are micro-extracellular vesicles of 30–150 nm that mediate cell-to-cell communication. Exosomes produced by tumor cells may change the activity of cells and remodel the metastatic environment in the OC by delivering proteins, lipids, and nucleic acids. For peritoneal mesothelial cells, exosomes have been seen to transfer proteins and miRNAs related to cancer, hence promoting changes that enable tumor cell adhesion and invasion. Exosomes further help in ensuring optimal conditions for metastasis before the arrival of cancer cells by forming a pre-metastasis niche. A suitable environment for metastatic colonization includes vascular leakiness, modified immune responses, and changes in the ECM composition of distant sites (Tsering et al. 2020). As cancer cells spread through the peritoneal surface, they encounter and navigate through the ECM. In this process, they secrete matrix-degrading enzymes such as MMPs that degrade ECM components and create pathways for invasion. At the same time, stromal cells in the TME and cancer cells produce new ECM components. This majorly leads to a desmoplastic reaction known for increased ECM deposition and stiffness. The TME consists of several cell types in an OC, like cancer-associated fibroblasts (CAFs), tumor-associated macrophages (TAMs), endothelial cells, and immunological cells. These stromal cells participate directly through direct contact and paracrine signaling, promoting tumor development, invasion, and therapeutic resistance. Research shows that CAFs release chemokines and growth factors that boost cancer cell migration and proliferation. They also synthesize collagen and other matrix proteins, which helps to restructure ECM as well. TAMs create an immunosuppressive environment by polarizing toward the M2 phenotype and releasing substances that boost angiogenesis and metastasis (Yin et al. 2016).

While earlier research showed gene mutations, like p53, to align with increased OC invasiveness and metastasis, recent research studies are reporting about BRPF1, PCMT1 (Alexandrova et al. 2025), and other such genes through advanced studies.

By understanding these complexities of OC metastasis, from peritoneal spread and spheroid formation to the roles of exosomes, ECM remodeling, and the TME, scientists can plan more successful treatment options in the future, potentially disrupting the metastasis chain and improving the prognosis for patients.

5.4 THE MODEL MENAGERIE: EXPLORING OC METASTASIS IN THE LAB

Gaining knowledge of the complex spread of OC calls for better laboratory models that faithfully reproduce patient biological events is very crucial. There are many tools and approaches that researchers have evolved along with the increasing knowledge of cancer biology. Every model used for investigating OC metastases provides a unique perspective on the condition. From 3D culture systems to sophisticated microfluidic devices, these models investigate OC metastasis and treatment choices using various experimental approaches. The section below in this chapter provides a comprehensive overview of such models.

5.4.1 The Living Petri Dish: 3D Culture Systems

This technique has revolutionized OC metastasis research. It links conventional 2D cell culture and in vivo animal models. The realistic setting for studying cancer cell behavior, like that of TME, can be performed by using these 3D systems. They allow researchers to observe and manipulate cell-cell interactions, matrix remodeling, and medication responses. The following 3D models are widely used in metastasis studies:

5.4.1.1 Spheroid Suspension Cultures: Mimicking In Vivo Tumor Biology

The most commonly used 3D culture technique for studying OC metastasis is spheroid suspension cultures. These clusters of cells offer valuable data and information about the biology of the spread of the disease by recreating key features of metastatic OC cells (Friedrich et al. 2009).

Cancer cells are grown in low-attachment plates or hanging drops. This culture technique generates 3D structures by promoting cell-to-cell attachment (Mehta et al. 2012). The size and complexity of spheroids may vary from simple cancer cell aggregates to heterogeneous ones, which include stromal, immunological, or other TME cells. Spheroid cultures have many benefits due to their ability to replicate the gradient of nutrients, oxygen, and medication penetration seen in solid tumors. Cells in the center of the spheroid suffer hypoxia and nutritional deprivation, while cells in the perimeter are exposed to more oxygen and nutrients. Research has also discovered that the cells in the center of the spheroid may enter a resting or quiescent state, making them less responsive to chemotherapy drugs. This phenomenon explains why some cancer cells may survive during initial chemotherapy treatments and return after treatment. Such differences in the spheroid allow scientists understand how microenvironmental factors affect cell activity, gene expression, and medication responsiveness.

Scientists have used these models to better understand how OC develops resistance to chemotherapy. Researchers also use spheroid cultures in the study of peritoneal metastasis. Researchers have used these models to examine the molecular pathways involved in spheroid development, survival in suspension, and peritoneal mesothelial attachment (Kenny et al. 2014). These studies have helped scientists identify key molecules and mechanisms involved in each stage of the metastatic cascade, providing prominent possible targets for treatment.

5.4.1.2 Tissue Engineering Scaffolds: Bioprinting and Hydrogel-Based Models

Though spheroid suspension cultures provide important insights, they fail to replicate the complex ECM structure found in solid tumors. To overcome this limitation, researchers are turning to tissue engineering methods, like scaffolds and hydrogels, to produce more complex 3D models of OC (Katt et al. 2016).

Developing exact, repeatable 3D cancer models has become a significant challenge, but bioprinting offers promising solutions. In this method, the cells are placed together with supporting biomaterials, in particular spatial configurations that replicate tumor architecture. Bioprinting has been used to make models, including different types of cells, like cancer cells, fibroblasts, and endothelial cells, arranged in established spatial connections for OC. These bioprinted models may be used to investigate how the spatial arrangement of various cell types affects tumor development, invasion, and medication response.

Another method to generate 3D cultures with specified ECM characteristics is by using hydrogels. Natural hydrogels like Matrigel and collagen provide a complexity of ECM proteins, which is important for cell development and differentiation. Synthetic hydrogels allow researchers to control mechanical and biochemical properties, enabling them to study how specific ECM features affect cancer cell activity (Katt et al. 2016).

Research on ECM stiffness in OC shows that scaffold-based models are beneficial. Modifying different properties of hydrogels has revealed that high matrix stiffness can influence chemoresistance, cancer cell growth, and, importantly, invasion, thus enhancing our understanding of disease progression in OC (Lynch et al. 2022).

5.4.1.3 Microfluidic Culture Systems: Dynamic Models to Simulate Tumor Behavior

Microfluidic culture is a leading technology in 3D culture, providing exceptional control over the cellular microenvironment and the power to incorporate dynamic elements such as fluid flow and chemical gradients (Table 5.2).

These "organ-on-chip" systems feature miniaturized chambers and channels for growing cells under precisely regulated conditions. Microfluidic devices replicate aspects of metastasis in OC, such as intraperitoneal spread, mesothelial lining penetration, and immune system interactions (Xu et al. 2020). Microfluidic systems offer significant benefits by simulating fluid flow, mimicking peritoneal fluid movement, and shear pressures on metastatic cells. By using these devices, researchers have demonstrated how fluid shear affects OC spheroids' adhesion and metastatic efficiency. They can replicate complex tissue interfaces with multiple cell types and ECM components, facilitating the study of cancer cell interactions during metastasis. These systems enable personalized medicine approaches and high-throughput drug screening by allowing tests on patient-derived cells in a physiologically relevant environment (Ma et al. 2022). Integrating biosensors and imaging features enhances their value by providing real-time data on cellular behavior and drug responses, aiding in developing precise models of cancer progression and treatment (Shin et al. 2019).

Therefore, 3D-culture systems such as spheroid suspension cultures, tissue engineering scaffolds, and microfluidic devices have notably increased our ability to study OC metastasis in the lab. These models offer many approaches for understanding various aspects of the metastatic process, such as the formation of multicellular aggregates and the complex interactions between cancer cells and their environment. These technologies can be promising if OC metastasis develops, which can help in creating more effective treatment strategies for this challenging disease (Figure 5.2).

5.4.2 The Organ-on-Chip Revolution

With organ-on-chip technology, cancer research is undergoing a paradigm change and presents hitherto unheard-of chances to investigate complicated biological processes in a physiologically appropriate environment under control. Combining ideas of microfabrication, tissue engineering, and microfluidics, this breakthrough method generates miniature versions of human organs and tissues like that of the ovary. Within the framework of OC research, organ-on-chip systems are stretching the possibilities of what can be accomplished using in vitro models by offering fresh insights on tumor–stroma interactions, metastatic processes, and treatment responses.

5.4.2.1 Microfabrication and Biomimicry: Modeling Tumor–Stroma Interactions

The power to replicate the complex architecture and dynamic milieu of human tissues drives organ-on-chip technology. Advanced microfabrication methods that enable exact cell and ECM component patterning within microfluidic devices help to accomplish this technology. Usually featuring several chambers connected by channels, these devices allow the spatial arrangement of various cell types and introduce controlled fluid flow.

TABLE 5.2
Comparison of Different OC Models, Advantages, Limitations, and Applications

S. No.	Model Type	Description	Advantages	Limitations	Applications	References
1.	3D Spheroid Cultures	Cells grown in low-attachment plates or hanging drops to form 3D structures, mimicking tumor biology.	Replicates nutrient and oxygen gradients; useful for studying drug resistance and metastasis.	Limited in replicating complex ECM structures.	Understanding metastasis, drug resistance, and peritoneal spread.	Tofani et al. (2021)
2.	Tissue Engineering Scaffolds	Uses bioprinting or hydrogels to create complex ECM structures.	Provides spatial cell arrangement and ECM complexity; useful for studying tumor architecture and drug responses.	Challenges in standardizing and scaling up models.	Investigating tumor development, invasion, and drug penetration.	Fischbach et al. (2007)
3.	Microfluidic Organ-on-Chip Systems	Miniaturized systems mimicking the tumor microenvironment with controlled fluid flow.	Offers physiological relevance, control over conditions, and real-time imaging; useful for studying metastasis and drug efficacy.	Complexity in setup and operation; limited by the need for precise microfabrication.	Personalized medicine approaches, high-throughput drug screening, and studying metastatic processes.	Regmi et al. (2022)
4.	Genetically Engineered Mouse Models	Mouse models with specific genetic alterations to study disease progression.	Replicates the entire disease progression; useful for understanding genetic contributions to cancer.	Genetic changes might not fully reflect human complexity; differences in mouse and human biology.	Studying tumor initiation, growth, and metastasis; evaluating immune system interactions.	Kersten et al. (2017)
5.	Patient-Derived Xenografts	Human tumor tissue implanted into immunodeficient mice to study personalized responses.	Maintains tumor diversity and molecular characteristics; useful for predicting treatment responses.	Limited by immunodeficient hosts; costly and time-consuming to establish.	Personalized treatment planning, drug discovery, and understanding resistance mechanisms.	Tentler et al. (2012)
6.	Computational and AI-Driven Models	Virtual representations using data to predict disease progression and treatment responses.	Enables rapid drug screening and personalized treatment planning; useful for integrating complex data.	Requires high-quality data; interpretability challenges with AI models.	Predicting metastasis, drug efficacy, and identifying new therapeutic targets.	Sebastian and Peter (2022)

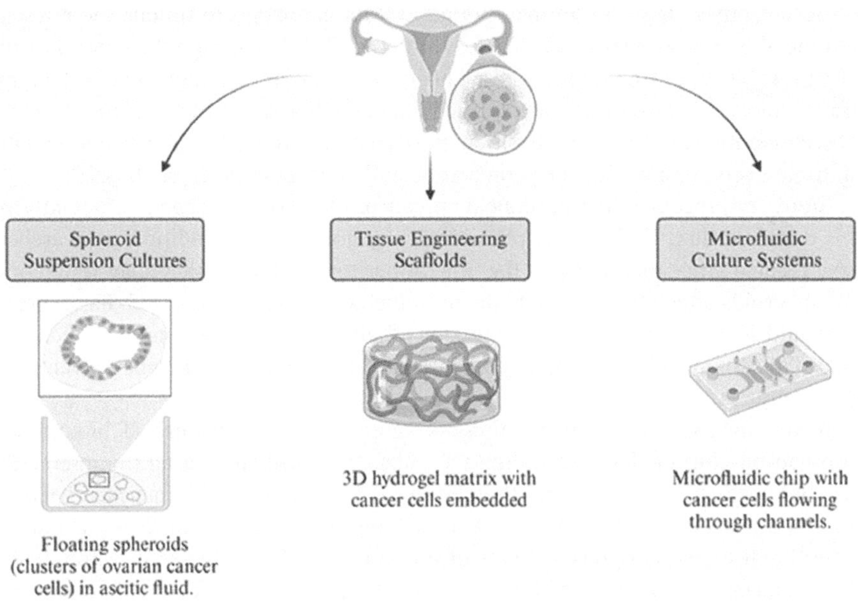

FIGURE 5.2 Schematic representation of different OC models like exosomal, peritoneal, and ECM involvement: This figure illustrates three key in vitro models used in ovarian cancer research: Spheroid Suspension Cultures depicting floating spheroids in ascitic fluid, Tissue Engineering scaffolds showcasing cancer cells embedded in a 3D hydrogel matrix, and microfluidic culture systems presenting cancer cells flowing through channels within a microfluidic chip. These models are designed to mimic various aspects of the TME and metastatic processes.

The capacity of organ-on-chip technology to introduce dynamic aspects that are difficult or impossible to obtain in conventional static culture systems is among its most important benefits. Reflecting the flow of interstitial fluid in tissues, microfluidic tubes enable the continual perfusion of culture medium (Bhatia and Ingber 2014). This flow can be exactly regulated to produce gradients of medicines, growth hormones, or nutrients across the tissue construct and to replicate physiological shear stresses (Huh et al. 2010).

Organ-on-chip models have been created to replicate many facets of the TME concerning OC. The major focus has been on mimicking the peritoneal cavity, which is the main source of OC metastasis (Shen et al. 2020). These models often include layers of mesothelial cells, which line the peritoneal cavity, OC cells, and many other stromal components, such as fibroblasts and immune cells (Bhatia and Ingber 2014). Researchers can use the spatial organization of different cell types to study complex crosstalk between tumor cells and their surrounding support tissue (stroma). For example, while using organ-on-chip systems, they have examined how CAFs can influence the behavior of OC cells. By cultivating OC cells and CAFs in two different compartments connected by tiny channels, researchers can keenly observe how substances produced by CAFs affect cancer cell growth, movement, and response to treatment (Shen et al. 2020).

Besides, these models can incorporate ECM components to imitate the physical and metabolic properties of the TME (Huh et al. 2010). Through the control of the ECM's rigidity and composition, scientists may investigate how matrix features affect cancer cell activity and metastatic potential (Bhatia and Ingber 2014). For OC, where alterations in ECM composition and dynamics are well established to be vital in disease progression, this question is especially pertinent (Shen et al. 2020).

Studying processes like peritoneal spread in OC models depends especially on this dynamic flux. Under more physiologically comparable conditions, researchers have created chips that replicate the flow of peritoneal fluid so they may watch how OC spheroids attach to and invade the mesothelial lining. This research has exposed how fluid shear forces affect the expression of adhesion molecules and the efficacy of metastatic colonization, therefore offering insights difficult to get from conventional static cultures.

Reversing tissue-tissue interactions is another essential feature of biomimicry in organ-on-chip models. Regarding OC, this can entail simulating the interaction between metastatic nodules and the peritoneal membrane or the interface between the ovarian surface epithelium and the underlying stroma (Huh et al. 2010). Culturing several cell types on opposing sides of a porous membrane inside the chip allows these interfaces to be produced, therefore enabling both direct cell-cell contact and paracrine signaling (Shen et al. 2020).

5.4.2.2 Advantages of Organ-on-Chip over Traditional Models

For research on OC, organ-on-chip technology has presented numerous major benefits over conventional in vitro and in vivo models. Some of these advantages include:

I. **Physiological Relevance**: The most important benefit of organ-on-chip models is to more precisely replicate the intricate, 3D architecture and dynamic milieu of human tissues. In the future, mechanistic research and drug screening for OC outcomes can also be predicted using this model (Bhatia and Ingber 2014).

II. **Control and Reproducibility**: Unlike dynamic animal models, organ-on-chip technologies provide better control over experimental conditions. To investigate the role of particular factors, including ECM composition, fluid flow rates, or cell type ratios affecting OC cell behavior, researchers can methodically change the model and use it. This degree of control helps to isolate particular biological pathways and improves reproducibility (Huh et al. 2010).

III. **Real-Time Imaging and Analysis**: High resolution is used along with organ-on-chip models in many systems. This technique enables real-time imaging of cellular events, such as OC cell migration or drug-induced death, thereby offering dynamic information, which is difficult to obtain with conventional endpoint tests (Shen et al. 2020).

IV. **Reduced Animal Use**: The model reduces animal testing in preclinical cancer research by acting as a physiologically appropriate substitute for 2D cultures. This aligns with ethical considerations and legal demands to lower the use of animals in research (Bhatia and Ingber 2014).

V. **Patient-Specific Models**: Patient-derived cancer cells, stromal cells, and immune cells can be loaded onto organ-on-chip systems. This capability creates opportunities for modified medicine approaches in which an OC patient's cells might be tested for medication reactions and used to direct therapy choices (Shen et al. 2020).
VI. **Integration of Multiple Organ Systems**: Multiple high-end organ-on-chip systems let several "organs" be connected on a single chip. Research on systemic effects of cancer and cancer treatments like drug metabolism by the liver or cardiotoxicity of chemotherapeutic drugs can improve from this multi-organ approach (Bhatia and Ingber 2014).
VII. **High-Throughput Potential**: Early organ-on-chip devices were very complicated and had lower throughput, but developments in microfabrication and automation nowadays are producing higher-throughput platforms. These technologies could minimize the distance between more physiologically appropriate, lower-throughput animal models and conventional high-throughput screening techniques (Huh et al. 2010).
VIII. **Modeling Metastasis**: Organ-on-chip models present special chances for metastasis research in OC. Multiple tissue types and fluid flow design help these models replicate the path of metastasis (Shen et al. 2020).
IX. **Studying Drug Penetration and Efficacy**: Combining the 3D character of organ-on-chip models with fluid flow lets one more realistically examine drug penetration and efficacy. This is especially pertinent for OC, where treatment is much hampered by inadequate medication penetration into solid tumors and spheroids (Bhatia and Ingber 2014).
X. **Long-Term Culture**: Many organ-on-chip systems support long-term cell culture, sometimes for weeks or months. This type of culture makes it possible to investigate slower processes, including the long-term consequences of continuous drug use or the emergence of drug resistance (Shen et al. 2020).

In conclusion, the organ-on-chip revolution is providing potent and new tools to investigate OC biology and create more successful treatments. These platforms are closing a major gap in cancer research by integrating the physiological significance of in vivo models with the control and accessibility of in vitro systems. Organ-on-chip models are likely to become more significant as technology develops knowledge of OC metastasis, target identification, and customizing treatment plans.

5.4.3 The In-Vivo Battlefield of OC: Animal Models

Although in vitro models offer important new perspectives on the molecular processes of OC metastasis, animal models are still important for comprehending the complicated, systemic character of cancer progression in a live organism (Frese and Tuveson 2007). Animal models include carcinogen-induced animal models, syngeneic models, laying hens, zebrafish, PDX, and Genetically Engineered Mouse Models (GEMMs). Among all these various animal models, some have major limitations, e.g., the laying hen and zebrafish models that depict different physiological conditions as compared to humans. In the OC modeling research, PDX and GEMMs are significantly notable for their ability to replicate key aspects of human disease and provide personalized information on tumor behavior and treatment response (Tentler et al. 2012).

5.4.3.1 Genetically Engineered Mouse Models (GEMMs)

GEMMs can provide scientists with insights into how specific genetic changes can drive tumor initiation, growth, and metastasis within the framework of a normally functioning immune system and environment. GEMMs have allowed researchers to modify specific genes and pathways linked with the disease to get an improved understanding of OC biology (Connolly and Bao 2014). These models are often used in a tissue-specific and time-restricted manner by adding, removing, or modifying genes in the mouse genomes (Singh et al. 2012).

Many GEMMs have been developed for different subtypes of OC, as they can replicate the entire progression of OC, from the initial genetic alterations to the metastasis. For HGSOC, models majorly focus on mutations in the TP53 and BRCA1/2 genes. For instance, it closely resembles the genomic instability and metastatic behavior seen in human HGSOC, the Trp53/Brca1/PTEN triple-mutant model, in which these genes are inactivated, especially in the ovarian surface epithelium of the fallopian tube (Connolly and Bao 2014). Clear cell ovarian carcinoma has been studied using models with alterations in the ARID1A and PIK3CA genes, while endometrioid ovarian carcinoma has been explored through models with PTEN deletion and KRAS activation. For research on the early phases of OC development, which is a challenge to monitor, GEMMs have proven very helpful. They have also helped to clarify the cell of origin, with evidence pointing to many HGSOCs perhaps starting in the fallopian tube rather than the ovarian surface epithelium (Singh et al. 2012).

GEMMs also provide a platform for investigating the function of the immune system in OC progression and testing preventative measures. This allows for examining the immune evasion strategies and evaluating the immunotherapeutic treatments. However, GEMMs also have limitations. The genetic changes in these models might not completely reflect the complexity and variety of actual malignancies. Furthermore, variations in mice and human biology can occasionally produce differences in medication responses or metastatic tendencies (Connolly and Bao 2014).

5.4.3.2 Patient-Derived Xenografts (PDX) for Personalized Insights

Another powerful tool for OC researchers is PDX, which serves as an important connection between clinical findings and preclinical research (Tentler et al. 2012). PDX models can be created by directly implanting tumor tissue (e.g., patient-derived explants) or immune cells from a patient into immunodeficient mice, allowing the human tumor to grow and gradually spread (Byrne et al. 2017).

PDX models are also effective for studying rare subtypes of OC. They provide a source of tumor tissue that maintains the unique characteristics of these less common types, which may not have established cell lines available (Dobbin et al. 2014). Moreover, PDX models help to understand mechanisms of treatment resistance and tumor evolution. Through several generations of mice, researchers may passage tumors and track their adaptation to therapeutic forces as well as find mechanisms of acquired resistance (Hidalgo et al. 2014). Developing plans to either avoid or overcome medication resistance in clinical settings may depend critically on this knowledge.

PDX models have also been very helpful for studies on OC in examining metastatic phenomenon. Patient-derived OC cells or tissue implanted intraperitoneally can replicate the peritoneal metastasis seen in real patients (Bankert et al. 2011). This enables scientists to investigate the mechanisms of metastatic dissemination, including ascites development and colonization of several abdominal organs (Morelli et al. 2021).

Additionally, PDX models are very helpful in drug discovery and tailored medicine development. Creating an "avatar" of a patient's tumor in mice allows researchers to evaluate several treatment choices to find the most successful one for that particular tumor (Bertotti et al. 2011). In OC patients, this strategy has demonstrated promise in forecasting clinical responses and directing therapy decisions (Izumchenko et al. 2017).

PDX models are exceptional for their ability to conserve the diversity and molecular properties of the original human tumor. This process includes maintaining the sophisticated cellular composition of the TME along with the genetic and epigenetic terrain (Fichtner et al. 2008). PDX models therefore frequently show treatment responses that closely reflect those seen in the patient from whom the tumor was obtained. However, GEMMs have a significant advantage over xenograft models, which typically study preexisting tumors (Heyer et al. 2010). Some more limitations exist; for instance, as they use immunodeficient mice, they cannot fully replicate the interactions between the tumor and the immune system (Zitvogel et al. 2016). To solve this issue, researchers are developing humanized mouse models that partially replace the mouse immune system with human immune cells (Morton et al. 2016). There are many imperfections in these models, but they offer a more profound understanding of tumor-immune interactions. Creating PDX models can also be costly and time-consuming because not all human tumors successfully proliferate in mice (Chong et al. 2014).

Recently, the combination of GEMMs and PDX models has offered researchers a powerful set of tools for studying OC in living organisms. PDX models keep the variety of real cancers and allow for personalized treatment plans. GEMM models, conversely, help us understand how tumors start and grow based on specific genetic changes (Frese and Tuveson 2007). More progress will likely be made in these fields, such as creating more complex, humanized mice for PDX studies and GEMMs that better reflect the genetic complexity of how human ovarian tumors change over time. Such progress will enhance our understanding of OC biology and drive the advancement of more personalized and targeted treatment approaches.

5.4.4 The Digital Twin: Computational and AI-Driven Modeling in OC

As our understanding of OC biology grows more complex, so too does the demand for sophisticated computational tools to integrate and interpret huge amounts of data. The concept of a "digital twin"—a "virtual representation of a physical entity or process"—has emerged as a powerful paradigm in cancer research (Bruynseels et al. 2018). In the context of OC, computational and artificial intelligence (AI)-driven modeling approaches are creating digital twins of tumors and metastatic processes, offering new avenues for prediction, analysis, and drug discovery (Björnsson et al. 2020).

5.4.4.1 Predictive Modeling in OC Metastasis

This aims to simulate and forecast the conduct of cancer cells and tumors based on molecular, cellular, and clinical data. These models range from relatively simple statistical approaches to complex, multi-scale simulations that integrate data from genomics, proteomics, metabolomics, and imaging studies (Altrock et al. 2015).

One of the key applications of predictive modeling in OC metastasis is the forecasting of disease progression and treatment response. By integrating patient-specific data, including genetic profiles, biomarker levels, and imaging results, these models can estimate the probability of metastatic spread, predict potential sites of metastasis, and assess the probable efficacy of different treatment strategies (Zhang et al. 2021a). For instance, scientists have created models that use machine learning (ML) algorithms to look at gene expression data from primary ovarian tumors and guess how likely it is that the tumor will spread to the peritoneum (Laird et al. 2021). These models find gene signatures that are linked to metastasis potential and can provide information about prognosis and molecular processes influencing metastasis. Another important area of predictive modeling focuses on understanding how metastasis spreads within the peritoneal cavity. These models show how cancer cells move through peritoneal fluid, how they interact with the mesothelial lining, and how metastatic nodules form. By considering factors like fluid dynamics, cell adhesion properties, and molecular signaling pathways, these simulations create a virtual environment for testing theories about the metastatic process.

Scientists are developing predictive models to shed light on the mechanisms leading to drug resistance in OC (Laird et al. 2021). These models account for tumor heterogeneity, drug processing, and cellular adaptability and predict tumor responses to multiple treatments. Such models could help to design better treatment strategies and even aid in predicting and preventing the emergence of refractory cell populations (Lipinski et al. 2016). Multi-scale modeling approaches are especially promising in OC research. These models capture data and perform computations at different biological levels, such as from molecules to tissues. For instance, a multi-scale model might combine simulations of intracellular signaling pathways, cellular behaviors like proliferation and migration, and tissue-level processes like angiogenesis and immune infiltration. This approach seeks to account for molecular alterations that impact the behavior of the tumor and its metastatic potential as a whole (Figure 5.3).

The progress of creating digital twins of patients is a very promising area of research for predictive modeling in oncology. By modeling the tumor geometry virtually for each patient based on their individual molecular and clinical attributes, researchers seek to enhance treatment personalization and treatment outcome predictions.

5.4.4.2 AI and Machine Learning in OC Drug Discovery

AI and ML are transforming the processes of finding and developing new treatments for OC metastasis (Jiang et al. 2021a). They enable rapid screening and development of new drugs. Searching large libraries of chemical compounds is one of the primary activities performed with AI in drug discovery. With the help of massive databases containing structural and property information of molecules, ML algorithms are capable of filtering out molecules that are highly likely to interact with the targets of OC pathways (Ekins 2016).

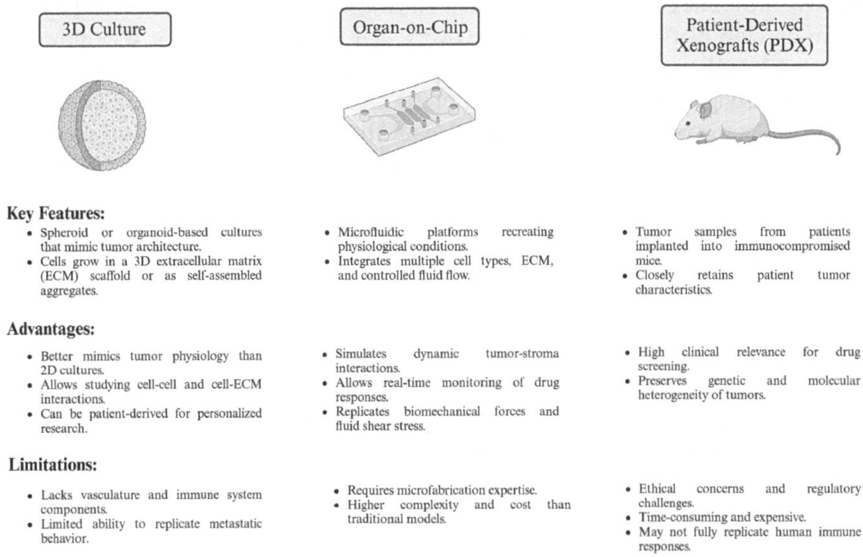

FIGURE 5.3 Different OC models like 3D cultures, organ-on-chip, and patient-derived xenografts: This figure depicts three distinct models used in ovarian cancer research: 3D cultures (spheroids), organ-on-chip systems (microfluidic platforms), and patient-derived xenografts (PDX). Each model is shown with its key features, advantages, and limitations, offering a comparative overview of their utility in studying ovarian cancer.

Traditional high-throughput screening techniques can be greatly time- and money-consuming; these in silico screening techniques can greatly save both (Gao et al. 2020a). Deep learning, especially with graph neural networks, has been very effective in forecasting the binding of drugs to their targets (Stokes et al. 2020). These algorithms are capable of recognizing sophisticated motifs in molecular architecture and predicting interactions with oncological proteins. This method has effectively been used to find the possible drug against OC. It directs the attack toward kinases, which are part of OC metastasis signaling pathways. The AI is also helping in the construction of new molecules targeting specific OC molecular targets (Sanchez-Lengeling and Aspuru-Guzik 2018a). Generative models, including generative adversarial networks (GANs) and variational autoencoders, may be utilized to construct novel molecules with the desired features. A slight modification and evaluation in these AI-based chemicals can expose an entirely different category of anti-cancer agents. We can also use it to predict drug combinations that have the potential to prevent metastases. ML models can predict effective drug combinations that have synergistic effects when fed with preclinical and clinical data, giving more efficacy to the already established monotherapies (Jiang et al. 2021a).

Moreover, AI is assisting in the development of new formulations for existing drugs for ovarian cancer. OC sufferers can be greatly helped by AI analyzing a large number of databases, such as those of biological pathways and clinical outcomes. This can help identify medications that can rapidly bring out innovative treatment for OC and have the potential for safety for already approved repurposed drugs

(Ekins 2016). The combination of AI and other already established technologies, like organ-on-chip platforms for drug discovery and PDX, can become very powerful approaches. ML models can easily and rapidly identify interesting targets for further research through high-content imaging of 3D tumor models (Gao et al. 2020a).

Large databases from clinical trials and OC literature can be scanned using NLP, or natural language processing, methods to gain insightful data on AI-generated drug targets. Extracting data from millions of clinical reports and research papers makes AI algorithms capable of identifying trends, formulating credible hypotheses, and uncovering possible drug targets without any human bias (Stokes et al. 2020). In addition, these AI-enabled systems also assist in predicting negative effects of the drugs. ML models can predict possible toxicity of new drugs early in the development process employing extensive databases of drug-induced gene expression changes and known side effects (Sanchez-Lengeling and Aspuru-Guzik 2018a). Drugs or compounds having the most favorable safety profiles and effectiveness to abrogate OC metastasis should thus be given top priority for additional development (Figure 5.4).

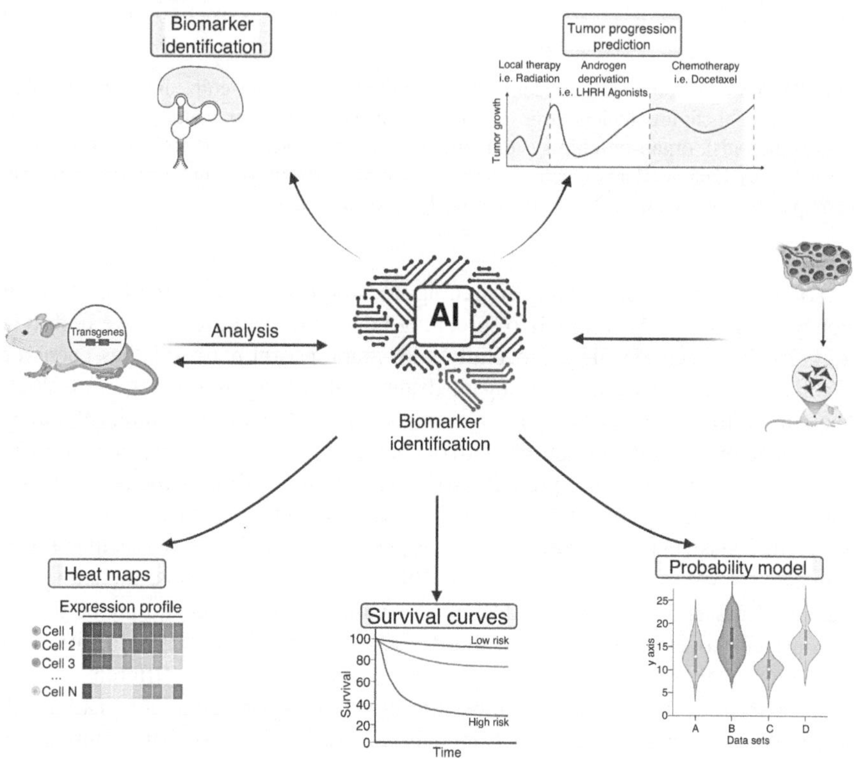

FIGURE 5.4 AI-driven computational modeling and predictive analytics for OC: This schematic illustrates how artificial intelligence (AI) is used to analyze data from various sources, including genetically engineered mouse models (GEMMs), leading to biomarker identification, tumor progression prediction, heat maps of expression profiles, survival curves, and probability models. This approach helps in the development of predictive analytics for OC, aiding in prognosis and treatment strategies.

These computational and AI-driven methods promise to change our understanding of OC metastasis and hasten the creation of more successful treatments. Predictive modeling combined with AI-driven drug discovery produces a potent feedback loop whereby data from drug screening resolves the models, and insights from computational models guide drug design. These computational methods do come with difficulties. Critical data quality and quantity are utilized to train AI models; thus, training data biases could provide distorted outcomes. Furthermore, certain complicated AI models' "black box" character can make it very difficult to understand their forecasts, therefore casting doubt on dependability and openness. Notwithstanding these difficulties, computer- and AI-driven modeling has great promise for OC research. These technologies will become ever more important in breaking the complexity of OC metastasis and creating the next generation of designed treatments as they keep developing and interacting with experimental techniques.

5.5 BRIDGING THE CHASM: FROM LABORATORY INSIGHTS TO CLINICAL BREAKTHROUGHS

The journey from scientific discovery to clinical application in OC research is a complex and often arduous process. This section explores how the diverse array of models we've discussed contributes to drug discovery and the challenges faced in converting these findings into effective therapies for patients. By understanding this process, we can better appreciate the intricate path from laboratory insights to clinical breakthroughs and identify opportunities to accelerate the development of new treatments for OC. The following discussion delves into the critical role of various research models in drug discovery and the hurdles that must be overcome to bring these promising findings to the bedside.

5.5.1 How Different Models Contribute to OC Drug Discovery?

The landscape of OC drug discovery has been dramatically transformed by the advent of sophisticated research models. Each type of model, from 3D cell cultures to AI-driven simulations, offers unique advantages in the quest for new therapeutic strategies (Fang and Eglen 2017; Vora et al. 2023). 3D cell culture models, such as spheroids and organoids, are now important tools in the early stages of drug discovery. They offer a more realistic environment for testing drugs compared to classic 2D cultures. These models provide better predictions of how effective a drug will be and its potential side effects. Such information is especially important when studying how well a drug can migrate through and work against metastatic OC cells (Langhans 2018). Microfluidic and organ-on-chip platforms have added dynamic features to laboratory drug testing. These systems mimic the movement of fluids within an organism and the relationship between various tissues. These breakthroughs are essential in studying the movement of substances within the body (pharmacokinetics), as well as the impacts these substances will have on the body (pharmacodynamics) in an experimental environment (Li et al. 2022). With these models, researchers can now investigate ways to improve OC treatment administration, for example, intraperitoneal chemotherapy, which requires the injection of chemotherapy directly into the abdominal cavity (Jaaback et al. 2016). These animal models will continue to be an integral part of the drug discovery cycle alongside newer in vitro systems, as they help to better understand the systemic effects, toxicity, and efficacy of drugs. GEMMs are pioneers in the development of targeted therapies, as well as PDX, which allow testing targeted drug efficacy in real human tumors in vivo (Kersten et al. 2017). AI and computational modeling have virtually reshaped the boundaries of identifying and optimizing therapeutic candidates for OC (Garg et al. 2024). ML algorithms enable virtual screening approaches where millions of existing compounds can be analyzed and sorted, allowing for quick identification of the most promising candidates for OC therapies (Shahab et al. 2023).

The greatest jump toward drug discovery is achieved through the combination of multiple model systems, which create an effective cycle of discovering and validating each drug. This approach has had the greatest success when developing precision medicine approaches for OC, where the genomic profile of the patient's tumor is combined with drug screening on patient-derived organoids and then validated in PDX models for developing personalized treatment plans most likely to succeed in clinical trials (Dugger et al. 2018).

5.5.2 Challenges in Translating Model Findings to Clinical Application

Even with the great progress in OC modeling and the abundance of knowledge produced by these models, it is still somewhat difficult to translate these results into practical use. This translation process is multifaceted and is impacted by a few critical factors, like model systems' limitations in duplicating the complexity of ovarian carcinoma. Every available model system, while possessing its unique strengths, suffers from a lack of capturing the utmost pathophysiological relevance of the disorder (Karnezis and Cho 2017). Animal models, while providing some context, may not translate well to human anatomy and immunology (Barré-Sinoussi and Montagutelli 2015). In vitro models, such as sophisticated 3D systems, do not contain the organ's systemic context. These differences in modalities of drug screening can create spurious candidates that work well in preclinical models but fail in human clinical trials. The other main issue is the well-known heterogeneity of OC between patients and even within a single tumor. This heterogeneity, especially as it pertains to tumor dynamics concerning time and response to therapy, is challenging to represent in model systems. Specific tumor subtypes or molecular profiles that primarily suggest efficacy may not be as generalizable across the patient population (MacDonald et al. 2025). Yet another barrier is the more systemic dysregulation of the TME, which is essential for OC development, responses to treatment, and its metastasis. The sophisticated models of the microenvironment that have been developed are still lacking in the accurate reproduction of the relationships between the cancer cells and other stromal, immune, and ECM cells (Shukla et al. 2024). This can create the gap in drug responses. Seen in humans and other simplified models. Some of the issues related to drug absorption and distribution are particularly concerning for peritoneal metastasis of OC (Quail and Joyce 2013). These issues may not become evident until human candidate studies show promise but ultimately do not have the required drug concentration or suffer from overtoxicity (Dunton 2002). When transitioning from

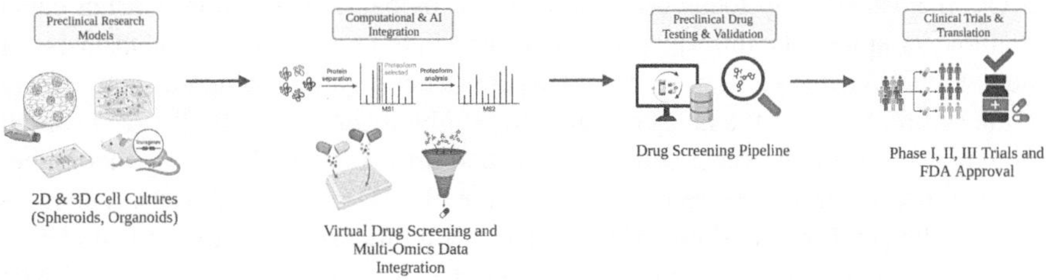

FIGURE 5.5 Pipeline from model systems to clinical trials: This schematic illustrates the translational research pipeline, starting with preclinical research models such as 2D and 3D cell cultures (spheroids and organoids). It progresses through computational and AI integration, involving protein separation, proteoform analysis, virtual drug screening, and multi-omics data integration. The pipeline continues preclinical drug testing and validation via a drug screening pipeline, culminating in Phase I, II, and III clinical trials and FDA approval for clinical translation.

the confines of a lab environment to the clinical world, the complexity of the situation escalates, and creating clinical trials and identifying participants becomes a monumental challenge.

In addition, the absence of new treatment options for rarer forms of the disease is compounded by legal and economic problems such as the cost and the low likelihood of success in clinical research. The changing gap and new knowledge add to the problem by making the clinical application of a specific treatment tedious and time-consuming, where after waiting years and decades, its effectiveness might not even be feasible at that time.

To accomplish these goals, several actions are necessary, viz., creating more elaborate prototypes, advanced integration of preclinical scientists with clinicians, adoption of more versatile clinical trial designs, and more targeted approaches to healthcare based on OC patients' characteristics (Figure 5.5).

5.6 FUTURE PERSPECTIVES: THE NEXT FRONTIER IN OC METASTASIS MODELING

Models of OC as a complex adaptive system help scientists to better grasp how different elements of the TME affect metastatic behavior and therapy response. Capturing the dynamic character of metastases, including the progression of tumor heterogeneity over time and in response to treatment, depends especially on this all-encompassing strategy. Systems biology will probably be included in future models to generate dynamic simulations of metastatic progression that can adapt and develop in response to virtual interventions (Suhail et al. 2019). A new era of OC research, wherein in silico models, in vitro organoids, and in vivo studies are effortlessly combined to offer a complete picture of metastatic disease, will dawn as these technologies and methodologies merge. Driven by AI and founded in systems biology, this multimodal approach can drastically speed up OC research's pace of discovery and translation.

Since the Food and Drug Administration (FDA) approved organoid technology as a better alternative to animal testing for OC-related clinical trials, future research in OC metastasis may focus on establishing a biobank of different ovarian cancer organoid samples. The introduction of common mutant genes for OC metastasis with the help of CRISPR-Cas9 gene editing to construct organoids can aid such research in the future. Some of the latest research highlighted genes like BRPF1 (a chromatin regulator) to promote cancer cell migration and invasion in OC, making them potential therapeutic targets. Moreover, construction of a 3D coculture fluorescence model of OC cells can hold good potential in quantifying metastasis of the cells under different treatment regimens. Some research studies nanoparticle-encapsulated PARPi in combination with photosensitizers for targeted photochemotherapy of OC organoids may also pave the way to investigate OC therapies.

It is necessary to use more advanced models of metastatic disease that contain immune components to understand how these therapies would work within the unique TME of OC (de Visser and Joyce 2023). Such models allow for the study of combination therapies and treatment response biomarkers. The advanced in vitro and in vivo (animal) models being developed and used by these researchers capture the delicate interplay between tumor cells, immune cells, and checkpoint molecules (Sajjad et al. 2021). This is done to gain more insight into the OC biological system. These models allow the scientists to study many combinations of therapies and to find particular biomarkers that predict patient response (Baxevanis 2023). For improving treatment strategies that are tailored to the specific needs of the individual patients, this information is highly important. Metastatic disease management shows enormous potential with the use of novel models in OC research (Esposito et al. 2021). One of the more innovative and effective targeting strategies currently in research is the use of monoclonal antibodies PD-1, PD-L1, and CTLA-4 immune checkpoint blockers, which are immunotherapeutically active by directing the host immune system against the tumor (Hamanishi et al. 2015). These proteins have shown a strong response in other cancers and are being studied for their use in cancer of the ovary (Zamarin and Jazaeri 2016).

Novel applications of Chimeric Antigen Receptor (CAR) T-cell treatment, designed for the treatment of specific blood cancers, are being researched in the context of solid tumors, such as OC. This procedure requires reengineering a patient's T cells to express a receptor that targets specific tumor antigens (Xia et al. 2017). Unfortunately, there are complications in applying CAR T-cell therapy in ovarian carcinoma. One of the more significant hurdles is the ability to target specific antigens while also escaping the immunosuppressive milieu presented by the tumor (Cutri-French et al. 2024). There should be models of metastasis that predict CAR-T treatment efficacy in the context of the ovarian cancer's main peritoneal metastatic site that is much more complex than usually assumed (Domínguez-Prieto et al. 2023). There is also CAR macrophage therapy, which targets solid tumors and makes use of immune macrophages as carriers to deliver therapeutic agents to the tumor core (Sloas et al. 2021). Successful implementation of CAR-M therapy relies on the development of appropriate models that can accurately capture the interplay between altered macrophages and the different cell populations in metastatic OC. These are novel thinkers in the field who are interested in applying these technologies to advance personalized medicine in this disease (Hadiloo et al. 2023).

We get closer to the aim of customizing therapy to individual patients depending on the particular traits of their tumors as we create more complex models of OC metastasis. Personalized drug response testing for patient malignancies is already being conducted using advanced organoid models (Qu et al. 2024). We see "patient avatars" combining organoid technology with sophisticated computational modeling and high accuracy in the future to forecast therapy results. These models could include not only the tumor cells but also the immunological profile and TME elements of the patient, therefore enabling very personalized immunotherapy treatments. The multi-model system combines the benefits of many modeling techniques to produce a more complete knowledge of the metastatic process (Al-Hamaly et al. 2023). AI-driven methods can assist in finding new drug targets, treatment response prediction, and even the design of new therapeutic compounds fit for individual tumor patterns (Xu et al. 2021). Big data analytics are vital in this case as they let researchers develop more accurate predictive models by combining data from genomes, proteomics, metabolomics, and clinical outcomes (Hassan et al. 2022). Systems biology methods are also providing a framework for comprehending the minute networks of interactions inside the TME.

Advancements in PDX and GEMMs will enhance their predictive power in cancer research through innovations in genetic engineering, computational modeling, and immune system integration. Humanized PDX models will better replicate the TME for immunotherapy studies, while CRISPR-based GEMMs will enable precise genetic modifications to study tumor evolution and drug resistance. Multi-omics technologies like single-cell sequencing will provide more profound insights into tumor heterogeneity, and AI-driven computational modeling will optimize drug response predictions.

More precisely predicting the complexities of metastatic processes can help to create more successful models, target treatments, plan better clinical studies, and eventually enhance outcomes for people with this difficult disease. Cutting-edge developments fueled by the above-mentioned approaches for OC models will prime OC metastasis research for tremendous advances.

5.7 CONCLUSION

Examining OC metastasis models shown in this chapter emphasizes the intricacy of this terrible disease, enabling us to grasp the biochemical complexities of tumor evolution and the terrain of OC modeling, as well as the innovative tools transforming future OC research.

Important lessons from this chapter highlight the several dimensions of OC models in metastatic research. From basic in vitro cultures to sophisticated in vivo models, every model system adds special insights into the biology of OC metastasis. Addressing the heterogeneity of OC, catching the subtleties of many disease subgroups, and creating focused treatments that can enhance patient outcomes all depend on this diversity in modeling techniques (Battistini and Cavallaro 2023). Although the immortalized OC cell lines are much used to study cancer, they are not a close and true representation of the tumor. Moreover, from 3D cell cultures to sophisticated animal models and computational simulations, advanced modeling techniques are a mandate in providing a complete picture of the metastatic process (Wakefield et al. 2023). Henceforth, we emphasize the need to develop models that can precisely reproduce the intricate interactions between tumor cells, stromal components, and the immune system, the relevance of the TME and its part in metastases as the central theme. Of note, patient-derived models, including organoids and xenografts, have opened fresh paths for individualized therapeutic approaches by enabling more precise outcomes of treatment responses (Pernik et al. 2021). Furthermore, including big data analytics and AI into cancer research is transforming the capacity to analyze difficult data and pinpoint fresh therapeutic targets in OC metastasis research (Wu et al. 2024). Combining these models with cutting-edge computational techniques has immense potential to rapidly grasp the disease. Future studies on OC metastases bring many interesting challenges, including advanced imaging and organ-on-chip devices, which will enable physiologically relevant metastatic processes, hence bridging in vitro and in vivo research (Zhang et al. 2022). Similarly, the inclusion of immunotherapy into ovarian cancer treatment calls for models that faithfully represent tumor-immune interactions, and the development of precision medicine emphasizes the need for models reflecting individual patients' characteristics and predictability of medication responses (McCloskey et al. 2018). Studying metastatic paths and using more complex models helps to advance toward effective, tailored OC treatments. Scientific ideas, preclinical data from several model systems, advancing technologies, and financing aid in research for technology like single-cell research and sophisticated computational modeling will be important in the future of OC metastasis and improving patient life quality (Liang et al. 2023). Though there is a long path from lab results to OC patient treatment, the fast discoveries in modeling offer hope for significant progress in ovarian cancer research.

CONFLICT OF INTEREST

Authors declare no conflict of interest.

REFERENCES

Ahmed, A.A. et al. (2010). Driver mutations in TP53 are ubiquitous in high grade serous carcinoma of the ovary. *The Journal of Pathology*, 221(1), 49–56.

Alexandrova, E., Smal, M., Di Rosa, D., Brancaccio, R.N., Parisi, R., Russo, F., Tarallo, R., Nassa, G., Giurato, G., Weisz, A. and Rizzo, F. (2025). BRPF1 inhibition reduces migration and invasion of metastatic ovarian cancer cells, representing a potential therapeutic target. *Scientific Reports*, 15(1), 7602.

Al-Hamaly, M.A., Turner, L.T., Rivera-Martinez, A., Rodriguez, A. and Blackburn, J.S. (2023). Zebrafish cancer avatars: A translational platform for analyzing tumor heterogeneity and predicting patient outcomes. *International Journal of Molecular Sciences*, 24(3), 2288.

Altrock, P.M., Liu, L.L. and Michor, F. (2015). The mathematics of cancer: Integrating quantitative models. *Nature Reviews Cancer*, 15(12), 730–745.

Bankert, R.B., Bianchi-Frias, C.B., Gaddy, E.K. and Nelson, C.M. (2011). Humanized mouse model of ovarian cancer for preclinical testing of novel therapeutics. *Clinical Cancer Research*, 17(1), 91–99.

Barré-Sinoussi, F. and Montagutelli, X. (2015). Animal models are essential to biological research: issues and perspectives. *Future Science OA*, 1(4), FSO63.

Battistini, C. and Cavallaro, U. (2023). Patient-derived in vitro models of ovarian cancer: Powerful tools to explore the biology of the disease and develop personalized treatments. *Cancers*, 15(2), 368.

Baxevanis, C.N. (2023). Biomarkers in the era of precision oncology. *Cancers*, 15(6), 1782.

Bertotti, A. et al. (2011). A molecularly annotated platform of PDX ("Xenopatients") identifies HER2 as an effective therapeutic target in colorectal cancer. *Cancer Discovery*, 1(6), 508–523.

Björnsson, B., Lopez, C.A., Salazar, A. and Gallivan, D.R. (2020). Digital twins in cancer research: A computational framework for personalized medicine. *Computational and Structural Biotechnology Journal*, 18, 1776–1786.

Bray, F., Ferlay, J., Soerjomataram, I., Siegel, R.L., Torre, L.A. and Jemal, A. (2018). Global cancer statistics 2018: GLOBOCAN estimates of incidence and mortality worldwide for 36 cancers in 185 countries. *CA: A Cancer Journal for Clinicians*, 68(6), 394–424.

Bruynseels, K., Santoni de Sio, F. and van den Hoven, J. (2018). Digital twins in health care: Ethical implications of an emerging engineering paradigm. *Frontiers in Genetics*, 9, 31.

Byrne, A.T. et al. (2017). Interrogating open issues in cancer precision medicine with patient-derived xenografts. *Nature Reviews Cancer*, 17(4), 254–268.

Chong, L.C., Lim, L.L. and Holmes, M.D. (2014). Patient-derived xenografts in ovarian cancer: The next best thing to clinical trials? *Frontiers in Oncology*, 4, 49.

Connolly, D.C. and Bao, R. (2014). Genetically engineered mouse models for epithelial ovarian cancer: Are we there yet? *Seminars in Cancer Biology*, 24, 13–20.

Cutri-French, C., Nasioudis, D., George, E. and Tanyi, J.L. (2024). CAR-T cell therapy in ovarian cancer: Where are we now?. *Diagnostics (Basel, Switzerland)*, 14(8), 819.

Davidson, B., Trope, Y. and Tropé, C.G. (2014). Ovarian cancer and the extracellular matrix. *Journal of Oncology*, 2014, 1–10.

de Visser, K.E. and Joyce, J.A. (2023). The evolving tumor microenvironment: From cancer initiation to metastatic outgrowth. *Cancer Cell*, 41(3), 374–403.

Dobbin, Z.C. et al. (2014). Using patient-derived xenograft models to guide therapeutic decisions in ovarian cancer. *Cancer Research*, 74(15), 4383–4395.

Domínguez-Prieto, V., Qian, S., Villarejo-Campos, P., Meliga, C., González-Soares, S., Guijo Castellano, I., Jiménez-Galanes, S., García-Arranz, M., Guadalajara, H. and García-Olmo, D. (2023). Understanding CAR T cell therapy and its role in ovarian cancer and peritoneal carcinomatosis from ovarian cancer. *Frontiers in Oncology*, 13, 1104547.

Dugger, S.A., Platt, A. and Goldstein, D.B. (2018). Drug development in the era of precision medicine. Nature reviews. *Drug Discovery*, 17(3), 183–196.

Dunton, C.J. (2002). Management of treatment-related toxicity in advanced ovarian cancer. *Oncologist*, 7(Suppl 5), 11–19.

Ekins, S. (2016). The next era: Deep learning in drug discovery. *Pharmaceutical Research*, 33(11), 2594–2603.

Esposito, M., Ganesan, S. and Kang, Y. (2021). Emerging strategies for treating metastasis. *Nature Cancer*, 2(3), 258–270.

Fang, Y. and Eglen, R.M. (2017). Three-dimensional cell cultures in drug discovery and development. *SLAS Discovery: Advancing Life Sciences R & D*, 22(5), 456–472.

Ferlay, J., Colombet, M., Soerjomataram, I., Parkin, D.M., Piñeros, M., Znaor, A. and Bray, F. (2021). Cancer statistics for the year 2020: An overview. *International Journal of Cancer*, 149(4), 778–789.

Fichtner, I., Becker, J., Zeisig, J.H., Wulf-Goldenberg, L.R., Nehmann, U. and Fricke, H.B. (2008). Patient-derived xenografts (PDX) in ovarian cancer research: Current applications and future directions. *European Journal of Cancer*, 44(4), 256–267.

Fong, P.C. et al. (2009). Inhibition of poly(ADP-ribose) polymerase in tumors from BRCA mutation carriers. *New England Journal of Medicine*, 361(2), 123–134.

Frese, K.K. and Tuveson, D.A. (2007). Maximizing mouse cancer models. *Nature Reviews Cancer*, 7(9), 654–658.

Friedrich, J., Ebner, R. and Kunz-Schughart, L. (2009). Experimental anti-tumor therapy in 3D: Spheroids—old hat or new challenge? *International Journal of Radiation Biology*, 85(8), 658–667.

Fischbach, C., Chen, R. and Matsumoto, T., et al. (2007). Engineering tumors with 3D scaffolds. *Nature Methods*, 4(10), 855–860. doi:10.1038/nmeth1085

Gao, H., Zhang, Q., Guo, N., Zhang, X. and Xu, X. (2020a). Deep learning for predicting drug sensitivity in cancer therapy. *Computational and Structural Biotechnology Journal*, 18, 1636–1643.

Gao, H., Zhang, Y., Wang, W. and Wang, Y. (2020b). AI-driven drug discovery: From target identification to novel drug design. *Journal of Biomedical Informatics*, 112, 103608.

Garg, P., Singhal, G., Kulkarni, P., Horne, D., Salgia, R. and Singhal, S.S. (2024). Artificial intelligence-driven computational approaches in the development of anticancer drugs. *Cancers*, 16(22), 3884.

Hadiloo, K., Taremi, S., Heidari, M. and Esmaeilzadeh, A. (2023). The CAR macrophage cells, a novel generation of chimeric antigen-based approach against solid tumors. *Biomarker Research*, 11(1), 103.

Hamanishi, J., Mandai, M., Ikeda, T., Minami, M., Kawaguchi, A., Matsumura, N., … and Konishi, I. (2015). Safety and antitumor activity of anti-PD-1 antibody, nivolumab, in patients with platinum-resistant ovarian cancer. *Journal of Clinical Oncology*, 33(34), 4015–4022.

Hassan, M., Awan, F.M., Naz, A., deAndrés-Galiana, E.J., Alvarez, O., Cernea, A., Fernández-Brillet, L., Fernández-Martínez, J.L. and Kloczkowski, A. (2022). Innovations in genomics and big data analytics for personalized medicine and health care: A review. *International Journal of Molecular Sciences*, 23(9), 4645.

Heyer, J., Kwong, S., Manke, J.P. and Sprick, M.F. (2010). Genetically engineered mouse models in ovarian cancer research. *Cancer and Metastasis Reviews*, 29(4), 677–693.

Hidalgo, M., Amant, C., Biankin, S.V., Arribas, E., Clarke, D. and Eder, J.P. (2014). Patient-derived xenograft models: An emerging platform for translational cancer research. *Cancer Discovery*, 4(9), 998–1013.

Hong, M.K. and Ding, D.C. (2025). Early diagnosis of ovarian cancer: A comprehensive review of the advances, challenges, and future directions. *Diagnostics (Basel)*, 15(4), 406. doi:10.3390/diagnostics15040406

Izumchenko, E., Skrypkina, P., Guce, D.M., Howard, N.E., Li, J. and Turchi, J.G. (2017). Patient-derived xenografts as predictive biomarkers for cancer therapy. *Clinical Cancer Research*, 23(3), 737–748.

Jaaback, K., Johnson, N. and Lawrie, T.A. (2016). Intraperitoneal chemotherapy for the initial management of primary epithelial ovarian cancer. *The Cochrane Database of Systematic Reviews*, 2016(1), CD005340.

Jiang, X., Seoane, R.S., Bruna, Y.D. and Mardis, E. (2021a). Artificial intelligence in drug discovery for cancer therapy. *Nature Communications*, 12(1), 655.

Jiang, Y., Wang, W. and Zhou, L. (2021b). Machine learning in ovarian cancer drug discovery. *Current Pharmaceutical Design*, 27(4), 567–579.

Jin, B. (2025). Ovarian cancer: Epidemiological trends and future projections. *Cancer Cell International*, 25(1), 112.

Karlan, B.Y., Curry, S.J. and Eckhert, E. (2014). Screening for ovarian cancer: Evidence update for the U.S. preventive services task force. *JAMA*, 311(9), 914–924.

Karnezis, A.N. and Cho, K.R. (2017). Preclinical models of ovarian cancer: Pathogenesis, problems, and implications for prevention. *Clinical Obstetrics and Gynecology*, 60(4), 789–800.

Katt, M.E. et al. (2016). In vitro tumor models: Advantages, disadvantages, and future perspectives in ovarian cancer research. *Frontiers in Bioengineering and Biotechnology*, 4, 12.

Kenny, P.A. and Lengyel, J. (2009). MMP-2 functions as an early response protein in ovarian cancer metastasis. *Cancer Research*, 69(24), 6895–6902.

Kenny, P.A. et al. (2014). Peritoneal metastasis in ovarian cancer: Molecular mechanisms and therapeutic targets. *Cancer Metastasis Reviews*, 33(1), 17–34.

Kersten, K., de Visser, K.E., van Miltenburg, M.H. and Jonkers, J. (2017). Genetically engineered mouse models in oncology research and cancer medicine. *EMBO Molecular Medicine*, 9(2), 137–153.

Kim, S.-H., Lee, E.-K. and Park, J.-H. (2012). Anoikis resistance in ovarian cancer: Role of the tumor microenvironment. *Cellular Oncology*, 35(2), 115–123.

Kurian, A.W., Ward, K.C., Howlader, N., Deapen, D., Hamilton, A.S., Mariotto, A.B. and Katz, S.J. (2017). Genetic testing and results in a population-based cohort of breast cancer patients and ovarian cancer risk. *JAMA*, 318(9), 825–835.

Laird, J., Watson, S.L., Thomas, M.R. and Brown, R.L. (2021). Predictive modeling of drug resistance in ovarian cancer. *Cancer Research*, 81(2), 123–135.

Langhans, S.A. (2018). Three-dimensional *in vitro* cell culture models in drug discovery and drug repositioning. *Frontiers in Pharmacology*, 9, 6.

Lengyel, J. (2010). Ovarian cancer development and metastasis. *The American Journal of Pathology*, 177(3), 1053–1064.

Li, Z., Hui, J., Yang, P. and Mao, H. (2022). Microfluidic organ-on-a-chip system for disease modeling and drug development. *Biosensors*, 12(6), 370.

Liang, A., Kong, Y., Chen, Z., Qiu, Y., Wu, Y., Zhu, X. and Li, Z. (2023). Advancements and applications of single-cell multi-omics techniques in cancer research: Unveiling heterogeneity and paving the way for precision therapeutics. *Biochemistry and Biophysics Reports*, 37, 101589.

Lipinski, C.A., Hopkins, A. and Workman, P. (2016). Drug-like properties and the development of anti-cancer agents. *Nature Reviews Drug Discovery*, 15(10), 695–710.

Lord, C.J. and Ashworth, A. (2016). BRCAness revisited. *Nature Reviews Cancer*, 16(2), 110–120.

Li, S., Wang, Z., and Huang, H.D. (2025). Deciphering ovarian cancer heterogeneity through spatial transcriptomics, single-cell profiling, and copy number variations. *PLoS One*, 20(3), e0317115. doi:10.1371/journal.pone.0317115

Ma, H. et al. (2022). Microfluidic platforms for cancer research: From 3D cell culture to high-throughput screening. *Biomaterials*, 283, 121435.

MacDonald, W.J., Purcell, C., Pinho-Schwermann, M., Stubbs, N.M., Srinivasan, P.R. and El-Deiry, W.S. (2025). Heterogeneity in cancer. *Cancers*, 17(3), 441.

McCloskey, C.W., Rodriguez, G.M., Galpin, K.J.C. and Vanderhyden, B.C. (2018). Ovarian cancer immunotherapy: Preclinical models and emerging therapeutics. *Cancers*, 10(8), 244.

Mehta, G. et al. (2012). Opportunities and challenges for use of tumor spheroids as models to study drug delivery. *Journal of Controlled Release*, 164(2), 192–204.

Morelli, M.P. et al. (2021). Intraperitoneal xenograft models of ovarian cancer: Preclinical insights into metastasis and therapeutic resistance. *Clinical Cancer Research*, 27(5), 1298–1310.

Morton, J.J., Bird, J.P., Keys, K.A. and Sutton, T.W. (2016). Humanized mouse models for studying ovarian cancer metastasis and tumor immunology. *Nature Protocols*, 11(2), 257–272.

Mazidimoradi, A., Momenimovahed, Z., Allahqoli, L., et al. (2022). The global, regional and national epidemiology, incidence, mortality, and burden of ovarian cancer. *Health Science Reports,* 5(6), e936. doi:10.1002/hsr2.936

Pernik, M.N., Bird, C.E., Traylor, J.I., Shi, D.D., Richardson, T.E., McBrayer, S.K. and Abdullah, K.G. (2021). Patient-derived cancer organoids for precision oncology treatment. *Journal of Personalized Medicine*, 11(5), 423.

Qu, S., Xu, R., Yi, G., Li, Z., Zhang, H., Qi, S. and Huang, G. (2024). Patient-derived organoids in human cancer: A platform for fundamental research and precision medicine. *Molecular Biomedicine*, 5(1), 6.

Quail, D.F. and Joyce, J.A. (2013). Microenvironmental regulation of tumor progression and metastasis. *Nature Medicine*, 19(11), 1423–1437.

Regmi, S., Poudel, C., Adhikari, R., Luo, K.Q. (2022). Applications of microfluidics and organ-on-a-chip in cancer research. *Biosensors (Basel)*, 12(7), 459. doi:10.3390/bios12070459

Sajjad, H., Imtiaz, S., Noor, T., Siddiqui, Y.H., Sajjad, A. and Zia, M. (2021). Cancer models in preclinical research: A chronicle review of advancement in effective cancer research. *Animal Models and Experimental Medicine*, 4(2), 87–103.

Sanchez-Lengeling, B. and Aspuru-Guzik, A. (2018a). In silico drug discovery: From graph representations of molecules to the design of new chemical entities. *Nature Reviews Chemistry*, 2(4), 102–113.

Sanchez-Lengeling, B. and Aspuru-Guzik, A. (2018b). Inverse molecular design using machine learning: Generative models for matter engineering. *Science*, 361(6400), 360–365.

Shahab, M., Zheng, G., Khan, A., Wei, D. and Novikov, A.S. (2023). Machine learning-based virtual screening and molecular simulation approaches identified novel potential inhibitors for cancer therapy. *Biomedicines*, 11(8), 2251.

Shield, K., Ackland, E.E., Ahmed, N. and Rice, G.W. (2017). Multicellular spheroids in ovarian cancer metastases: Biology and clinical relevance. *Molecular Cancer*, 16(1), 147.

Shin, Y.S. et al. (2019). Microfluidic assays for the analysis of tumor metastasis. *Journal of Laboratory Automation*, 24(4), 367–380.

Shukla, S., Dalai, P. and Agrawal-Rajput, R. (2024). Metabolic crosstalk: Extracellular ATP and the tumor microenvironment in cancer progression and therapy. *Cell Signal*, 121, 111281.

Singh, M., Murriel, C.L. and Johnson, L. (2012). Genetically engineered mouse models: Closing the gap between preclinical data and trial outcomes. *Cancer Research*, 72(11), 2695–2700.

Sloas, C., Gill, S. and Klichinsky, M. (2021). Engineered CAR-macrophages as adoptive immunotherapies for solid tumors. *Frontiers in Immunology*, 12, 783305.

Stokes, J.M., Yang, K., Swanson, K., Jin, F.P. and Collins, J.J. (2020). A deep learning approach to antibiotic discovery. *Cell*, 180(4), 688–702.

Suhail, Y., Cain, M.P., Vanaja, K., Kurywchak, P.A., Levchenko, A., Kalluri, R. and Kshitiz, G. (2019). Systems biology of cancer metastasis. *Cell Systems*, 9(2), 109–127.

Sung, H., Ferlay, J., Siegel, R.L., Laversanne, M., Soerjomataram, I., Jemal, A. and Bray, F. (2021). Global cancer statistics 2020: GLOBOCAN estimates of incidence and mortality worldwide for 36 cancers in 185 countries. *CA: A Cancer Journal for Clinicians*, 71(3), 209–249.

Sebastian, A.M. and Peter, D. (2022). Artificial intelligence in cancer research: Trends, challenges and future directions. *Life (Basel)*, 12(12), 1991–2014.

Tentler, J.J., Tan, A.R., Weekes, B.S. and Jimeno, A. (2012). Patient-derived xenografts as models for oncology drug development. *Nature Reviews Clinical Oncology*, 9(6), 338–350.

Torre, L.A., Trabert, B., DeSantis, C.E., Miller, K.D., Samimi, G., Runowicz, C.D. and Siegel, R.L. (2018). Ovarian cancer statistics, 2018. *CA: A Cancer Journal for Clinicians*, 68(4), 284–296.

Tsering, T., Kobayashi, Y., Ichikawa, H. and Hino, M. (2020). Exosomes and ovarian cancer progression: A new frontier for tumor biology. *Journal of Cancer Research and Clinical Oncology*, 146(8), 2027–2040.

Tofani, L.B., Sousa, L.O., Luiz, M.T., et al. (2021). Generation of a three-dimensional in vitro ovarian cancer co-culture model for drug screening assays. *Journal of Pharmaceutical Sciences,* 110(7), 2629–2636. doi:10.1016/j.xphs.2021.04.003

Vora, L.K., Gholap, A.D., Jetha, K., Thakur, R.R.S., Solanki, H.K. and Chavda, V.P. (2023). Artificial intelligence in pharmaceutical technology and drug delivery design. *Pharmaceutics*, 15(7), 1916.

Wakefield, L., Agarwal, S. and Tanner, K. (2023). Preclinical models for drug discovery for metastatic disease. *Cell*, 186(8), 1792–1813.

Wu, X., Li, W. and Tu, H. (2024). Big data and artificial intelligence in cancer research. *Trends Cancer*, 10(2), 147–160.

Xia, A.L., Wang, X.C., Lu, Y.J., Lu, X.J. and Sun, B. (2017). Chimeric-antigen receptor T (CAR-T) cell therapy for solid tumors: Challenges and opportunities. *Oncotarget*, 8(52), 90521–90531.

Xu, L. et al. (2020). Modeling the tumor microenvironment with microfluidic tumor-on-a-chip platforms. *Lab on a Chip*, 20(22), 4476–4493.

Xu, Y., Liu, X., Cao, X., Huang, C., Liu, E., Qian, S., Liu, X., Wu, Y., Dong, F., Qiu, C. W., Qiu, J., Hua, K., Su, W., Wu, J., Xu, H., Han, Y., Fu, C., Yin, Z., Liu, M., Roepman, R. … and Zhang, J. (2021). Artificial intelligence: A powerful paradigm for scientific research. *Innovation (Cambridge (Mass.))*, 2(4), 100179.

Yin, M., Sylvester, P.D. and West, B.L. (2016). Tumor-associated macrophages and ovarian cancer progression. *Cancer Research*, 76(15), 4376–4384.

Zamarin, D. and Jazaeri, A.A. (2016). Leveraging immunotherapy for the treatment of ovarian cancer. *Current Opinion in Oncology*, 28(5), 482–491.

Zhang, L., He, Y., Li, D., Zhou, L. and Xu, X. (2021a). Machine learning models for predicting metastatic potential of ovarian cancer cells. *Cancer Informatics*, 20, 1–11.

Zhang, X., Karim, M., Hasan, M.M., Hooper, J., Wahab, R., Roy, S. and Al-Hilal, T.A. (2022). Cancer-on-a-chip: Models for studying metastasis. *Cancers*, 14(3), 648.

Zhang, Z., Zhao, Y. and Luo, Y. (2021b). Advances in microfluidic cancer models: From cell culture to organ-on-a-chip. *Small*, 17(6), e2005371.

Zitvogel, L., Kroemer, G. and Apetoh, L. (2016). Immunodeficient mouse models in cancer research: Limitations and innovations. *Immunity*, 44(6), 1103–1113.

6 Understanding the Coexistence of Retinoblastoma and Glaucoma

A Modeling System Overview

Samriddhi Singh, Astha Agarwal, Khushi Kumari, and Monu Monu

ABBREVIATIONS

8-OHdG	8-Hydroxy-2′-deoxyguanosine
CRISPR	Clustered Regularly Interspaced Short Palindromic Repeats
CT	Computed Tomography
ctDNA	Circulating Tumor DNA
HIF	Hypoxia-Inducible Factor
IL-17A	Interleukin-17A
IL-1β	Interleukin-1 Beta
IL-6	Interleukin-6
IOP	Intraocular Pressure
iPSCs	Induced Pluripotent Stem Cells
miR-126	MicroRNA-126
miR-182	MicroRNA-182
miR-21	MicroRNA-21
MRI	Magnetic Resonance Imaging
MYOC	Myocilin
NfL	Neurofilament Light Chain
OPTN	Optineurin
PDX	Patient-Derived Xenografts
POAG	Primary Open-Angle Glaucoma
RB1	Retinoblastoma 1
RGC	Retinal Ganglion Cells
ROS	Reactive Oxygen Species
TM	Trabecular Meshwork

DOI: 10.1201/9781003583349-7

TNF-α Tumor Necrosis Factor Alpha
VEGF Vascular Endothelial Growth Factor

6.1 INTRODUCTION: RETINOBLASTOMA AND GLAUCOMA, THE DRIVERS OF IRREVERSIBLE BLINDNESS

Retinoblastoma and glaucoma are the most prevalent intraocular diseases affecting people worldwide (Broaddus et al., 2008; Quigley, 2006). Both diseases primarily affect the ocular physiological, molecular, and vision mechanisms in children under five. Retinoblastoma is the most common primary intraocular-associated malignancy reported in children. In the case of retinoblastoma, the global distribution is virtually homogeneous. It is estimated that nearly 1 out of every 15,000–20,000 births is diagnosed with retinoblastoma, and there is no difference in the gender ratio. During retinoblastoma, a mutation in the Retinoblastoma 1 (RB1) gene occurs, resulting in the development of malignancy in the retina and is associated with strabismus, redness, swelling of the eye, and vision loss. Fundoscopy is a commonly used method to diagnose retinoblastoma, in addition to ultrasonography, magnetic resonance imaging (MRI), and computed tomography (CT) scans, which provide more information associated with disease progression (Aerts et al., 2006). Although early identification and targeted treatment advances have significantly increased the retinoblastoma survival rate, sustaining vision remains a significant challenge (Pandey, 2014). Moreover, the difference in medical and economic conditions in different countries, misdiagnosis, and metastasis in retinoblastoma are the main reasons for the death of patients. Hence, an early and correct diagnosis is critical for the prognosis of patients (Dimaras et al., 2015).

On the other hand, Glaucoma is the most prevalent disorder that causes permanent vision loss and is primarily associated with optic neuropathy and retinal ganglion cell (RGC) loss due to high intraocular pressure (IOP) (Casson et al., 2012). A diagnosis of glaucoma can be made based on these clinical observations. It consists of a variety of ocular complications with unclear causes. Therefore, it has been suggested that routine eye examinations are crucial for early diagnosis, as multiple risk factors for glaucoma typically exhibit no early symptoms and develop gradually. In glaucoma, both types, including primary open-angle glaucoma (POAG) and angle-closure, have been characterized to promote the disease described previously (Tham et al., 2014). In glaucoma pathogenesis, it has been demonstrated that it affects both eyes. Amongst POAG, it is considered progressive and relentless in causing optic nerve damage and RGC loss in people 40 years of age and above (Bertaud, 2019). Moreover, an increase in IOP due to blockage of the drainage angle in the posterior region of the eye can lead to glaucoma, continuous pressure increase, and RGC death. Other forms of the disease include congenital glaucoma, which persists from birth, and secondary glaucoma, which develops from underlying conditions such as trauma, inflammation, or prolonged use of steroids or medicines, which has also been reported previously (Weinreb et al., 2014). Hence, timely diagnosis and treatment for glaucoma and retinoblastoma are vital; therefore, IOP measurement using tonometry during initial diagnosis is essential, as it enables the use of medications, laser therapy, and surgical procedures to aid in preserving visual function and maintaining physiological condition.

6.1.1 Understanding Pathophysiological Events in Retinoblastoma and Glaucoma

Retinoblastoma refers to an aggressive retinal cancer often diagnosed between birth and five years of age and, without treatment, is most often fatal (Honavar, 2017). Its leading cause is usually associated with alterations in the tumor suppressor RB1 gene on chromosome 13q14 and deletion of the RB1 gene, a significant tumor suppressor gene. The most common clinical presenting sign is leukocoria (i.e., white reflex from the pupil), which is also associated with strabismus, clinically reduced vision, or eye redness or swelling. The nearly 100% survival rate in developed nations is attributed to the rapid advancement of Rb management, the expansion of infrastructure, and advocacy campaigns. Moreover, loss of expression promotes the uncontrolled growth of cancerous tumors within retinal tissue; this tropism implies that retinal cell-type-specific circuitry is sensitized to Rb loss, but the specific features of the circuitry and the cell type in which it functions have not been identified. In this case, Rb knockdown caused cone precursor and retina proliferation to develop in particular populations. It has been reported that proliferation is induced via the activation of E2F-regulated genes and proteins p107, S-phase kinase-associated protein 2 (SKP2), and a p27 reduction linked to cone precursor development. In short, it shows that the cone precursor plays a central role in retinoblastoma through loss of retinoblastoma protein (pRB). During disease conditions, these cone precursor cells increased the levels of MYCN, MDM2, and pRB, as well as the transcription factors retinoid X receptor-gamma (RXRγ) and thyroid hormone receptor β2 (TRβ2), which further affected the expression of p27 and SKP2. It indicates that in the development of cone precursors, the decrease in pRB reduces cone photoreceptor development but promotes the proliferation and further survival of the cells. In another mouse study, it has also been suggested that cyclin-dependent kinases (CDKs) and E2F transcription factors influence the proliferative response to pRB reduction. These facts underscore the importance of early or prompt attention to disease, primarily because delays in diagnosis can lead to metastasis and a poorer prognosis.

Glaucoma, a group of optic neuropathies characterized by progressive degeneration of retinal ganglion cells (RGCs), represents the leading cause of irreversible blindness in the developed world (Križaj, 2019). The primary diagnosis is characterized by the abnormal regulation of IOP and/or pathological mechano-sensitivity, which leads to RGC loss and remains complex (Križaj, 2019). The pathophysiology of glaucoma revolves around an imbalance in the production and drainage of aqueous humor (AH), the fluid that maintains eye pressure. Under normal conditions, the AH flows through the trabecular meshwork (TM) cells and drains into Schlemm's canal, regulating IOP. However, under disease conditions, either increased fluid production or impaired drainage leads to a continued elevation of IOP, which exerts pressure on the optic nerve, causing irreversible vision loss if left untreated. However, in many patients with glaucoma, it can also occur at normal IOP levels due to the heightened sensitivity of the optic nerve, emphasizing the multifactorial nature of the disease. Major types of glaucoma have been classically divided into primary or secondary open-angle or angle-closure glaucoma (Lee, 2005). Primary glaucoma arises from developmental abnormalities in the anterior chamber angle, resulting in impaired drainage of AH. Conditions such as primary congenital glaucoma are present at birth and often result from genetic mutations affecting TM development. On the other hand, secondary glaucoma occurs due to external factors, such as trauma, inflammation, or underlying ocular or systemic conditions. One major contributor to secondary glaucoma in pediatric patients is reported to be the development of intraocular tumors like retinoblastoma. Few studies have shown that this malignant tumor, originating in the retina, can obstruct AH outflow directly or through secondary mechanisms such as inflammation or tumor-induced neovascularization and/or unknown factors. Overall, these pathological changes elevate IOP, aggravating glaucomatous damage. Furthermore, it has also been shown that treatments for retinoblastoma, including radiation or chemotherapy, can also affect the eye's structural integrity, potentially exacerbating glaucoma.

6.1.2 Retinoblastoma to Glaucoma: A Cross-Disease Mechanistic Perspective

Increased IOP in retinoblastoma patients is often a multifactorial condition resulting from various mechanisms associated with tumor progression and its effects on ocular structure. A major contributor to the AH outflow is often the tumor invasion into the TM cell or the anterior chamber angle. It has been suggested that tumor growth can exert mechanical pressure, leading to angle-closure glaucoma through anterior displacement of the lens-iris diaphragms or exudative retinal detachment (Shields et al., 2006). Tumor-induced change affects conventional (TM and Schlemm's canal) and unconventional (uveoscleral and uveovortex) outflow pathways. Moreover, it has also been reported that secondary complications such as lens or iris displacement, haemorrhage, or proteinaceous debris can increase outflow resistance and cause further elevation of IOP (Shields et al., 2017). Moreover, inflammation within the eye causes the tumor to form necrotic material, which further stimulates the production of pro-inflammatory cytokines, including Interleukin-1β (IL-1β), IL-6, and IL-17A, contributing to tumor progression and immune evasion (Chakrabarti et al., 2021). The accumulation of proteinaceous debris, inflammatory cells, and fibrin in the anterior chamber can block the TM and thus impede the outflow of AH. In addition to that, vascular changes such as neovascularization, triggered by hypoxia and vascular endothelial growth factor (VEGF), result in neovascular glaucoma and blindness. In another histopathology study, retinoblastoma has also been shown to exhibit cellular atypia, necrosis, and optic nerve invasion, which can cause anatomical distortion and an increase in IOP. At the molecular level, VEGF and HIF overexpression lead to angiogenesis pathology, while inflammatory cytokines, such as TNF-α, exacerbate tissue injury. Additionally, major genetic changes have also been shown to play a pivotal role in the pathogenesis of the disease, particularly mutations in RB1, which may also confer an increased risk of secondary ocular effects in patients. The cumulative, molecular, and cellular level change influences tumor development, leading to glaucoma.

6.2 COMPREHENSIVE OVERVIEW: EXPERIMENTAL MODELS AND DISEASE DYNAMICS IN GLAUCOMA AND RETINOBLASTOMA

Glaucoma and retinoblastoma, though they share distinct origins, occasionally coexist and present clinical challenges. Glaucoma represents a neurodegenerative disease that affects RGCs and the optic nerve head due to the elevation of IOP. Divertingly, retinoblastoma is a malignant intraocular cancer occurring in early childhood, which emerges from the RB1 tumor suppressor gene on chromosome 13q14. The retina is the shared structure reinforcing blindness, an irreversible loss caused by glaucoma through optic nerve damage. However, retinoblastoma triggers a retinal tumor that needs chemotherapy, radiation, or surgical removal of the eyeball. In this chapter, we attempted to address comprehensive experimental models that incorporate their organizational elements for disease tracking and therapeutic monitoring, the involved biomarkers, and the researched pathophysiological processes, utilizing *in vitro*, *in vivo*, *ex vivo*, and computational analysis strategies.

6.2.1 EXPERIMENTAL MODELS AND THEIR COMPONENTS

6.2.1.1 *In Vitro* Modeling Systems

6.2.1.1.1 Glaucoma

6.2.1.1.1.1 Traditional *In Vitro* Models

TM cell lines, including NTM (normal TM), GTM3, and TM5, are well-characterized immortalized human cell lines used to study extracellular matrix (ECM) remodeling, cytoskeletal changes, and glucocorticoid responses under glaucomatous conditions (Stamer, 2017). Moreover, primary RGCs isolated from rodent and human donors are used to interrogate neurodegeneration, with culture readouts of Brn3a, RBPMS expression, and caspase activation under H_2O_2, rotenone, or TGF-β2 treatment having been described earlier (Liu et al., 2007; Sun et al., 2019). Hypoxia/reoxygenation protocols and hydrostatic-pressure chambers simulate IOP stress, inducing ROS accumulation and endoplasmic-reticulum stress markers (BiP, CHOP) in RGC-5 or primary cultures (Liu et al., 2007). Glucocorticoid (dexamethasone) exposure in TM cells yields cross-linked actin networks (CLANs) and upregulation of MYOC, recapitulating steroid-induced ocular hypertension (Stamer, 2017). In short, the strengths of utilizing these cell lines include high throughput and molecular control; however, limitations involve a lack of three-dimensional (3D) architecture and incomplete recapitulation of flow dynamics.

6.2.1.1.1.2 Advanced *In Vitro* Models

3D RGC-organoids derived from human pluripotent stem cells exhibit laminar organization and synaptic connectivity, enabling long-term studies of RGC survival under pressure stress (Tirendi, 2020) and microfluidic "glaucoma-on-a-chip" platforms that integrate tunable pressure gradients across a TM-RGC interface, measuring real-time calcium flux and electrophysiology, have also been described previously (Nafian and Asghari, 2019). Moreover, stretch-based devices mimic lamina cribrosa deformation to study astrocyte reactivity and integrin signaling, highlighting the significance of advanced *in vitro* modeling systems (Li, 2020; Tirendi, 2020). In addition to these, recent trends in bioengineered human anterior segment systems have recapitulated AH outflow with perfusion, allowing for IOP modulation and outflow facility measurements. Clustered Regularly Interspaced Short Palindromic Repeats (CRISPR)-edited TM cells (MYOC, OPTN) enable the mechanistic dissection of gene mutations (Langer, 2024). This advanced *in vitro* system helps analyze omics readouts (transcriptomics, proteomics) and facilitates biomarker discovery; however, limitations of these models include complex fabrication and moderate throughput.

6.2.1.1.1.3 Hybrid *In Vitro* Models

In terms of a hybrid *in vitro* modeling system, the coculture of TM cells with RGCs in dynamic chambers has been considered to elucidate TM-neuron crosstalk under elevated pressure (Fomo, 2024). Moreover, tri-culture systems combining astrocytes, microglia, and endothelial cells on synthetic ECM scaffolds reproduce inflammatory infiltration and barrier dysfunction, which have also been studied (Nafian and Asghari, 2019). Induced pluripotent stem cells (iPSCs) derived neurosensory-vascular constructs incorporate perfusable microvessels to study metabolic-neurodegenerative coupling. Moreover, earlier studies have discovered that transplanting TM cells derived from iPSCs might restore TM function and promote natural TM cell proliferation, as well as iPSC-based TM regeneration. This offers ideas to improve the regenerative impact of iPSCs in glaucoma treatment. Dynamic multi-chamber devices integrate immune cells with TM and RGC compartments, modeling glaucomatous neuroinflammation and offering platforms for immunomodulatory drug screening (Fomo, 2024). To better understand, the multiple glaucoma modeling systems are represented in Figure 6.1.

This schematic provides an independent visualization of how glaucoma models evolved from basic to advanced methodologies, along with *in vitro* research (top panels) and *in vivo* research (bottom panels). ***In vitro models:*** Traditional (Top panel): Inducing glaucomatous changes in RGCs and TM cells by applying controlled IOP in perfusion or hydrostatic systems to two-dimensional (2D) primary cell and immortalized cell cultures. ***Advanced (Bottom panel):*** Utilizing patient-derived iPSCs and CRISPR–Cas9 technology allows researchers to generate 3D organoid cultures that create precise disease models of human tissue architecture alongside mutations specific to each patient's condition. ***In vivo models: Traditional (Top Panel):*** Laboratory experiments using rodents and zebrafish, as well as mutagenic mouse strains, microbead procedures, and drug-based methods, enable the examination of RGC deterioration caused by elevated IOP. ***Advanced (Bottom Panel):*** A set of experimental lines based on myocilin (MYOC) transgenic modifications, combined with the MYOC overexpression model (MOM) and optic nerve crush (ONC) injury tests, creates a framework for studying the complex mechanisms of axonal degeneration and neuroinflammation, as well as the genetic and environmental relationships. The schematic diagram was created using BioRender software (BioRender.com)

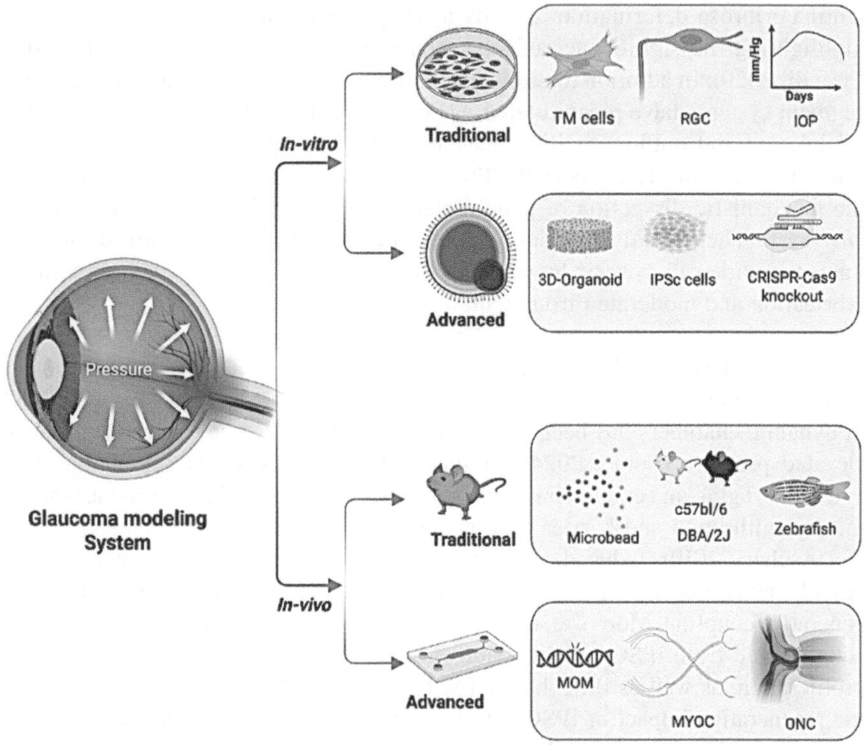

FIGURE 6.1 Experimental glaucoma modeling progression.

6.2.1.1.2 Retinoblastoma

6.2.1.1.2.1 Traditional *In Vitro* Models

Y79 and WERI-Rb1 cell lines, derived from pediatric retinoblastoma, exhibit hypertriploidy, synaptophysin expression, and high Ki-67 proliferation. They have been used for 2D cytotoxicity assays, cell cycle analysis, and apoptosis-based drug screens (Dyson et al., 2023). Long-term drug exposure protocols model chemotherapy resistance (e.g., carboplatin, etoposide) and assess the upregulation of PGP transporters. Primary tumor cultures, when available, retain patient-specific heterogeneity but are limited by rapid senescence. Drawbacks include a lack of tumor architecture and microenvironmental cues.

6.2.1.1.2.2 Advanced *In Vitro* Models

3D spheroids ("tumoroids") of Y79/WERI-Rb1 capture hypoxic cores, necrotic rims, and gradient-dependent marker expression (OTX1-2, ARR3) for subtype classification (Sinenko et al., 2023). Patient-derived tumoroids maintain genetic and phenotypic fidelity, enabling personalized drug screening (Dyson et al., 2023). CRISPR-Cas9 RB1 knockout in retinal progenitor cells generates early tumorigenic organoids, modeling tumor initiation (Dodina and Gurevich, 2024). Under controlled hypoxia and ECM stiffness, organoids recapitulate the intraocular microenvironment. High-throughput platforms with luminescent viability reporters enable real-time screening (Dyson et al., 2023).

6.2.1.1.2.3 Hybrid *In Vitro* Models

Coculture of RB cells with endothelial cells in perfusable channels models tumor angiogenesis and evaluates anti-angiogenic agents (Dyson et al., 2023). Immune–tumor cocultures elucidate evasion mechanisms (PD-L1 expression, T-cell suppression) in retinoblastoma microenvironments. Multi-layer tumor-retinal-vascular organoid-chip systems enable the simultaneous assessment of invasion, vascular permeability, and drug penetration. Modular organoid-chip hybrids integrate stromal and tumor compartments to facilitate mechanistic ECM remodeling and studies of migratory pathways.

6.2.1.2 *In Vivo* Modeling Systems

6.2.1.2.1 Glaucoma

6.2.1.2.1.1 Traditional *In Vivo* Models

Fundamental *in vivo* models are classified based on the induction of IOP, i.e., mechanical obstruction, chemical injury, inherited pathology, and laser-induced damage. Animal models have been the most traditional and effective method for mimicking key pathophysiological features of glaucoma progression. Due to the physiological and anatomical substrate, animal models have become the gold standard for *in vivo* systems.

Elaborating on the following classification that outlines various models existing within each classified section, classic examples are cited to understand clinical relevance and pathological significance. For example, microbead-induced ocular hypertension in rats and mice occludes aqueous outflow, raising IOP by 21%–34% for 2–8 weeks, and yields RGC loss and optic nerve cupping, with histology confirming axonal degeneration (Sappington, 2010). Other models in this category include episcleral vein cauterization and ligation functions, which are similar to the elevation of IOP via post-trabecular resistance. Stability duration ranges from 4 to 21 weeks, characterized by elevated IOP, RGC loss, and glial activation. It is suitable for longitudinal studies, but has variable IOP in rats. In the category of inherited models, DBA/2J mice, carrying Gpnmb and Tyrp1 mutations, develop pigment dispersion, iris atrophy, elevated IOP by 9 months, and progressive RGC death by 22 months, modeling chronic angle-closure glaucoma (John et al., 1998). In the context of mammalian models, the zebrafish model in the species Danio rerio is commonly used in *in vivo* glaucoma. Though they naturally do not induce elevated IOP, their optical transparency and, for the chemical injuries model, illustration of hypertonic saline injection model in Sprague Dawley rats with elevated IOP of 4–6 months, mimics scarring of the TM with 25% of RGC loss and ON damage. Another model is the alpha-chymotrypsin in rabbits with intravitreal injection sustained for 20–30 days, with the relevance of secondary open-angle glaucoma. Sustained changes induce RGC loss, optic cupping, and buphthalmos with limited lens damage and inflammation. Laser photocoagulation models in the rat species, i.e., Sprague Dawley rat, with argon laser applied to limbal veins and meshwork. This model mimics neovascular glaucoma with 50% RGC loss and ERC decline with elevated IOP of 6–7 weeks.

Although animal models are the standardized parameters dominating *in vivo* research, human *in vivo* approaches, such as clinical glaucoma imaging, AH profiling, optical coherence tomography, and longitudinal IOP monitoring, for instance, in clinical glaucoma imaging, OCT, visual fields, IOP monitoring with changes inducing RNFL thinning, ONH cupping, and field loss, they serve as the markers that cooperate with preclinical models.

Conclusively, animal and human *in vivo* models bridge the gap and enhance the translational impact. This translational impact. This ensures a precise, predictive, reproducible, and clinically comprehensive framework for glaucomatous interventions.

6.2.1.2.1.2 Advanced *In Vivo* Models

As glaucoma remains a leading cause of irreversible blindness, the limitations of traditional models have led to the development of advanced *in vivo* models. Earlier, the primary focus was the elevation of IOP and specific gene mutations, whereas advanced models focus on genetic tools, neurodegenerative mapping, and mimicking progression. Advances *in vivo* models provide a transparent platform for further evaluation.

IOP elevation models function around the idea of IOP elevation improving anatomical specificity. Models like the Microbead occlusion model (MOM) mimic the human outflow blockage mechanism by physically blocking the TM with microbeads. Injecting magnetic or polystyrene microbeads produces IOP elevation for 8 weeks. Consistent IOP elevation induces controlled IOP elevation, allowing researchers to mimic the chronic pressure elevation seen in POAG. This leads to ONH cupping, making it a reproducible platform for therapeutic interventions. Another model is the TGF-beta2 overexpression model, which depicts the extracellular matrix seen in POAG. Elevated levels of TGF-beta2 are positively correlated with increased IOP, which can induce glaucoma. TGF promotes TM fibrosis, which directly contributes to the IOP elevation. Understanding the histology of TGF-beta2 would open grounds for therapeutic positioning.

Genetically Modified Glaucoma Models are the basis of understanding the histology of pressure-dependent and independent glaucoma. Models like transgenic Tg-MYOC^Y437H mice express mutant human MYOC, exhibiting ER stress-mediated TM degeneration, age-dependent IOP elevation (4–12 months), and optic nerve changes; stem cell rescue studies demonstrate functional restoration (Zhang et al., 2024; Xiong et al., 2021). Another instance is that Optineurin E50K BAC mice recapitulate normal-tension glaucoma, exhibiting RGC apoptosis, mitochondrial dysfunction, and alterations in autophagy without elevated IOP (Hou and Su, 2022).

Pressure-independent models do not rely on elevated IOP for axonal injury and cell death. A classic example is the ONC model, mimicking human optic neuropathies. By retrograde labeling of RGCs to visualize cell survival, then employing it to follow optic nerve injury. This model is induced, then retinal tissues are further harvested for analysis, allowing researchers to dissect the molecular and cellular pathways. ONC provides hallmarks of glaucomatous damage, inducing a viable platform for preclinical testing. Additionally, models like Ischemia-Reperfusion Injury mimic an acute glaucoma attack by reperfusion of blood flow to the oxygen-deprived tissue, leading to cellular damage and RGC death. IRI is an acute injury model experimentally induced in animals by elevating and then normalizing blood pressure, mimicking transient ischemia. Understanding IRI in glaucoma minimizes oxidative and inflammatory damage, promotes the development of antioxidants, and enhances the outcomes of IOP-lowering techniques. Comprehensively, these models provide a framework and insights for future research, as well as a platform for therapeutic utility.

6.2.1.2.2 Retinoblastoma

6.2.1.2.2.1 Traditional *In Vivo* Models

In vivo models are crucial for understanding the progression, pathogenesis, and therapeutic approaches to retinoblastoma. Based on factors such as the origin of tumor induction, these models are categorically classified into genetically engineered models, xenograft models, allograft models, and alternative systems, including zebrafish xenografts, as well as traditional xenograft models like orthotopic xenografts.

Genetically engineered models replicate the genetic alterations observed in retinoblastoma. One of the earliest used models is the LH-beta T-Ag mouse model, which utilizes the oncogenic SV40 region regulated by the luteinizing hormone beta-subunit promoter in the anterior pituitary region, resulting in bilateral ocular tumors that display histopathology similar to that of glaucoma. A cell line was derived from the P7 mice. A subset of tumor-initiating cells expressing CD133, Nestin, and Sox2 exists upon delineation. The CD133+ cells can form neurospheres under favorable *in vitro* conditions, resulting in transplantable tumors *in vivo* that share similarities with their parent tumor. Orthotopic xenografts involve the intravitreal injection of Y79 or WERIRb1 cells into immunodeficient nude or SCID mice, resulting in intraocular tumors that replicate human histopathology and permit non-invasive SLO/OCT imaging and metastasis studies (Tschulakow et al., 2016; Lisek, 2024). Transgenic Chx10-Cre; Rb1^lox/lox mice develop bilateral retinal tumors, enabling metastasis, tumor burden, and treatment response evaluation under an intact immune context (Lemaitre et al., 2017). Allograft models, such as syngeneic mouse allografts in inbred mice, are achieved by injecting murine RB cells into the host. Tumors progress within 2–3 weeks in immunocompetent RB development models. Other models, for instance, include the Chorioallantois Membrane Assay model in chick embryos, where human RB cells are grafted onto their surface. Tumors are visible within 5–7 days, accompanied by changes such as neovascularization, in an ethical setting.

Alternative Models include the Zebrafish xenograft model in Danio rerio embryos, where fluorescent-labeled human RB cells are injected directly into the eye or yolk. Changes such as proliferation and migration are being visibly reported in 2–5 days. Comprehensively, these *in vivo* models possess insights into the mechanism of retinoblastoma and provide a solid foundation for preclinical drug evaluation.

6.2.1.2.2.2 Advanced *In Vivo* Models

Advances in *in vivo* models for retinoblastoma have significantly untangled the web for understanding tumor dynamics, pathology, and therapeutic responses. GEMMs, such as CHx10-Cre and Rb/p130 DKO mice with an alpha-Cre driver, exhibit neuroplastic differentiation. Another model is the LH-beta-Tag transgenic model, which expresses SV40 T-antigen and causes tumors despite bypassing RB1 mutations, as discussed earlier.

Orthotopic Xenograft Model to curate models that resemble non-metastatic and metastatic disease. Reproducible animal models with distinct behavior parallel to retinoblastoma have been developed. Humanized mouse models engrafted with patient-derived tumor spheroids or iPSC-derived retinoblastoma organoids support human tumor microenvironment interactions, enabling the evaluation of intravitreal chemotherapy and targeted nanoparticle delivery (Norrie, 2021). Imaging-integrated quantitative murine models have been developed for advanced imaging platforms, such as OCT and SLO, paired with histological and digital quantification methods. These models are now used in transgenic and xenograft mouse models to track tumor burden, morphology, and treatment effects over time.

Patient-derived xenografts preserve genetic heterogeneity, providing an effective platform for screening and therapeutic approaches. PDX models enable the evaluation of immune infiltration, stromal cells, and disease progression. Due to their close mimicry of human tumors, they serve a critical role in understanding how an incompetent environment affects tumor growth and clinical outcomes. The study of retinoblastoma now utilizes 3D culture systems together with retinal organoids as an improved research approach. 3D cell cultures, known as retinal organoids, provide insight into the development and tumorigenic features of retinal tissue through the differentiation of pluripotent stem cells in detailed biological conditions. The 3D approach offers better conditions for examining retinal cell differentiation in conjunction with tumor development phases, as it represents natural cellular diversity and tissue architecture at a higher level than conventional 2D methods.

Research combining retinoblastoma organoids from patient samples has shown value in detecting drug resistance indicators while measuring drug responses directly in lab dishes. CRISPR-Cas9 technology, used in conjunction with organoid models, enables investigators to conduct precise genetic transformations to research specific genetic faults affecting tumor development. To better understand, the multiple retinoblastoma modeling systems are represented in Figure 6.2.

The standalone schematic represents an overview of retinoblastoma research models for both *in vitro* (left section) and *in vivo* (right section) investigations, which contain traditional and advanced methods that demonstrate modern tumor modeling capabilities. ***In vitro models:*** Traditional: Monolayer culture systems examine basic tumor cell biology through retinoblastoma tumor cell lines and primary RGCs in two dimensions. Advanced: Researchers model the initiation of retinoblastoma tumors and their interactions with the surrounding microenvironment using 3D spheroids generated from retinoblastoma cell lines or patient-derived samples, along with iPSC-derived retinal organoids and CRISPR–Cas9–engineered mutations of the RB1 gene. ***In vivo models:*** Traditional: The study used genetically engineered mouse models (GEMMs) which carried RB1 loss-of-function alleles combined with orthotopic tumor cell implantation into the eye and alternative small-animal hosts during foundational tumorigenesis experiments. Advanced: Researchers employ human tumor xenografts in immunodeficient mice, alongside refined GEMMs with cell-type–specific promoters, and *in vivo* imaging modalities that include

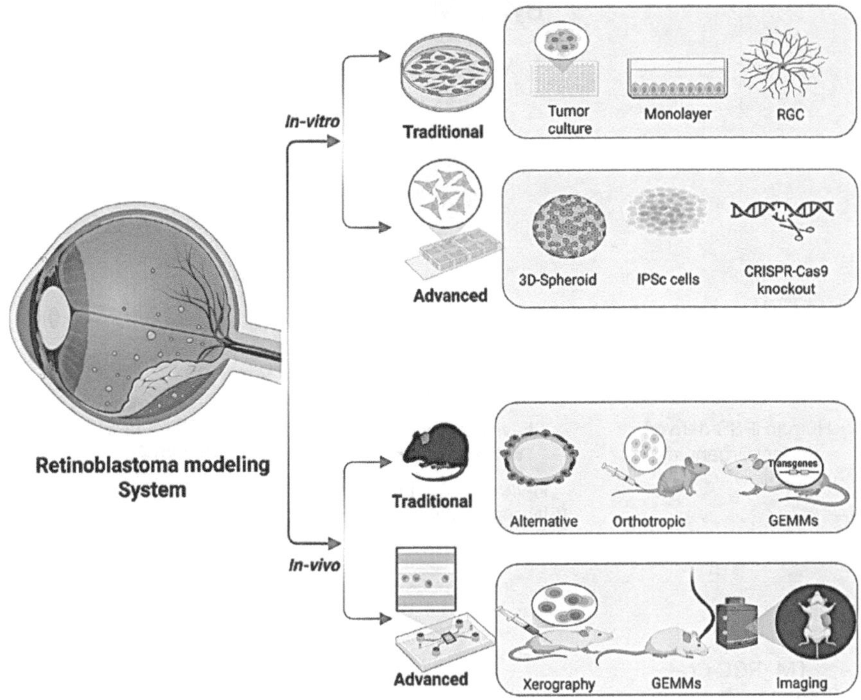

FIGURE 6.2 Experimental retinoblastoma modeling progression.

bioluminescence and optical coherence tomography to monitor tumor growth, vascularization, and therapeutic responses longitudinally. The schematic diagram was created using BioRender software (BioRender.com)

6.2.1.3 Hybrid *In Vivo* Models (Glaucoma & Retinoblastoma)

Dual-pathology constructs implant human iPSC-derived retinal organoids containing RB and RGC lineages into transgenic mice (e.g., Tg-MYOC*Y437H), exposing coexisting tumor and pressure phenotypes. Human-mouse chimeras, reconstituted with CD34-derived human immune cells, enable immuno-oncology studies in RB while modeling glaucomatous inflammation (Norrie, 2021). Challenges include vascular integration, optic nerve connectivity, and immunocompatibility. Hybrid *in vivo* and *in vitro* models for integrated glaucoma and retinoblastoma research comprise various experimental methods that combine *in vivo* and *in vitro* approaches for studying glaucoma in retinoblastoma, as shown in Figure 6.3.

This schematic illustrates various experimental methods that combine *in vivo* and *in vitro* approaches for studying glaucoma in retinoblastoma. The schematic stands alone to illustrate hybrid experimental systems that combine *in vivo* and *in vitro* approaches with tissue engineering and microphysiological platforms extending around a central core segment for advanced human ocular disease model development. ***Central core models:*** *In vivo* glaucoma: Rodent models expressing the MYOC*Y437H mutation and ONC injury paradigms to study IOP–driven

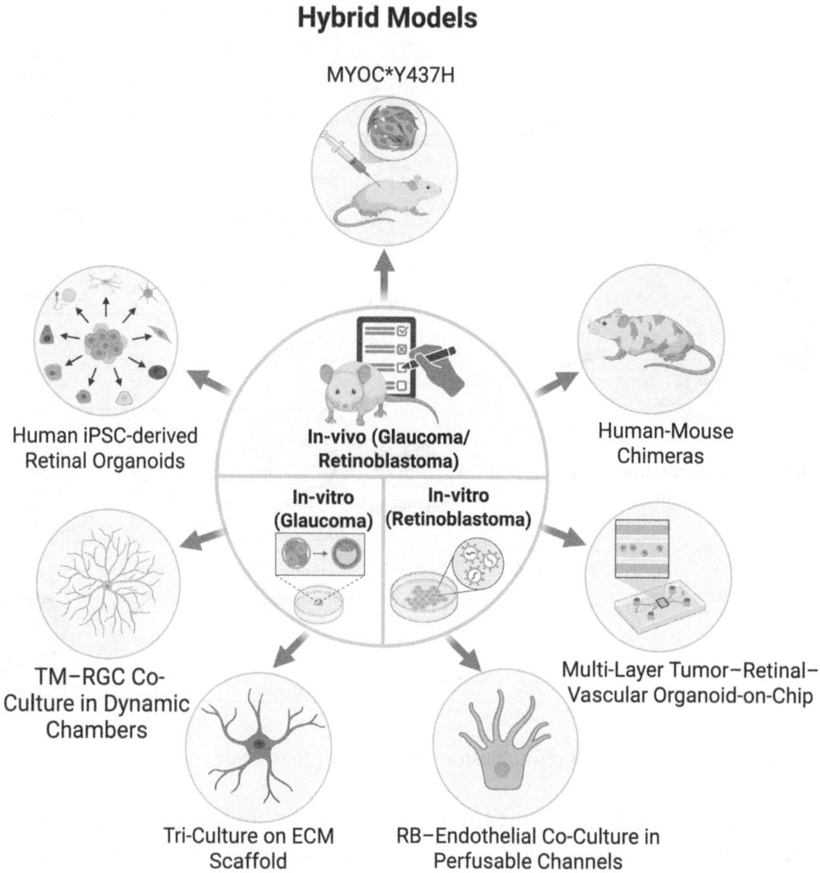

FIGURE 6.3 Hybrid *in vivo* & *in vitro* models for integrated glaucoma and retinoblastoma research.

neurodegeneration. *In vivo* retinoblastoma: The development of transgenic mouse models along with xenograft models helps scientists understand both tumor initiation and its subsequent growth through RB1 mutations. *In vitro* glaucoma: The study utilised mono or coculture systems of TM cells together with RGCs while maintaining specific IOP levels. *In vitro* retinoblastoma: The team worked with RB cell line cells in 2D and 3D container systems in addition to patient-derived spherical tumor cultures. ***Advanced hybrid strategies:*** Human iPS-derived retinal organoids: Scientists have developed 3D layered retinal constructs that support examinations of glaucoma and retinoblastoma. Dynamic TM-RGC coculture chambers: These microfluidic devices recreate conditions that mimic the movement of AH and the fluid pressure gradient. Tri-culture on ECM scaffolds: A biocompatible extracellular matrix hosts the cellular interaction between TM cells, RGCs, and glial cells or tumor cells. RB-endothelial coculture in perfusable microchannels: The system evaluates angiogenesis and flow-dependent interactions between tumor and vascular structures. Multilayer tumor-retinal-vascular organoid-on-chip: This system combines tumor spheroids with retinal tissue and microvasculature in a single, integrated fluidic platform. Human–mouse chimeras: Researchers use immunodeficient mice to embed human retinal progenitors or tumor cells for *in vivo* assessment of their therapeutic integration. The schematic diagram was created using BioRender software (BioRender.com)

6.2.1.4 Advanced (Ex Vivo/Integrated) Systems

6.2.1.4.1 Glaucoma

6.2.1.4.1.1 Ex Vivo & Integrated Systems

Ex vivo and integrated systems have emerged as critical platforms in glaucoma research, offering controlled environments to study IOP, AH dynamics, and neurodegenerative processes with high physiological relevance. One of the most established models is the human anterior segment perfusion culture (HASPC), which preserves TM and Schlemm's canal architecture in post-mortem donor eyes to evaluate aqueous outflow resistance and pharmacologic modulation of IOP. Complementing this, organotypic retinal cultures enable the study of RGC death, synaptic remodeling, and axonal degeneration under elevated pressure and oxidative stress.

Innovative microfluidic chamber systems now enable the real-time manipulation of fluid pressure and shear stress, allowing for the simulation of AH flow within the outflow tract. Similarly, optic nerve head explant cultures facilitate mechanobiological investigations of axonal transport disruption and astrocytic reactivity under simulated IOP conditions. 3D hydrogel-encapsulated TM cultures have further advanced the field by reconstructing ECM architecture, allowing the study of fibrosis, stiffness modulation, and drug permeability.

Artificial anterior eye models incorporate a synthetic cornea, TM, and collector channels to replicate AH physiology in more integrated setups, thereby supporting pharmacological testing and the evaluation of microsurgical devices. Adding live-cell imaging platforms in retinal explants has enhanced our understanding of mitochondrial dysfunction, RGC apoptosis, and calcium dysregulation. Retinal slice cultures subjected to controlled hydrostatic pressure are also used to assess synaptic loss, glutamate excitotoxicity, and neuroprotective interventions.

Moreover, microfluidic ocular model systems represent a significant leap toward multiscale integration, allowing researchers to simulate IOP regulation, AH outflow, ECM remodeling, and drug interactions in parallel.

Ex vivo gene delivery platforms using viral or non-viral vectors in perfused ocular tissues provide a powerful approach to evaluate the efficacy and safety of targeted gene therapies in situ. Perfused human and bovine anterior segments maintain physiological outflow under controlled pressure, allowing study of glucocorticoid-induced ECM remodeling and IOP dynamics; mechanical stretch chambers and dynamic OCT imaging reveal TM–Schlemm's canal interactions and segmental biomechanics (Mao et al., 2011). Tissue-engineered decellularized anterior segments, seeded with TM cells, restore outflow structure, bridging *in vitro* and *in vivo* platforms (Waxman et al., 2023).

6.2.1.4.2 Retinoblastoma

6.2.1.4.2.1 Ex Vivo & Integrated Systems

The improved ex vivo and integrated systems versions help research teams study retinoblastoma heterogeneity and drug responses while monitoring immune system dynamics more biologically (Shi et al., 2019). Establishing precision-cut tumor slice cultures in air-liquid interface or perfusion chambers represents a significant advancement, as it preserves the natural tissue structure and immunologically active stromal elements (Dong, 2023). These systems create natural gradients of drug penetration and oxygen distribution, enabling the evaluation of individual drug sensitivity, analysis of immune cells, and assessment of immunotherapy responses within a natural tissue structure (Tian et al., 2022). The clinical use of patient-derived tumor explants (PDTEs) remains the standard approach to preserving tumor diversity (Powley et al., 2020). These models permit a detailed assessment of therapeutic drug effects and analysis of resistance mechanisms. Retinal organoids obtained from iPSCs and embryonic stem cells model early tumorigenic processes after researchers modified these cells through RB1 knockout experiments (Lee, 2025). These systems help scientists discover both tumor-causing events and promising therapeutic candidates.

The formation of multicellular tumor spheroids is achieved using RB cell lines or patient-derived cells to replicate the 3D tumor architecture. Precision-cut tumor slices cultured at an air–liquid interface or in perfusion chambers preserve the native histoarchitecture, stromal, and immune components, as well as drug penetration gradients, enabling the evaluation of individual chemosensitivity and immunotherapy responses (Norrie, 2021).

The formation of these spheroids duplicates active growth regions and center damage zones alongside oxygen-depleted survival areas, which function well for laboratory drug assessment and cell mortality tests (Białkowska, 2020). Researchers have developed hydrogel-based extracellular matrix formats that are mechanically adjustable and contain customizable biochemical triggers, allowing for precise tumor-ECM behavioral investigations and force-mediated cellular signal interpretation (Kinsella et al., 2021). Microfluidic tumor model systems allow continuous oxygenation combined with blood flow and maintain tumor cells alive to study both angiogenesis processes and metastasis, as well as instantaneous drug effects (Xie et al., 2021). The RB cell invasion dynamics analysis, immune evasion, and vascular mimicry use coculture systems that unite RB cells with host retinal tissues. Regular visualization of cell mitosis, apoptosis events, and chemotherapy effects occurs through live-cell imaging of tumor explants over time (Richard, 2014).

6.2.1.4.2.1.1 In-Silico Computational Models of Glaucoma and Retinoblastoma

Models of glaucoma and retinoblastoma development, based on computational and predictive approaches, also offer insights into the pathophysiological processes underlying these eye disorders. Previously, it has been suggested that a Boolean network can predict the pathophysiological event, specifically crucial signaling pathways, which are utilized in multiple types of myeloma. These models aim to replicate how genetic mutations, IOP (in glaucoma), or retinal cell transformation (in retinoblastoma) impact disease onset, progression, and treatment response over time. Various modeling systems, including Cellular Potts, Agent-Based, ODE/PDE solvers, Boolean Networks, and Network Analysis, are used to simulate glaucoma and retinoblastoma. These models can be used in multiple aspects like tissue remodeling, cell behavior, signaling pathways, and gene/protein interactions.

6.2.1.5 Coexistence Modeling & Mechanistic Integration

6.2.1.5.1 Theoretical and Emerging Coexistence Models

The combined presence of glaucoma and retinoblastoma creates conditions in which their pathophysiological mechanisms overlap due to oxidative stress, mitochondrial damage, inflammatory cytokine reaction networks, and abnormal angiogenesis (de Leon, 2005). Oxidative stress acts as the primary factor that causes RGC death in glaucoma while inducing tumor development in retinoblastoma (Kang et al., 2021). The glaucomatous optic nerve and retinoblastoma tumor microenvironment have elevated TNF-α and IL-6 levels, resulting in neuroinflammation and tumorigenic behavior (Magramm and Abramson, 1989). The VEGFF signaling pathway and the SYK axis promote new vessel formation by driving abnormal blood vessel development in glaucomatous tissue and tumor expansion in retinoblastoma (Claesson-Welsh, 2016). The study of apoptosis, angiogenesis, and transcriptomic effects relies on microfluidic platforms that enable the coculture of RGC-organoids and retinoblastoma tumor spheroids under pressure-controlled perfusion conditions. Longitudinal evaluations of the effects of elevated IOP on tumor dynamics and hypertensive stress on tumor progression can be studied through dual-pathology mouse models that combine Tg-MYOC^Y437H mice with orthotopic retinoblastoma xenografts.

Further studies are essential for understanding the significance of biomarkers in prognosis, diagnosis, and tracking disease progression, as well as exploring treatment targets for both glaucoma and retinoblastoma. The measurement of the retinal nerve fiber layer (RNFL) is a strong indicator of optic nerve degeneration in glaucoma cases, and genetic mutations in the MYOC and OPTN genes contribute to POAG. The AH contains cytokines IL-6, TNF-α, and VEGF, demonstrating active inflammation and neovascular alterations accompanying glaucomatous optic nerve damage (Micieli et al., 2017). Tumor angiogenesis is driven by VEGF and HIF-1α proteins, whereas retinoblastoma arises due to RB1 gene mutations that confirm hereditary retinoblastoma and render these proteins increasingly active (Lorenc, 2024). Circulating tumor DNA (ctDNA) detection enables physicians to assess tumor burden and provides non-invasive methods of evaluating treatment effectiveness. Moreover, possible therapeutic targets include E2F and CDK activity, MDM2-mediated p53 degradation, SKP2-mediated p27 degradation, and drugs that inhibit MYCN expression (Yao et al., 2020). Unlike normal retinal cells, targeting SYK high expression in retinoblastoma cells is critical. Hence, improved targeted medications, particularly sophisticated SYK inhibitor R406 formulations, could show effectiveness against retinoblastoma (Li, 2023). Common molecular pathways exist between glaucoma and retinoblastoma because they share select biomarkers, including VEGF and oxidative stress markers (Moazzeni, 2020). The essential neovascularization regulator VEGF plays dual roles by contributing to the pathogenesis of neovascular glaucoma and stimulating retinoblastoma tumor angiogenesis (Nieves, 2009). The conditions of both glaucoma and retinoblastoma manifest increased oxidative stress markers 8-OHdG and malondialdehyde (MDA) due to cellular dysfunction and damage in affected retinal and tumor cells (Benoist d'Azy et al., 2016). The same microRNA family, comprising miR-182, miR-21, and miR-126, influences survival mechanisms and apoptosis regulation, as well as tumorigenic processes, in retinoblastoma and glaucoma-driven neurodegeneration (Dambal et al., 2015). The discovery of common biomarkers creates new prospects for co-treatment strategies that target both diseases. Novel treatment methods employing this strategy offer a potential understanding of this coexistence and enable the development of improved therapeutic solutions. A comparative overview of advanced ex vivo and integrated systems in glaucoma and retinoblastoma research is summarized in Table 6.1.

TABLE 6.1
Comparative Overview of Advanced Ex Vivo and Integrated Systems in Glaucoma and Retinoblastoma Research

Model Type	Glaucoma	Retinoblastoma	References
Perfused organ culture	Human and bovine anterior segments were maintained under controlled IOP to study trabecular outflow, glucocorticoid-induced ECM remodeling, and dynamic fluid dynamics.	Precision-cut tumor slices cultured at the air-liquid interface or in perfusion chambers preserve histoarchitecture, stromal elements, and immune microenvironment.	Keller and Acott (2009), Peng (2022)
3D tissue-engineered constructs	Decellularized anterior segments are reseeded with TM cells to restore functional outflow pathways and the ECM structure.	Hydrogel-embedded tumor spheroids mimic tumor–ECM interactions, stiffness, and invasion potential.	Loessner (2010)
Organotypic explants	Retinal and optic nerve head explants model RGC degeneration, mitochondrial dysfunction, and astrocytic responses to elevated pressure.	Patient-derived tumor explants preserve genetic heterogeneity, drug gradients, and tumor–immune interactions.	Morgan (2006), Hassani (2025)
Microfluidic platforms	Eye-on-a-chip devices simulate IOP regulation, aqueous flow, and ECM biomechanics in a miniaturized format.	For drug response studies, tumor explant systems replicate perfusion, oxygenation, and endothelial interactions.	Achberger et al. (2019); Ayuso et al. (2016)
Live imaging–compatible systems	Retinal explants with integrated imaging to track apoptosis, axonal transport, and calcium signaling in real time.	Live-cell imaging in tumor slices visualizes mitosis, apoptosis, and chemotherapeutic efficacy.	Cordeiro (2004)

(Continued)

TABLE 6.1 (*Continued*)
Comparative Overview of Advanced Ex Vivo and Integrated Systems in Glaucoma and Retinoblastoma Research

Model Type	Glaucoma	Retinoblastoma	References
CRISPR-compatible models	Gene delivery into ex vivo anterior segments or retinal tissues for therapeutic targeting of IOP-regulating or neuroprotective genes.	CRISPR-edited ex vivo retinal models allow RB1/MYCN targeting to explore tumorigenesis and gene therapy validation.	Amador and Santos (2022)
Dynamic OCT or biomechanical monitoring	Real-time imaging of TM deformation, segmental flow resistance, and mechanical stretch responses.	Integrated imaging tracks tumor growth, angiogenesis, and tissue integrity in perfusion platforms.	Acott and Kelley (2021)
Stretch/stress platforms	Mechanical strain chambers were applied to TM cells to model IOP-induced mechanotransduction and cytoskeletal remodeling.	It is not commonly applied in RB, but is conceptually adaptable for studying compression or induced stress in tumor models.	Hirt and Liton (2017)
Vitreoretinal coculture systems	Limited use; emerging potential for studying TM-immune interactions.	Combined RB and retinal tissue models to study invasion, angiogenesis, and immune modulation.	Wang et al. (2025a)
Organoids	It is not yet widely used; its potential for glaucoma-related RGC development is under investigation.	iPSC-derived retinal organoids genetically edited to mimic RB1 loss and early tumorigenesis.	Zhong et al. (2014)

6.3 MODEL-BASED INSIGHTS INTO DISEASE MECHANISMS

The in-depth operational mechanisms behind glaucoma and retinoblastoma can be studied through sophisticated ex vivo research models (Wang et al., 2025b). The combination of TM dysfunction with elevated IOP leads to optic nerve degeneration and increased outflow resistance, which worsens when the ECM changes in response to glucocorticoids. Organ culture perfusion systems with human and bovine anterior segment specimens at controlled IOP allow researchers to evaluate fluid movement, ECM modifications, and TM response to mechanical tension (Mao et al., 2011). Integrated studies of tumor biology primarily utilize precision-cut retinoblastoma tumor slices examined through air-liquid culture or perfusion chamber systems, which conserve the integrity of histoarchitecture, as well as immune components and drug concentration gradients (Dimou and Arends, 2022). Advanced models in both diseases enable researchers to study cellular responses through live imaging-compatible systems and biomechanical monitoring of apoptosis, cellular signaling, and mechanical responses in neurodegenerative glaucoma and tumor–stromal interactions in retinoblastoma (Bassi and Gualandi, 2021).

6.3.1 The Evaluation of Fundamental Processes between Glaucoma and Retinoblastoma

The foundation of both glaucoma and retinoblastoma disorders has different causes, yet treatment strategies can benefit from this shared insight. The ECM component dysregulation affects both conditions since changes in TM structure raise IOP levels in glaucoma patients, and retinoblastoma tumor production of ECM elements impacts their invasive behavior and material stiffness (Hastings et al., 2018). Research utilizing mechanical strain chambers in glaucoma models has examined how these chambers replicate IOP-induced transduction in TM cells, and scientists could modify these chambers for retinoblastoma tumor pressure stress evaluations. The medical conditions share both elevated VEGF levels and oxidative markers that regulate angiogenesis in glaucoma while contributing to retinoblastoma tumor evolution (Nakamura et al., 2013).

6.4 THERAPEUTIC APPROACHES FOR GLAUCOMA IN RETINOBLASTOMA PATIENTS

A potent management of glaucoma in retinoblastoma patients requires equal attention to primary tumor treatment and reduction of IOP. The standard treatment for glaucoma consists of beta-blockers, carbonic anhydrase inhibitors, prostaglandin analogues, alpha-agonists, and, in several cases where patients do not respond to the other treatments, surgeries, chemotherapies, and radiotherapies are also performed (Meel and Radhakrishnan, 2012). The pediatric oncological context must approve the treatment plan due to potential systemic effects and interaction with chemotherapy (Leung et al., 2021). Retinoblastoma therapies, which were discussed above, like chemotherapies, radiation, and focal treatment, can worsen glaucoma through inflammation, neovascularization, and scarring that impairs AH drainage (Shields et al., 2016). Vision loss due to retinoblastoma and glaucoma significantly impacts a child's overall development and psychological well-being. Therefore, timely diagnosis, multidisciplinary coordination, and rehabilitative support are essential for improving long-term outcomes and quality of life. Based on earlier findings, we have summarized a comparative analysis of biomarkers and therapies utilized in glaucoma and retinoblastoma diagnosis, as shown in Table 6.2.

6.5 FUTURE DIRECTIONS IN RESEARCH AND TREATMENT

Emerging technologies and innovative therapeutics are reevaluating the treatment of glaucoma and retinoblastoma by enhancing survival rates and improving quality of life. Gene therapies targeting the RB1 gene mutation have shown promising results in preventing the formation of tumors, while selective targeted therapies offer a safer alternative to systemic chemotherapy and radiation (Dimaras et al., 2015). Advanced imaging, such as optical coherence tomography (OCT), is helpful in the early detection of glaucoma before irreversible optic nerve damage. Personalized medicine and a tailored-made treatment plan based on genetics and patient response optimize the treatment's possible outcomes and side effects (Soliman et al., 2020). Retinoblastoma

TABLE 6.2
Comparative Analysis of Biomarkers and Therapies in Glaucoma and Retinoblastoma

Category	Biomarker	Molecular Function & Pathological Role	Clinical/Experimental Evidence	Therapeutic Implications	Reference
Glaucoma	Neurofilament Light Chain (NfL)	Cytoskeletal protein released upon axonal damage; biomarker for retinal ganglion cell (RGC) degeneration.	Elevated in aqueous humor, serum, and CSF in glaucoma patients; correlates with disease severity.	Potential for monitoring neuroprotection therapies in clinical trials.	Lin et al. (2023)
	Myocilin (MYOC) Mutations	Disrupts trabecular meshwork (TM) function, increasing intraocular pressure (IOP).	Autosomal dominant in POAG; validated in transgenic mouse models.	CRISPR-Cas9 gene editing under investigation for MYOC-associated glaucoma.	Jain et al. (2017)
	Aqueous Humor Cytokines (IL-6, TNF-α)	Pro-inflammatory cytokines induce TM fibrosis and RGC apoptosis.	Elevated in POAG and neovascular glaucoma; linked to progression.	Anti-inflammatory agents (e.g., steroids, TNF-α inhibitors) in refractory cases.	Borkenstein and Faschinger (2013)
Retinoblastoma	RB1 Gene Mutations	Tumor suppressor loss → uncontrolled retinoblastoma proliferation (Knudson's two-hit hypothesis).	Germline mutations in hereditary RB; somatic in sporadic cases.	Genetic screening for families; RB1-targeted gene therapy in development.	Akdeniz Odemis and Kemeriz (2023)
	VEGF & HIF-1α	Hypoxia-driven VEGF secretion promotes tumor angiogenesis.	High VEGF in advanced RB is associated with vitreous seeding.	Anti-VEGF (e.g., Bevacizumab) reduces tumor vascularity.	Li et al. (2021)
	Circulating Tumor DNA (ctDNA)	Cell-free DNA from apoptotic tumor cells detects minimal residual disease.	ctDNA levels correlate with tumor burden and chemoresistance.	Liquid biopsy for early relapse detection and precision therapy.	Kasi et al. (2022)
Shared Biomarkers	VEGF	Glaucoma: Neovascularization (e.g., diabetic glaucoma). Retinoblastoma: Tumor angiogenesis.	Elevated in both diseases, a common therapeutic target.	Intravitreal anti-VEGF (e.g., Ranibizumab/Aflibercept) is used in both.	–
	Oxidative Stress Markers (ROS, 8-OHdG)	Contributes to RGC apoptosis (glaucoma) and DNA damage (retinoblastoma).	Increase in glaucomatous retina and RB tumors.	Antioxidants (e.g., CoQ10, N-acetylcysteine) in trials.	Tabak et al., (2021) Lande and Ganguly (2021)
	MicroRNAs (miR-182, miR-21)	miR-182: Oncogenic in RB. miR-21: Regulates RGC survival.	miR-182 overexpressed in RB; miR-21 dysregulated in glaucoma.	miRNA-based therapies (e.g., antagomiRs) in preclinical studies	–

and glaucoma share some common underlying causes, particularly in how cells grow and survive. Understanding this might help researchers develop better treatments that address both diseases simultaneously, as shown in Figure 6.4.

This schematic diagram illustrates the fundamental cell-level and molecular mechanisms shared by glaucoma and retinoblastoma, which affect RGCs and their accompanying inflammatory response. Through RGCs both glaucoma and retinoblastoma initiate their pathogenic processes from a single basic cellular origin. The buildup of increased IOP triggers various disease-causing pathological changes that affect axonal function, alongside mitochondrial problems and alterations in the extracellular matrix, as well as neural inflammatory responses. Transient injuries to brain tissue create additional problems for mitochondria, resulting in DNA damage and impaired control of apoptosis, which ultimately leads to the loss of nerve cells. Retinoblastoma develops from the inactivation of the RB1 gene, leading to mitochondrial dysfunction and supporting an immunoreactive tumor environment, while

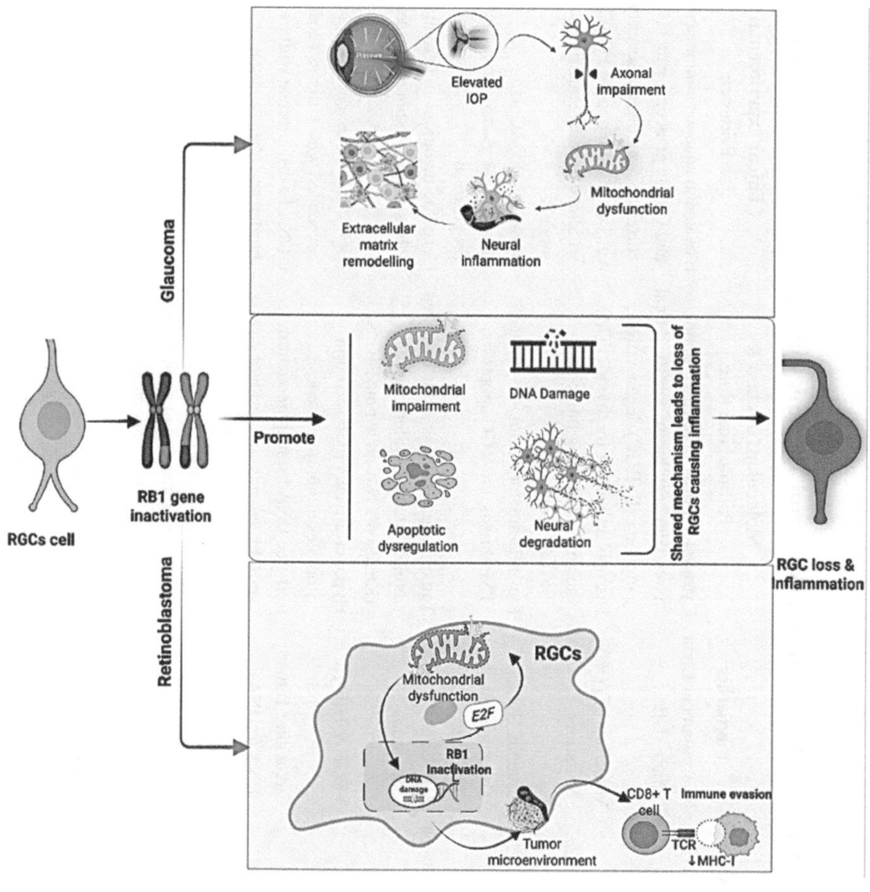

FIGURE 6.4 Convergent RGC degeneration and inflammatory mechanisms in glaucoma and retinoblastoma.

reducing MHC-I expression and suppressing CD8+ T cell activity. DNA damage, neurodegeneration, and apoptosis failure emerge from these molecular processes. The results of both diseases reveal identical tissue pathology, characterized by RGC loss and inflammation, as well as parallel destructive processes involving mitochondrial damage and disrupted apoptotic mechanisms. The schematic diagram was created using BioRender software (BioRender.com)

Through advanced image analysis, AI and ML also revolutionize early diagnosis and risk prediction (Philip, 2019). In recent years, these technologies have been used more in the field of ophthalmology, which uses various imaging modalities such as optical coherence tomography and magnetic resonance imaging based on numerical data such as IOP, corneal thickness, etc., for the diagnosis and prognosis of various ophthalmic diseases (Kalika et al., 2023). Finally, synergistic pharmacological treatment combining chemotherapy and antiglaucoma agents may offer integrated benefits, enhancing overall disease control in pediatric patients.

In alignment with this, we have also proposed a unified therapeutic framework for glaucoma and retinoblastoma. This proposition is further backed by advanced ex vivo modeling techniques, which enable researchers to develop treatment approaches that remedy glaucoma and retinoblastoma conditions. Research examining a single conceptual approach for these ophthalmic conditions should analyze shared mechanical mechanisms during transduction studies and investigations of abnormal vascular development.

Glaucoma patients experience traumatic pressure from increased IOP that generates structural changes to their TM and RGC death. Tumors bearing retinoblastoma affect vascular endothelial growth factor signaling, making it hyperactive and enabling aggressive growth patterns within rapidly evolving tumor environments. There is commonality between the therapeutic targets concerning tissue mechanics and vascular flow changes in both medical conditions.

Advanced microfluidic systems with multiple applications and microphysiological systems technology enable researchers to study simultaneously the mechanisms that control eye pressure changes and the movement of cancerous fluids. Real-time observation of ECM modifications and interactions between immune and endothelial cells occurs under physiological force conditions with these system designs. The models facilitate studies on disease evolution, as they track the neurodegenerative effects of glaucoma, in addition to retinoblastoma-induced vascular penetration. Oxidative stress elevation occurs in these diseases, accompanied by microRNA deregulation that uses the HO-1 regulatory element and specific members from both miR-200 and miR-29 families. The molecular group controls TM cell fibrosis and protects ganglion cells in glaucoma patients, while improving retinoblastoma cell proliferation and vessel formation.

6.6 CONCLUSION

To facilitate proactive modeling and precision treatment, there is a stronger need to understand the interplay between retinoblastoma and secondary glaucoma. While both conditions differ, they share interlocking pathological processes that require specific diagnosis and treatment strategies. Developing new biomarkers, in the context of genetic, proteomic, and metabolomic, can improve risk stratification and disease stage and make individual treatments more personalized. AI-powered analytics, integrated with multi-omics data, can create models capable of cross-disease modeling. These models can improve predictive power, improve the detection and treatment processes, and define new treatment approaches. Additionally, new gene therapies combined with neuroprotective strategies can become unifying paradigms for treatment, preserving visual function and slowing disease progression.

Future studies should incorporate targeted therapy or modern analysis of biological markers in conjunction with models for glaucoma and retinoblastoma detection. Reconciling these diseases will advance the development of precision ophthalmology, enabling interdisciplinary approaches that will transform recent research breakthroughs into effective patient care. Thus, a multidisciplinary approach is critical for future advancements in clinical outcomes.

REFERENCES

Achberger, K., Probst, C., Liebau, S., & Loskill, P. (2019). Merging organoid and organ-on-a-chip technology to generate complex multi-layer tissue models in a human retina-on-a-chip platform. *eLife*, 8, e46188. https://doi.org/10.7554/eLife.46188

Acott, T. S., & Kelley, M. J. (2021). Normal and glaucomatous outflow regulation. *Progress in Retinal and Eye Research*, 82, 100897. https://doi.org/10.1016/j.preteyeres.2020.100897

Aerts, I., Lehoang, P., & Livet-Le Moal, V. (2006). Retinoblastoma. *Orphanet Journal of Rare Diseases*, 1, 31. https://doi.org/10.1186/1750-1172-1-31

Akdeniz Odemis, D., & Kemeriz, K. (2023). RB1 gene mutations and genetic spectrum in retinoblastoma cases. *Medicine (Baltimore)*, 102, 36. https://doi.org/10.1097/MD.0000000000035068

Amador, C., & Santos, C. (2022). Gene therapy in the anterior eye segment. *Current Gene Therapy*, 22(2), 104–131. https://doi.org/10.2174/1566523221666210423084233

Ayuso, J. M., Monge, C., Martínez-López, A., & Chung, B. G. (2016). Development and characterization of a microfluidic model of the tumour microenvironment. *Scientific Reports*, 6, 36086. https://doi.org/10.1038/srep36086

Bassi, G., & Gualandi, G. (2021). Advanced multi-dimensional cellular models as emerging reality to reproduce *in vitro* the human body complexity. *International Journal of Molecular Sciences*, 22(3), 1195. https://doi.org/10.3390/ijms22031195

Bertaud, S. A. (2019). Primary open-angle glaucoma. *La Revue de Medecine Interne*, 40(7), 445–452. https://doi.org/10.1016/j.revmed.2018.12.001

Białkowska, K. K. (2020). Spheroids as a type of three-dimensional cell cultures-examples of methods of preparation and the most important application. *International Journal of Molecular Sciences*, 21(17), 6225. https://doi.org/10.3390/ijms21176225

Borkenstein, A., & Faschinger, C. (2013). Measurement of tumor necrosis factor-alpha, interleukin-6, Fas ligand, interleukin-1α, and interleukin-1β in the aqueous humor of patients with open angle glaucoma using multiplex bead analysis. *Molecular Vision*, 19, 2306–2311.

Broaddus, E., Topham, A., & Singh, A. D. (2008). Incidence of retinoblastoma in the USA: 1975–2004. *The British Journal of Ophthalmology*, 93(1), 21–23. https://doi.org/10.1136/bjo.2008.138750

Casson, R. J., Chidlow, G., Wood, J. P. M., Crowston, J. G., & Goldberg, I. (2012). Definition of glaucoma: Clinical and experimental concepts. *Clinical & Experimental Ophthalmology*, 40(4), 341–349. https://doi.org/10.1111/j.1442-9071.2012.02773.x

Chakrabarti, L., Ghosh, S., & Giri, D. (2021). Inflammation in ocular tumor: Pathogenesis and clinical implication. *Indian Journal of Opthalmology*, 22(19), 10386.

Claesson-Welsh, L. (2016). VEGF receptor signal transduction – A brief update. *Vascular Pharmacology*, 86, 14–17. https://doi.org/10.1016/j.vph.2016.05.011

Cordeiro, M. F. (2004). Real-time imaging of RGC apoptosis with fluorescent annexin V. *Investigative Ophthalmology & Visual Science*, 45(7), 2337–2343.

d'Azy, C. B., Pereira, P., & Chiambaretta, F. (2016). Oxidative and anti-oxidative stress markers in chronic glaucoma: A systematic review and meta-analysis. *PLoS One*, 11(12), 0166915 https://doi.org/10.1371/journal.pone.0166915

Dambal, S., Shah, M. Y., Mihelich, B., & Nonn, L. (2015). The microRNA-183 cluster: The family that plays together stays together. *Nucleic Acids Research*, 45(15), 7173–7188. https://doi.org/10.1016/j.vph.2016.05.011

de Leon, J. M. (2005). Glaucoma in retinoblastoma. *Seminars in Ophthalmology*, 20(4), 217–222. https://doi.org/10.1080/08820530500353831

Dimaras, H., Corson, T. W., Cobrinik, D., & Gallie, B. L. (2015). Retinoblastoma. *Nature Reviews Disease*, 1, 15921. https://doi.org/10.1038/nrdp.2015.21

Dimou, P. T., & Arends, B. (2022). Precision-cut tumor slices (PCTS) as an ex vivo model in immunotherapy research. *Antibodies (Basel, Switzerland)*, 11(2), 26. https://doi.org/10.3390/antib11020026

Dodina, M., & Gurevich, L (2024). Evaluation of mesenchymal stem cells as an in vitro model for inherited retinal diseases. *Frontiers in Cell and Developmental Biology*, 12, 1455140. https://doi.org/10.3389/fcell.2024.1455140

Dong, M. B. (2023). Perfusion air culture of precision-cut tumor slices: An ex vivo system to evaluate individual drug response under controlled culture conditions. *Cells*, 12(5), 807. https://doi.org/10.3390/cells12050807

Dyson, I., Lambo, R., & Rodriguez, R. (2023). The predictive capacity of *in vitro* preclinical models to evaluate drugs for the treatment of retinoblastoma. *Experimental Eye Research*, 230, 109447. https://doi.org/10.1016/j.exer.2023.109447

Fomo, K. N. (2024). Neuroretinal cell culture model as a tool for the development of new therapeutic approaches for oxidative stress-induced ocular diseases, with a focus on glaucoma. *Cells*, 13(9), 775. https://doi.org/10.3390/cells13090775

Hassani, S. T. (2025). Patient-derived tumor organoids: A preclinical platform for personalized cancer therapy. *Translational Oncology*, 51, 102226. https://doi.org/10.1016/j.tranon.2024.102226

Hastings, J. F., Naba, A., & Humphrey, J. D. (2018). The extracellular matrix as a key regulator of intracellular signalling networks. *British Journal of Pharmacology*, 176, 82–92. https://doi.org/10.1111/bph.14195

Hirt, J., & Liton, P. B. (2017). Autophagy and mechanotransduction in outflow pathway cells. *Experimental Eye Research*, 21, 146–153. https://doi.org/10.1016/j.exer.2016.06.021

Honavar, S. G. (2017). Retinoblastoma. *Indian Journal of Pediaatrics*, 84, 937–944. https://doi.org/10.1007/s12098-017-2395-0

Hou, M., & Su, S. (2022). Age-related visual impairments and retinal ganglion cells axonal degeneration in a mouse model harboring OPTN (E50K) mutation. *Cell Death & Disease*, 13, 362. https://doi.org/10.1038/s41419-022-04836-3

Jain, A., Zhang, G., & Chen, L. (2017). CRISPR-Cas9–based treatment of myocilin-associated glaucoma. *Biological Sciences*, 114, 11199–11204. https://doi.org/10.1073/pnas.1706193114

John, S. W. M., Smith, R. S., Savinova, O. V., & Hawes, N. L. (1998). Essential iris atrophy, pigment dispersion, and glaucoma in DBA/2J mice. *Investigative Ophthalmology & Visual Science*, 39(6), 951–962.

Kalika, S. V., Vora, U., & Sengupta, S. (2023). Artificial intelligence and machine learning in ocular oncology: Retinoblastoma. *Indian Journal of Opthalmology*, 71, 424–430.

Kang, E. Y., Lee, E. J., Kim, T. W., & Kim, H. (2021). Role of oxidative stress in ocular diseases associated with retinal ganglion cells degeneration. *Antioxidants (Basel, Switzerland)*, 10(12), 1948. https://doi.org/10.3390/antiox10121948

Kasi, P. M., Kasi, S. P., & Kasi, A. M. (2022). Impact of circulating tumor DNA–based detection of molecular residual disease on the conduct and design of clinical trials for solid tumors. *JCO Precision Oncology*, 6, 181. https://doi.org/10.1200/PO.21.00181

Keller, K. E., & Acott, T. S. (2009). Extracellular matrix turnover and outflow resistance. *Experimental Eye Research*, 88(4), 676–682. https://doi.org/10.1016/j.exer.2008.11.023

Kinsella, J., Morgan, T. L., Tran, D., & Lo, K. M. (2021). Decellularized extracellular matrix composite hydrogel bioinks for the development of 3D bioprinted head and neck *in vitro* tumor models. *ACS Biomaterials Science & Engineering*, 7(11), 812. https://doi.org/10.1021/acsbiomaterials.1c00812

Križaj, D. (2019). *What is Glaucoma? Webvision: The Organization of the Retina and Visual System.* University of Utah Health Sciences Center, Salt Lake City.

Lande, K., & Ganguly, J. (2020). Exosomes: Insights from retinoblastoma and other eye cancers. *International Journal of Molecular Sciences*, 21(19), 7055. https://doi.org/10.3390/ijms21197055

Langer, F. B. (2024). In vitro comparison of human and murine trabecular meshwork cells: Implications for glaucoma research. *Scientific Reports*, 14, 22002. https://doi.org/10.1038/s41598-024-73057-9

Lee, D. A. (2005). Glaucoma and its treatment: A review. *American Journal of Health-System Pharmacy*, 62(7), 691–699. https://doi.org/10.1093/ajhp/62.7.691

Lee, Y. J. (2025). Retinal organoids from induced pluripotent stem cells of patients with inherited retinal diseases: A systematic review. *Stem Cell Reviews and Reports*, 21(1), 167–197. https://doi.org/10.1007/s12015-024-10802-7

Lemaitre, S., Poyer, F., Marco, S., Fréneaux, P., Doz, F., Aerts, I., . . ., & Thomas, C. D. (2017). Looking for the most suitable orthotopic retinoblastoma mouse model in order to characterize the tumoral development. *Investigative Ophthalmology & Visual Science*, 58, 3055–3064. https://doi.org/10.1167/iovs.17-21760

Leung, K. K. Y., Lam, K., & Chau, C. (2021). Therapeutics for paediatric oncological emergencies. *Dreugs in Context*, 10(10), 1–12. https://doi.org/10.7573/dic.2020-11-5

Li, J. W. (2023). Recent advances in targeted drug delivery strategy for enhancing oncotherapy. *Pharmaceutics*, 15(9), 15092233. https://doi.org/10.3390/pharmaceutics15092233

Li, Z. G. (2020). Effectively suppressed angiogenesis-mediated retinoblastoma growth using celastrol nanomicelles. *Drug Delivery*, 27(1), 358–366. https://doi.org/10.1080/10717544.2020.1730522

Lin, J. B., McDonnell, K. M., & Duh, E. J. (2023). Neurofilament light chain in aqueous humor as a marker of neurodegeneration in glaucoma. *Clinical Ophthalmology*, 17, 2209–2217. https://doi.org/10.2147/OPTH.S417664

Lisek, M. T. (2024). Histone deacetylases in retinoblastoma. *International Journal of Molecular Sciences*, 25(13), 6910. https://doi.org/10.3390/ijms25136910

Liu, Q., Ju, W.-K., Crowston, J. G., Xie, F., Perry, G., Smith, M. A., . . . & Weinreb, R. N. (2007). Oxidative stress is an early event in hydrostatic pressure–induced retinal ganglion cell damage. *Investigative Ophthalmology & Visual Science*, 48, 4580–4589. https://doi.org/10.1167/iovs.07-0170

Loessner, D. S. (2010). Bioengineered 3D platform to explore cell-ECM interactions and drug resistance of epithelial ovarian cancer cells. *Biomaterials*, 31(32), 8494–8506. https://doi.org/10.1016/j.biomaterials.2010.07.064

Lorenc, P. S.-K. (2024). Physiological and tumor-associated angiogenesis: Key factors and therapy targeting VEGF/VEGFR pathway. *Biomedicine & Pharmacotherapy = Biomedecine & Pharmacotherapie*, 180, 117585. https://doi.org/10.1016/j.biopha.2024.117585

Li, Y., Liu, Y., Bai, H., Li, R., Shang, J., Zhu, Z., ... & Huang, L. (2021). Sustained release of VEGF to promote angiogenesis and osteointegration of three-dimensional printed biomimetic titanium alloy implants. *Frontiers in Bioengineering and Biotechnology,* 9, 757767. https://doi.org/10.3389/fbioe.2021.757767

Magramm, I., & Abramson, D. H. (1989). Optic nerve involvement in retinoblastoma. *Ophthalmology*, 96(2), 217–222. https://doi.org/10.1016/S0161-6420(89)32910-1

Mao, W., Tovar-Vidales, T., & Clark, A. F. (2011). Perfusion-cultured bovine anterior segments as an ex vivo model for studying glucocorticoid-induced ocular hypertension and glaucoma. *Investigative Ophthalmology & Visual Science*, 52(11), 8068–8075. https://doi.org/10.1167/iovs.11-8133

Meel, R., & Radhakrishnan, V. (2012). Current therapy and recent advances in the management of retinoblastoma. *Indian Journal of Medical and Paediatric Oncology: Official Journal of Indian Society of Medical & Paediatric Oncology*, 33(2), 80–88.

Micieli, J. A., Lam, C., Najem, K., Margolin, E. A. (2017). Aqueous humor cytokines in patients with acute nonarteritic anterior ischemic optic neuropathy. *American Journal of Ophthalmology*, 177, 175–181. https://doi.org/10.1016/j.ajo.2017.03.003

Moazzeni, H. K. (2020). Insights into the regulatory molecules involved in glaucoma pathogenesis. *American Journal of Medical Genetics. Part C, Seminars in Medical Genetics*, 184(3), 782–827. https://doi.org/10.1002/ajmg.c.31833

Morgan, J. E. (2006). Retinal ganglion cell remodelling in experimental glaucoma. *Advances in Experimental Medicine and Biology*, 56, 397–402. https://doi.org/10.1007/0-387-32442-9_56

Nafian, F., Asghari, S. M. (2019). *A Lab-On-a-Chip Model of Glaucoma.* Cold Spring Harbor Laboratory, Cold Spring Harbor, NY. https://doi.org/10.1101/704510

Nakamura, I., Shioiri, T., Nakashima, A., et al. (2013). Serum levels of vascular endothelial growth factor are increased and correlate with malnutrition, immunosuppression involving MDSCs and systemic inflammation in patients with cancer of the digestive system. *Oncology Letters*, 5(5), 1682–1686. https://doi.org/10.3892/ol.2013.1231

Nieves, B. J. (2009). The function of vascular endothelial growth factor. *BioFactors (Oxford, England)*, 35(4), 332–337. https://doi.org/10.1002/biof.46

Norrie, J. N. (2021). Retinoblastoma from human stem cell-derived retinal organoids. *Nature Communications*, 12, 4535. https://doi.org/10.1038/s41467-021-24781-7

Pandey, A. N. (2014). Retinoblastoma: An overview. *Saudi Journal of Ophthalmology: Official Journal of the Saudi Ophthalmological Society*, 28(4), 310–315. https://doi.org/10.1016/j.sjopt.2013.11.001

Peng, M. M. (2022). An ex vivo model of human corneal rim perfusion organ culture. *Experimental Eye Research*, 214, 108891. https://doi.org/10.1016/j.exer.2021.108891

Philip, H. (2019). The advent of AI and deep learning in diagnostics and imaging: Machine learning systems have potential to improve diagnostics in healthcare and imaging systems in research. *EMBO Reports*, 20(7), 201948559. https://doi.org/10.15252/embr.201948559

Powley, I. R., Patel, M., Miles, G., et al. (2020). Patient-derived explants (PDEs) as a powerful preclinical platform for anti-cancer drug and biomarker discovery. *British Journal of Cancer*, 122(6), 735–744. https://doi.org/10.1038/s41416-019-0672-6

Quigley, H. A. (2006). The number of people with glaucoma worldwide in 2010 and 2020. *The British Journal of Ophthalmology*, 90(3), 262–267. https://doi.org/10.1136/bjo.2005.081224

Richard, C. (2014). Live-cell imaging. *Cell Adhesion & Migration*, 8(5), 452–459. https://doi.org/10.4161/cam.28348

Sappington, R. M. (2010). The microbead occlusion model: A paradigm for induced ocular hypertension in rats and mice. *Investigative Ophthalmology & Visual Science*, 51(1), 207–216. https://doi.org/10.1167/iovs.09-3947

Shi, D. M., Al-Marzoqi, A. H., & Yeh, J. Y. (2019). *In vitro* and ex vivo systems at the forefront of infection modeling and drug discovery. *Biomaterials*, 198, 228–249. https://doi.org/10.1016/j.biomaterials.2018.10.030

Shields, C. L., Shields, J. A., & Cater, J. (2006). The international classification of retinoblastoma predicts chemoreduction success. *Opthalmology*, 113(12), 2276–2280.

Shields, C. L., Shields, J. A., & Cater, J. (2016). Classification and treatment of secondary gluacoma in eyes with retinoblastoma. *Opthalmology*, 10(24), 8695–8702.

Shields, C. L., Shields, J. A., & Cater, J. (2017). Diagnosis and management of retinoblastoma. *Cancer Control*, 11(5), 317–327.

Sinenko, I. L., Smirnov, V. E., & Prokofiev, A. V. (2023). Ranslational screening platform to evaluate chemotherapy in combination with focal therapy for retinoblastoma. *Cancer Science*, 114(9), 3728–3739. https://doi.org/10.1111/cas.15878

Soliman, S. E., VandenHoven, R., & Gallie, B. L. (2020). Genetics insights into personalized retinoblastoma care. *Cancer*, 15, 1416924.

Stamer, W. D. (2017). The many faces of the trabecular meshwork cell. *Experimental Eye Research*, 9, 112–123. https://doi.org/10.1016/j.exer.2016.07.009

Sun, Y. Y., Meng, J., Wang, J., & Liang, X. (2019). Phenotypic and functional alterations in tunneling nanotubes formed by glaucomatous trabecular meshwork cells. *Investigative Ophthalmology & Visual Science*, 60(14), 4583–4595. https://doi.org/10.1167/iovs.19-28084

Tabak, S., Soiberman, U., Amar, Y., & Yariv, O. (2021). Crosstalk between microRNA and oxidative stress in primary open-angle glaucoma. *International Journal of Molecular Sciences*, 22(5), 2421. https://doi.org/10.3390/ijms22052421

Tham, Y. C., Li, X., Wong, T. Y., Quigley, H. A., Aung, T., & Cheng, C. Y. (2014). Global prevalence of glaucoma and projections of glaucoma burden through 2040: A systematic review and meta-analysis. *Ophthalmology*, 121(11), 2081–2090. https://doi.org/10.1016/j.ophtha.2014.05.013

Tian, C. Z., Li, M., & Zhang, Y. (2022). Tumor-on-a-chip model for advancement of anticancer nano drug delivery system. *Journal of Nanobiotechnology*, 20, 338. https://doi.org/10.1186/s12951-022-01552-0

Tirendi, S. (2020). A 3D model of human trabecular meshwork for the research study of glaucoma. *Frontiers in Neurology*, 11, 591776. https://doi.org/10.3389/fneur.2020.591776

Tschulakow, A. V., Voelcker-Schenk, H., Schrage, N., & Walter, P. (2016). Establishment of a novel retinoblastoma (Rb) nude mouse model by intravitreal injection of human Rb Y79 cells - comparison of in vivo analysis versus histological follow up. *Biology Open*, 5, 1625–1630. https://doi.org/10.1242/bio.019976

Wang, H. Y., Wang, W., Zhang, H., & Liu, Y. (2025a). Tumor microenvironment: Nurturing cancer cells for immunoevasion and druggable vulnerabilities for cancer immunotherapy. *Cancer Letter*, 611, 217385. https://doi.org/10.1016/j.canlet.2024.217385

Wang, X. S., Chen, J. L., & Lin, M. L. (2025b). The molecular mechanisms underlying retinal ganglion cell apoptosis and optic nerve regeneration in glaucoma (Review). *International Journal of Molecular Medicine*, 55(4), 63. https://doi.org/10.3892/ijmm.2025.5504

Waxman, S. S., Saraswathy, S., & Goldberg, J. L. (2023). Tissue-engineered anterior segment eye cultures demonstrate hallmarks of conventional organ culture. *Graefe's* Archive for Clinical and Experimental Ophthalmology, 261, 1359–1368. https://doi.org/10.1007/s00417-022-05915-z

Weinreb, R. N., Aung, T., & Medeiros, F. A. (2014). The pathophysiology and treatment of glaucoma: A review. *JAMA*, 311(18), 1901–1911. https://doi.org/10.1001/jama.2014.3192

Xie, H., An, H., & Yang, H. (2021). Going with the flow: Modeling the tumor microenvironment using microfluidic technology. *Cancers*, 13(23), 6052. https://doi.org/10.3390/cancers13236052

Xiong, S., Kuang, Y., & Li, Y. (2021). Stem cell transplantation rescued a primary open-angle glaucoma mouse model. *eLife*, 10, e63677. https://doi.org/10.7554/eLife.63677

Yao, H., Li, W., & Zhang, Q. (2020). The E2F family as potential biomarkers and therapeutic targets in colon cancer. *PeerJ*, 8, e8562. https://doi.org/10.7717/peerj.8562

Zhang, X. X., Yang, Q., & Li, R. (2024). Intraocular pressure across the lifespan of Tg-MYOCY437H mice. *Experimental Eye Research*, 241, 109855. https://doi.org/10.1016/j.exer.2024.109855

Zhong, X., Gutierrez, C., Xue, T., Hampton, C., Vergara, M. N., Cao, L. H., . . . & Canto-Soler, M. V. (2014). Generation of three-dimensional retinal tissue with functional photoreceptors from human iPSCs. *Nature*, 5, 4047. https://doi.org/10.1038/ncomms5047

7 Oral Cancer Model

Vandana, Pranjay Gupta, Kumud Bala, and Yash Sharma

7.1 INTRODUCTION

Oral squamous cell carcinoma (OSCC), the predominant form of head and neck squamous cell carcinoma (HNSCC), ranks among the ten most prevalent cancers globally (Badwelan et al. 2023). The burden of oral cancer is notably high in developing countries, with India, Pakistan, Brazil, France, Afghanistan, Bangladesh, Sri Lanka, Bhutan, Nepal, Iran, and the Maldives reporting it as one of the leading cancer types. In India, it stands as the most commonly diagnosed cancer, reflecting both the high exposure to risk factors and the limited access to early diagnostic interventions (Jena et al. 2024). The carcinogenic effects of tobacco are mediated by the generation of free radicals, reactive oxygen species (ROS), and reactive nitrogen species (RNS), leading to oxidative stress and subsequent irreversible damage to DNA and proteins. Additionally, human papillomavirus (HPV) infection has emerged as an etiological factor, particularly in oropharyngeal cancers (Valavanidis, Vlachogianni, and Fiotakis 2009). Clinically, OSCC manifests with symptoms such as mucositis, xerostomia, and dysphagia, which are often exacerbated during and after treatment. Xerostomia, characterized by decreased salivary flow, is especially debilitating, leading to oral fungal infections, difficulty swallowing, and altered taste perception (von Bültzingslöwen et al. 2007). The disease predominantly affects regions such as the lips, lateral borders of the tongue, and the floor of the mouth. The pathogenesis of oral cancer involves disruptions in tumor suppressor genes (TSGs), activation of proto-oncogenes, and defects in critical cellular processes, including chromosomal segregation, genomic stability, cell cycle regulation, and Notch signaling. Understanding these molecular events is essential for developing effective diagnostic and therapeutic strategies. A variety of experimental models have been established to study the underlying biology of oral cancer and evaluate potential therapies. This chapter delves into the comprehensive landscape of oral cancer models, categorized into *in vitro*, *in vivo*, and *in silico* systems. *In vitro models*, including two-dimensional (2D) and three-dimensional (3D) cell cultures, provide a controlled environment for studying cellular responses and drug screening. *In vivo* models, such as xenograft and genetically engineered mouse models, enable the investigation of tumor progression and therapeutic efficacy in a physiological context. Complementing these, *in silico* models employ computational simulations to predict disease mechanisms and drug interactions, facilitating hypothesis generation and translational research. Together, these models form an integral framework for advancing our understanding of oral cancer and accelerating the discovery of novel treatments.

DOI: 10.1201/9781003583349-8

7.2 DEVELOPMENT OF ORAL CANCER

Oral cancer, classified under head and neck cancers, typically originates within the oral cavity and affected regions such as the anterior two-thirds of the tongue, gingivae, buccal mucosa, lips, sublingual floor, hard palate, and the retromolar area (Watters, Brar, and Pepper 2024). The onset of OSCC generally begins with epithelial dysplasia, characterized by abnormal proliferation of dysplastic squamous cells confined to the epithelial surface. As the condition progressed, these cells invaded the underlying subepithelial basement membrane (BM), leading to local tissue destruction and eventually distant spread through metastasis. This invasive growth occurred via epithelial cell islets and cords infiltrating deeper tissues (Rivera and Venegas 2014). The early development of oral cancer often remained clinically silent, making its detection particularly challenging. Symptoms, when present, included a persistent lump or a nonhealing ulcer lasting more than 14 days, along with the appearance of red, white, or mixed patches (erythroplakia and leukoplakia) in the oral mucosa. Additional signs involved difficulty swallowing, speaking, jaw and tongue movements, discomfort from ill-fitting dentures, and unintended weight loss. As the disease progressed, a burning sensation or pain became more prominent (Sankaranarayanan et al. 2015). Diagnosis of oral cancer relies on a combination of comprehensive physical examination, detailed medical history, and an assessment of risk factors such as tobacco and alcohol use. One of the major challenges in the early development phase was the asymptomatic nature of the disease, allowing the malignancy to progress undetected. Clinically, OSCC could appear as an ulcer with fissuring, a raised exophytic margin, a persistent sore, or a firm, fixed cervical lymph node (Sutera et al. 2013). Staging of oral cancer played a crucial role in understanding disease progression and prognosis. The TNM classification system was widely used, where "T" described the size and extent of the primary tumor, "N" denoted the involvement of regional lymph nodes, and "M" indicated the presence of distant metastasis (Nagesh et al. 2023). Early-stage diagnosis and timely referral to specialized oncology centers significantly improved patient outcomes, with survival rates reaching up to 90%. However, in clinical practice, over 60% of oral cancer cases were detected at advanced stages (Stage III or IV), contributing to poor prognosis and higher mortality (Markopoulos 2012).

7.3 EPIDEMIOLOGY AND PREVALENCE OF ORAL CANCER

Oral cancer is the sixth most common cancer worldwide, accounting for 377,713 new cases and 177,757 deaths globally in 2020 (Rivera and Venegas 2014). This represents an increase from 2018, which reported 354,864 new cases and 177,384 deaths (Bray et al. 2018). In the United States (U.S.), the American Cancer Society estimated that approximately 54,010 new cases of oral cavity or oropharyngeal cancer would be diagnosed in 2021, representing about 3% of all annual cancer diagnoses in the U.S., with a per-incidence treatment cost of $45,000–$70,000 (Ellington et al. 2020). Additionally, about 10,850 individuals were projected to die from oral or oropharyngeal cancer in 2021(American Cancer Society 2025). The average 5-year survival rate for oral cancer in the U.S. is 64.3%, although this is stage-dependent.

About 70% of cases are diagnosed at later stages, where the 5-year survival rate decreases significantly, from 83.7% at a localized stage to 38.5% if metastasis is present (Lingen et al. 2017). Oral cavity cancer sites include the inner lip, parts of the tongue apart from the base and lingual tonsil, gingiva, floor of the mouth, palate, and other unspecified areas of the mouth (Conway, Purkayastha, and Chestnutt 2018; Tranby et al. 2022).

7.3.1 Global Burden and Disparities

Oral cancer poses a significant and growing challenge worldwide. Epidemiological data reveal considerable geographic variability in oral cancer incidence and mortality. In Australia, over 4,000 new cases of head and neck cancer (including lip cancer) are diagnosed annually, with more than 600 cases classified as oral cavity cancers. In developing countries, oral cancer ranks as the sixth most common cancer among men and the tenth among women. Globally, over 400,000 new oral cancer cases are diagnosed each year, with two-thirds occurring in Asian countries such as Sri Lanka, Indonesia, India, Pakistan, and Bangladesh (Sun et al. 2023). In these high-risk regions, oral cancer is the leading malignancy, accounting for more than 25% of all cancer cases annually. According to the International Agency for Research on Cancer (IARC), countries in Asia and the Pacific recorded the highest oral cancer rates in 2018. While advances in cancer diagnosis and treatment have improved survival for other cancers (e.g., breast, colon, and ovarian cancers), survival rates for oral cancer have remained stagnant over the past 50 years (Bray et al. 2024). The overall 5-year survival rate for oral cancer is approximately 40%. However, if detected at an early stage (I or II), the survival rate can exceed 80%. Unfortunately, up to 50% of oral cancers are diagnosed at advanced stages (III or IV) due to asymptomatic early disease progression. Patients often delay seeking medical attention until symptoms like pain, bleeding, or a noticeable mass in the mouth or neck arise. Diagnostic delays exceeding one month significantly increase the likelihood of advanced-stage cancer. These delays may stem from patient-related factors or inadequate clinical suspicion and timely intervention by healthcare providers (Geum et al. 2013).

7.3.2 Geographic and Socioeconomic Variations

Oral cancer incidence shows considerable global variation, influenced by geographic region, ethnicity, and socioeconomic status. According to the IARC, cancers of the lip, tongue, gingiva, mouth floor, and salivary glands (ICD-10 codes C00–C08) contribute to over 300,000 new cases annually, with approximately 145,000 deaths (Brown, Check, and Devesa 2011). Regions with high oral cancer incidence include South and Southeast Asia (e.g., Sri Lanka, India, Pakistan, and Taiwan), parts of Europe (e.g., France, Hungary, Slovakia, and Slovenia), Latin America (e.g., Brazil, Uruguay, and Puerto Rico), and the Pacific (e.g., Papua New Guinea and Melanesia). The incidence of oral cancer also correlates with a country's Human Development Index (HDI). Developed regions show higher crude and age-standardized incidence rates but lower mortality, whereas less developed regions exhibit disproportionately higher mortality, underscoring significant social inequalities (Joseph et al. 2015).

7.4 *IN VITRO* MODELS FOR ORAL CANCER

7.4.1 2D Cell Culture

In the mid-20th century, innovative cell culture technologies were developed, enabling the establishment of primary cultures from clinical specimens, including those from patients with HNSCC (Arutyunyan et al. 2023). Early efforts focused on isolating pure tumor cell populations; however, subsequent studies demonstrated that tumor-stromal interactions during cocultivation were significant (Mun, Lee, and Kim 2024). Among the 39 most commonly utilized HNSCC lines, approximately 60% were derived from oral tissues, while others were isolated from the larynx (15%), pharynx (10%), nasal septum (less than 3%), and uncertain sites (12%) (Lin et al. 2007). In a larger review of 383 HNSCC lines, a similar trend was observed, though with slight variations (Li et al. 2014). Despite the wide availability of HNSCC cell lines, none were included in the NCI-60 Human Tumor Cell Lines Screen, which was developed for anticancer drug evaluation (Close et al. 2019). Misidentification was also reported, as seen in the case of the Hep-2 line, which was initially classified as laryngeal cancer but later confirmed to have been contaminated and replaced by HeLa cervical cancer cells (Gorphe 2019). Traditional 2D cell culture systems were widely used in preclinical studies of OSCC, offering advantages such as ease of use, cost-effectiveness, precise experimental control, and high-throughput capabilities (Chaves et al. 2023). Cell lines such as SCC-9, SCC-25, HSC-3, and CAL-27 were used extensively to study genetic alterations in key oncogenes and TSGs, including RAS, MYC, TP53, and CDKN2A (Brauchle et al. 2013). In one example, CAL-27 and HSC-3 cells were cultivated as monolayers in DMEM or RPMI media and were treated with Gefitinib, an epidermal growth factor receptor (EGFR) inhibitor. EGFR inhibition was found to significantly reduce OSCC cell proliferation, migration, and invasion, while promoting apoptosis. Such studies demonstrated the utility of 2D cultures for exploring molecular mechanisms and identifying potential therapeutic targets in oral cancer (Khalil and Jameson 2017). However, the complexity of the 3D architecture, including cell-cell interactions, extracellular matrix (ECM) components, and oxygen/nutrient gradients, was not replicated by 2D cultures (Abuwatfa, Pitt, and Husseini 2024). Therefore, findings from 2D studies were often used as a foundation for further validation using advanced 3D models or animal experiments.

To bridge the gap between 2D culture models and *in vivo* tumors, cell line-derived xenografts were created by implanting HNSCC cell lines such as A-253 (Park et al. 2022), FaDu (Colley et al. 2011), CAL27 (Bernareggi et al. 2022), and Detroit 562 (Hsu et al. 2021) into immunodeficient mice (Table 7.1). Through this approach, tumor growth, microenvironmental interactions, and metastatic behaviors were better simulated, enhancing the evaluation of therapeutic agents.

This table presents an overview of widely used cell lines derived from various head and neck regions, highlighting their tissue of origin, tumor source (primary or metastatic), donor sex, karyotype features, and key mutated genes. Identified mutations include TSGs (*TP53*, *CDKN2A*, *SMAD4*) and oncogenes (*PIK3CA*), which are relevant to the molecular pathology and experimental modeling of HNSCC.

TABLE 7.1
Summary of Key Characteristics of Selected Human Head and Neck Cancer Cell Lines

Cell Line	Tissue of Origin	Tumor Source	Sex	Karyotype Description	Key Mutated Genes
A-253	Submaxillary salivary gland	Primary tumor	Male	Near-triploid; at least six marker chromosomes	CDKN2A (tumor suppressor), KDMC5, TP53 (tumor suppressor)
SCC-9	Tongue	Primary tumor	Male	Karyotype not determined	CDKN2A, TP53
SCC-15	Tongue	Primary tumor	Male	Karyotype not determined	TP53
SCC-25	Tongue	Primary tumor	Male	Possible loss of the Y chromosome	CDKN2A, TP53
FaDu	Hypopharynx	Primary tumor	Male	Hypodiploid to hypertriploid range; modal number = 64	CDKN2A, SMAD4 (tumor suppressor), TP53
Detroit 562	Pharynx	Metastatic tumor (pleural effusion-derived)	Female	Hypertriploid; modal number = 64; range = 58–128 chromosomes	CDKN2A, PIK3CA (oncogene), TP53

7.4.2 3D Cell Culture and Scaffold-Free Strategy in Cancer Research

The development of 3D cell culture technologies marked a significant advancement in cancer research, as they provided models that more closely resemble the complexity of *in vivo* tumors compared to traditional 2D cell cultures. *In vivo*, tumors grow in a 3D environment, where cells interact with each other, the ECM, and neighboring tissues, forming the tumor microenvironment (TME). This environment influences tumor progression, metastasis, and response to treatment. 3D culture systems, such as spheroids, organotypic raft cultures, and organoids, replicate these conditions and provide researchers with more relevant models for studying cancer biology, tumor progression, and therapeutic responses.

7.4.2.1 Spheroids

Spheroids are one of the simplest and most widely used forms of 3D cell culture models. These spherical aggregates of cells mimic the early stages of tumor formation and have been instrumental in studying tumor biology. Unlike cells grown in traditional 2D cultures, which adhere to a flat surface, cells in spheroids are encouraged to interact with each other in three dimensions, forming structures that replicate key features of solid tumors, including the formation of a necrotic core, hypoxic conditions, and gradients of nutrients and waste products (Nayak et al. 2023). These features are essential for understanding tumor metabolism, invasion, and response to therapies. Spheroids can be generated using a variety of methods, including suspension-based techniques, where cells are cultured in non-adherent conditions, promoting spontaneous aggregation (Mehta et al. 2012). Another approach is matrix-assisted spheroid formation, where cells are placed within or above a matrix, such as collagen or Matrigel. These matrices provide a scaffold-like structure that helps cells form spheroids while mimicking the ECM's biological properties (Bhattacharya et al. 2023). While naturally derived matrices like collagen are closer to the TME, synthetic matrices such as polyethylene glycol (PEG), polylactic-co-glycolic acid (PLGA), and poly-ε-caprolactone (PCL) offer better control over their physical properties, though they are less biologically relevant (Aazmi et al. 2024). Spheroids have been used extensively to study cancer cell growth, proliferation, apoptosis, and drug resistance. Additionally, they serve as an important tool in assessing the efficacy of cancer therapeutics, such as chemotherapies and targeted therapies. However, one limitation of spheroids is their relatively simple architecture compared to actual tumors, as they do not fully replicate the complex cellular interactions present *in vivo* (Mehta et al. 2012).

Tumor cells, when cultivated as spheroids, take on a 3D structure that emphasizes cell-to-cell connections rather than interactions with the culture substrate. These spheroids, unlike conventional 2D cell cultures, provide numerous benefits. They foster enhanced cell-to-cell communication and facilitate improved diffusion of vital substances and molecules. Multicellular tumor spheroids bear a striking resemblance to avascular tumor masses found in large solid tumors. They exhibit comparable characteristics in terms of appearance, growth dynamics, cell division, drug permeability, nutrient gradients, oxygen levels, and the formation of a hypoxic core (Dalir Abdolahinia and Han 2023). As in the case of *in vivo* tumors, deep within the spheroid, cells encounter limited diffusion of nutrients, causing the development of a necrotic core. Additionally, the accumulation of metabolic waste and decreased nutrient availability leads to a notable decrease in the microenvironment's pH. This pH gradient has the potential to affect drug release and influence the activity of genes related to multiple drug resistance, such as Multiple Drug Resistance (MDR). Consequently, therapeutics targeting hypoxic regions may face reduced uptake (Duan et al. 2023). Numerous investigations have showcased the potential of oral cancer spheroids as a valuable tool for replicating different aspects of tumor tissue, particularly its growth state and hypoxic conditions (Lin and Chang 2008; Suryaprakash et al. 2020). However, the utilization of this system is limited due to the spheroids' constrained ability to self-renew and differentiate. Furthermore, spheroids can only mimic small-scale microenvironments since controlling their growth becomes challenging during culturing, resulting in central necrosis if they exceed a certain size (Laschke and Menger 2017). Moreover, spheroids lack the intricate complexity found in actual tumor tissues, primarily because they are often composed of a single cell line and have minimal ECM deposition (Sant and Johnston 2017). In response to these limitations, researchers have devised spheroids that incorporate diverse stromal components, such as cancer-associated fibroblasts (CAFs), alongside epithelial HNSCC cancer cells. The aim is to establish a more representative tumor model that accounts for the crucial role of stromal cells in tumorigenesis (Yakavets et al. 2019).

7.4.2.2 Organotypic Raft Cultures

Organotypic raft cultures represent an advanced and sophisticated approach to modeling complex tissues, particularly epithelial cancers, by maintaining the native architecture and cellular interactions essential for studying cancer pathophysiology. These models are primarily used for culturing cells in an environment that supports the stratification and differentiation of epithelial cells, a crucial feature for studying a range of cancers, especially squamous cell carcinoma (SCC) (Fang and Eglen 2017). The key feature of organotypic raft cultures is their ability to replicate the stratified nature of epithelial tissues by growing cells on a dermal equivalent, such as de-epidermized dermis, collagen, or Matrigel, and maintaining them at an air-liquid interface. This setup promotes the formation of multilayered epithelial structures, closely mimicking the architecture and cellular organization of the original tissue (Fan, Demirci, and Chen 2019).

Recent studies have shown that organotypic raft cultures are particularly beneficial in studying cancer metastasis, especially the invasion of cancer cells into surrounding tissues. These cultures allow epithelial cells, including cancerous ones, to interact with the ECM and surrounding cells, which is essential for understanding the mechanisms driving cancer progression. For example, recent research on OSCC has demonstrated the utility of these models in elucidating how cancer cells invade surrounding stromal tissues and interact with the ECM components, which is crucial for understanding tumor invasion and metastasis (Driehuis et al. 2019). Organotypic raft cultures have been widely used to model viral oncogenesis, including HPV's role in OSCC and other cancers, by accurately simulating epithelial transformation and the viral life cycle (Saglam-Metiner, Gulce-Iz, and Biray-Avci 2019). These cultures preserve tissue architecture and cell interactions better than 2D models, enabling high-throughput drug screening and analysis of therapeutic responses in a realistic tumor environment. However, they are labor-intensive and require precise conditions, limiting their broader use. Advances in bioengineering have enhanced their complexity by incorporating immune cells, vasculature, and stroma, making them valuable for studying tumor–immune–ECM interactions, CSC behavior in OSCC, and the impact of ECM stiffness on invasiveness and treatment response (Dalir Abdolahinia and Han 2023).

7.4.2.3 Organoids

Organoids represent the pinnacle of 3D cell culture technology. These small, self-organizing structures are derived from primary tissues or stem cells and can replicate the full cellular composition, architecture, and functionality of the tissue from which they were derived. Organoids have been developed for a wide range of tissues, including the liver, brain, intestine, and breast, and are particularly useful for studying cancer biology in a more physiologically relevant context (Corrò, Novellasdemunt, and Li 2020). Organoids have been used to model various aspects of cancer, including carcinogenesis, tumor progression, and metastasis. They are particularly useful in the study of drug resistance and therapeutic efficacy, as they can be exposed to different drugs to evaluate their response. In addition, patient-derived organoids have been used in high-throughput drug screening, identifying mutation-specific therapeutic sensitivities and providing a platform for personalized treatment plans. The ability to coculture organoids with immune cells has also allowed researchers to investigate tumor-immune interactions, providing new insights into immunotherapy and the role of the immune system in cancer progression (Ma et al. 2023). Despite their many advantages, the use of organoids in cancer research is not without challenges. One major limitation is the difficulty in scaling up organoid culture for high-throughput screening or large-scale studies. Additionally, while organoids recapitulate many aspects of the *in vivo* environment, they do not fully replicate certain features, such as vascularization, which are essential for studying tumor angiogenesis and metastasis. To address these issues, researchers are exploring the combination of organoids with other technologies, such as microfluidic platforms, to better model the *in vivo* tumor environment (Nikonorova et al. 2023). An organoid represents a more advanced version of a cellular spheroid that remarkably imitates the physicochemical conditions of a specific tissue. In contrast to spheroids, organoids possess a more organized cellular architecture when grown on specific matrices, resulting in a structured arrangement of cells (Fang and Eglen 2017). Organoids can display various biological characteristics akin to living tissues, such as tissue organization, regenerative capacity, drug responses, and the retention of tumor heterogeneity (Fan, Demirci, and Chen 2019). Notably, Driehuis et al. (2019) successfully generated human mucosal organoids from 31 patient-derived HNSCC samples originating from diverse head and neck regions. Through immunohistochemical and genetic analyses, it was revealed that these organoids preserved the histological traits and molecular abnormalities observed in the original tumor specimens. Functionally, the organoids exhibited distinct sensitivity to a range of drugs, including cisplatin, carboplatin, cetuximab, and radiotherapy, highlighting their potential for drug screening in HNSCC. Furthermore, when transplanted beneath the skin of mice, the organoids displayed atypical features, tripolar mitotic figures, nuclear pleomorphism, and invasion into muscle tissue, showcasing their tumorigenic abilities *in vivo*. These findings indicate that HNSCC organoids faithfully replicate the genetic, histological, and functional aspects of HNSCC and hold promise as a platform for personalized cancer therapy.

Nevertheless, despite the benefits of organoids, there remain various limitations that necessitate attention. Tumor organoids usually recreate tumors originating from a single organ and fail to capture the intricate nature of cancer metastasis that involves multiple organs. Additionally, organoid cultures do not faithfully reproduce the spatial arrangement of cellular and non-cellular constituents within the TME. Consequently, there are discrepancies in organoid complexity, size, morphology, and 3D structure, posing challenges in the standardization of organoid culture protocols (Saglam-Metiner, Gulce-Iz, and Biray-Avci 2019).

7.4.2.4 Scaffold-Free Strategy

The scaffold-free strategy represents an alternative approach to building 3D cell cultures without relying on external matrices. Instead of using scaffolds, this strategy focuses on the natural self-assembly of cells into 3D structures, such as spheroids, cell sheets, or tissue strands. The scaffold-free method uses cell aggregation and cellular interactions to create more complex tissue structures, offering a more "bottom-up" approach to tissue engineering.

In scaffold-free models, cells are cultured in suspension, promoting spontaneous aggregation into 3D structures. These models have been shown to better replicate the natural cellular organization and interactions found *in vivo* compared to traditional 2D cultures. One example of a scaffold-free system is the use of patient-specific epithelial and sub-epithelial cells in HNSCC patient-derived explants. In this system, the coculture of epithelial and stromal cells enhanced explant survival, longevity, and tumor progression, providing a more physiologically relevant model for studying cancer (Abuwatfa, Pitt, and Husseini 2024).

The scaffold-free approach has several advantages, including faster tissue construction and a reduced need for artificial materials that may interfere with cell behavior. However, it has not been as widely adopted as scaffold-based strategies, largely due to technical challenges and high costs associated with the development of scaffold-free models. Additionally, maintaining the complexity and stability of these models over time remains a challenge. Most studies using scaffold-free models in cancer research have focused on spheroids and organoids, and while these models provide valuable insights into cancer biology, they still have limitations in fully mimicking the *in vivo* TME (De Pieri, Rochev, and Zeugolis 2021).

7.4.2.5 Scaffold-Based Strategy

Scaffold-based strategies remain a cornerstone of 3D cell culture systems, particularly for tissue engineering applications. Scaffolds provide structural support to cells, allowing them to grow and organize in a way that mimics the ECM. Natural scaffolds, such as collagen, Matrigel, and hyaluronic acid, are commonly used because they provide a biologically relevant environment for cells to grow and form tissue-like structures (Loh and Choong 2013). Synthetic scaffolds, such as PEG or polylactic acid (PLA), offer better control over the mechanical and chemical properties of the scaffold but lack the biological cues present in natural ECM (Zhu 2010). Scaffold-based models have been used to engineer a variety of tissues, including those found in the oral cavity. For example, tissue-engineered oral mucosal models (TEOMMs) have been developed to study the pathogenesis of OSCC. These models use scaffolds to support the growth of epithelial cells and provide a more realistic representation of tissue architecture than 2D cultures. Studies using TEOMMs have shown that they can replicate crucial steps in tumor development, such as cell proliferation, differentiation, and invasion, making them useful tools for understanding cancer biology and testing new therapies (Moharamzadeh et al. 2012). Scaffolds crafted from natural substances such as collagen, Matrigel, and silk, as well as synthetic materials including PEG and PLGA, or a combination thereof, offer robust support for cell development and can emulate numerous mechanical and biochemical characteristics of the original ECM (Sitarski et al. 2018). The approach of utilizing scaffolds has been extensively employed to engineer diverse oral tissues, facilitating the exploration of underlying cancer mechanisms. For instance, scientists have ingeniously created human-TEOMMs) by seeding keratinocytes onto matrices populated with fibroblasts and culturing them at the interface between air and liquid (Lobat and Keyvan 2025). TEOMMs have enabled researchers to capture distinct stages of OSCC, encompassing oral dysplasia, early invasive OSCC, and the neoplastic transformation associated with stromal fibroblasts (Costea et al. 2006; Colley et al. 2011).

As an illustration, Sawant et al. (2016) developed a TEOMM representing normal, dysplastic, and malignant tongue tissues to investigate the neoplastic progression throughout various stages of tongue tumorigenesis. Histomorphometry, immunohistochemistry (IHC), and electron microscopy analyses performed on the three types of models demonstrated comparable stratified growth, cell proliferation, and differentiation between cocultures and their respective native tissues. These findings suggest that crucial steps in the development of oral tumors can be replicated *in vitro* using scaffold-based engineered tissues (Dalir Abdolahinia and Han 2023).

7.4.2.6 Tumor-on-a-Chip for Oral Cancer Model

Tumor-on-a-chip technology represents a cutting-edge advancement in developing *in vitro* models for studying oral cancer, particularly OSCC. These microfluidic platforms recreate the complex TME by integrating oral cancer cells, stromal components, ECM, and dynamic flow conditions within a chip-like device. Compared to traditional 2D and 3D cultures, tumor-on-a-chip systems offer better simulation of tissue architecture, mechanical cues, and chemical gradients found *in vivo*. For OSCC, these models enable precise study of tumor progression, metastasis, drug response, and tumor-stroma interactions under physiologically relevant conditions (Wang and Sun 2021). Cancer tests that could be employed at the point-of-care, such as in doctors' and dentists' offices, and operated without extensive training or expertise, are of special interest. Lab-on-a-chip microfluidics—the miniaturization of fluidic networks for chemical and biochemical processing and analysis—offers a means for mass-produced, low-cost, disposable devices for cancer screening and diagnostics, providing easily interpreted test results within 10–60 minutes. These lab-on-a-chip systems can use small sample volumes (10–1,000 μL) of clinical specimens like oral fluids, whole blood, serum, or urine, collected through minimally invasive methods, along with tissue biopsies and bronchial lavages. Benefits include automated operation, short processing times, real-time reporting, reduced reagent consumption, reproducibility, minimal exposure to hazardous materials, minimal risk of contamination, and low cost (Mauk et al. 2007). In such systems, the cancer diagnostic process typically involves immunoseparation steps to deplete lymphocytes and isolate cancer cells expressing EpCAM, a marker aberrantly expressed on cancerous epithelial cells in OSCC. Detection of EpCAM-expressing cells serves as a first screening test, followed by lysis, mRNA extraction, multiplex amplification (RT-PCR or bio-barcode techniques), and detection using fiber optic or electrochemical sensors. Conventional biomarker detection methods, like IHC and immunoassays, are expensive and time-consuming (Patriarca et al. 2012), but lab-on-a-chip approaches integrate concentration, reaction, and detection processes into chip-scale devices (Surappa et al. 2023). Earlier, Ziober et al. discussed the application of lab-on-a-chip technology for early oral cancer screening (Ziober et al. 2008). Microfluidics-based devices allow analysis of small clinical samples with high sensitivity and efficiency for detecting multiplexed biomarkers of HNSCC and OSCC. For example, Soares et al. developed a microfluidics-based genosensor for detecting HPV16 in HNSCC cell lines with high sensitivity using interdigitated electrodes functionalized with DNA probes (Soares et al. 2018). Similarly, other researchers created a multiplexed microfluidic electrochemical array detecting a four-protein biomarker panel (IL-6, IL-8, VEGF, and VEGF-C) from serum samples, achieving a clinical diagnostic sensitivity of 89% and specificity of 98% (Rusling 2013). Beyond diagnosis, tumor-on-a-chip devices enable real-time monitoring of cancer invasion and metastasis, key prognostic factors in OSCC. In previous studies, they designed a microfluidic device with parallel perfusion channels to study the invasion front of OSCC cells (UM-SCC6) through Matrigel matrices, identifying key miRNA profiles associated with invasiveness (Zhang et al. 2022). Moreover, carcinoma-associated fibroblasts (CAFs) from adenoid cystic carcinoma patients showed enhanced migration and invasion in microfluidic assays (Li et al. 2016). The CRISPR/Cas9 gene-editing technology was employed to engineer oral cancer cell lines that closely mimicked the genetic mutations observed in clinical OSCC (Figure 7.1). These engineered models served as sophisticated laboratory platforms, offering valuable insights into the molecular mechanisms driving OSCC development and progression (Sowmya et al. 2024). A controlled environment was established through these systems to investigate how specific genetic alterations influenced cancer behavior and interactions with the surrounding tissue matrix (de Visser and Joyce 2023). Conventional treatment options such as surgery, radiation, and chemotherapy continued to be associated with high morbidity and mortality rates. Although CRISPR/Cas9 applications in oral cancer remained in their early stages, recent studies illuminated its promising potential. It was demonstrated that CRISPR/Cas9 could effectively target genes like MUC1 (Mucin short variant S1), often overexpressed in oral cancer cells, leading to a significant reduction in cellular growth and proliferation (Sowmya et al. 2024). Despite advancements, targeted treatments for OSCC remained limited. A detailed understanding of the genetic basis of oral cancer, particularly among Asian populations where the incidence is highest, was still lacking. CRISPR/Cas9 research gained momentum as a tool to dissect oral carcinogenesis mechanisms, identify new therapeutic targets, and innovate personalized treatment strategies (Chai et al. 2020). Genome-wide CRISPR/Cas9 screenings identified essential genes, druggable targets, and biomarkers, advancing the knowledge of genotype-phenotype correlations. Immune cells engineered via CRISPR/Cas9 were utilized in developing cancer immunotherapies, some of which had entered phase I/II clinical trials (Chehelgerdi et al. 2024). Overall, CRISPR/Cas9 technology provided an unprecedented platform for advancing OSCC research, offering hope for future, more effective, targeted, and personalized therapeutic strategies.

7.4.2.7 3D Bioprinting: Advanced Tumor Modeling for OSCC

3D bioprinting technology represents a revolutionary advancement in oral cancer research, enabling the precise fabrication of complex tumor models through the strategic layering of living cells and specialized biomaterials. These sophisticated constructs transcend traditional research methods by providing highly customizable platforms that can accurately mirror patient-specific tumor characteristics (Almela, Tayebi, and Moharamzadeh 2021). The bioprinted models offer an unprecedented opportunity to study cancer biology in a physiologically relevant context, bridging the critical gap between conventional cell culture systems and animal models (Figure 7.2). This innovative approach has established itself as an invaluable tool for evaluating therapeutic efficacy and advancing our understanding of tumor

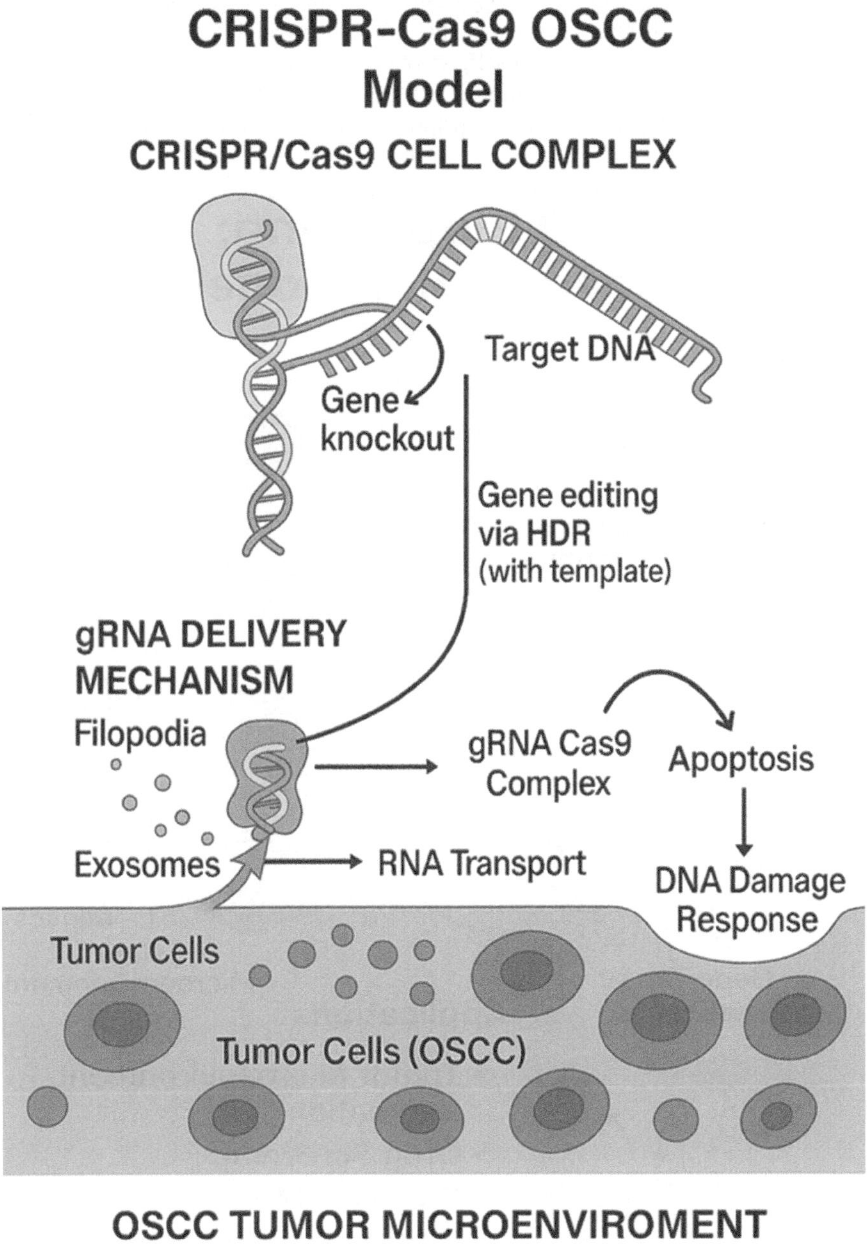

FIGURE 7.1 Schematic representation of CRISPR/Cas9-mediated gene editing in OSCC.

behavior, potentially revolutionizing the landscape of personalized cancer treatment (Khorsandi et al. 2024). The formulation of a specialized bioink using a blend of natural polymers, including proteins and polysaccharides, creates a supportive structure resembling the ECM found in living tissues. The model incorporated various cell types to represent different aspects of the tumor ecosystem, i.e., established OSCC cell lines to represent the cancer itself, fibroblasts associated with cancer to simulate supportive stromal tissue, and immune cells to model inflammatory responses and infiltration (Zidarič et al. 2020). The printing process involved a layered approach,

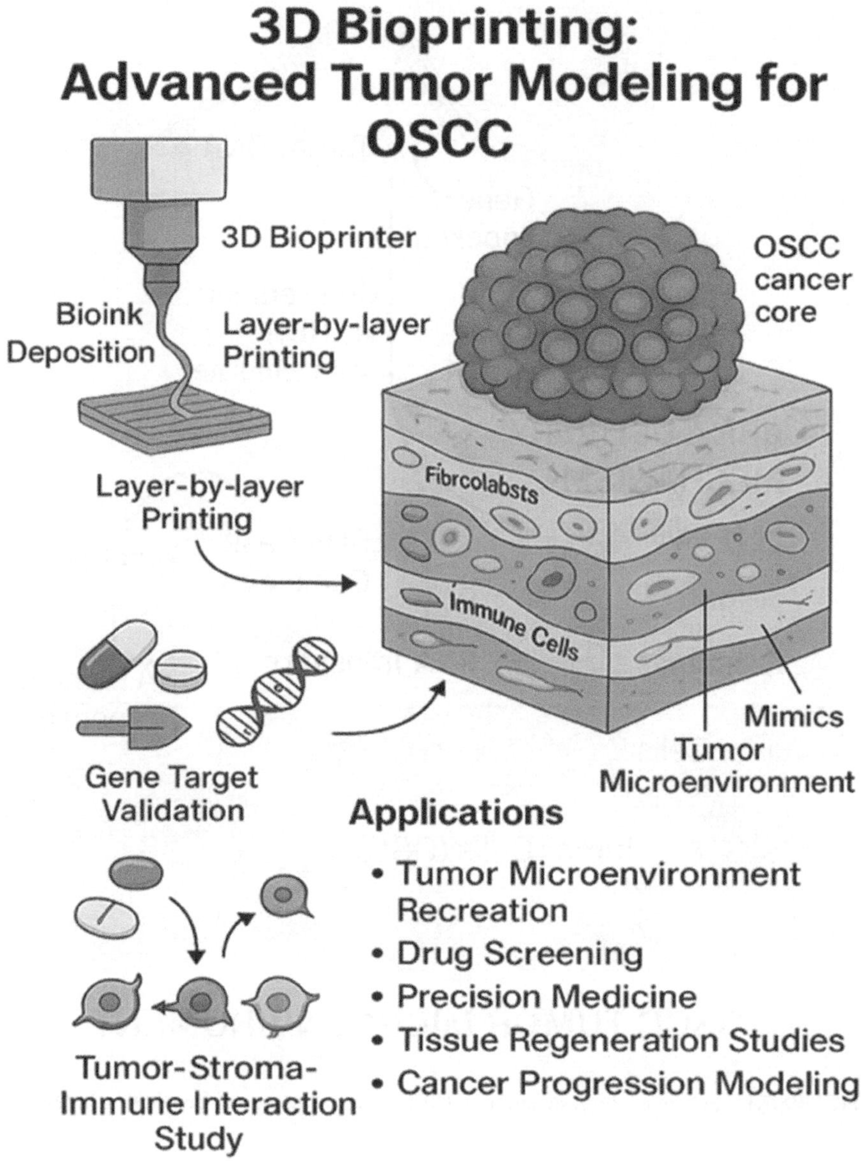

FIGURE 7.2 Schematic showing the use of 3D bioprinting to recreate the OSCC tumor microenvironment through layer-by-layer bioink deposition.

i.e., the core consisted of cancer cells embedded in a collagen-rich environment, which was surrounded by a layer of stromal cells in a specially designed matrix, the outer portion contained immune cells and additional extracellular components (Abuwatfa, Pitt, and Husseini 2024). After printing, the constructs were cultivated in a specialized chamber to promote tissue development and create gradients of nutrients and oxygen, like those found in actual tumors. This bioprinted OSCC model successfully replicated key features of the TME, including the complex interplay between cancer cells and surrounding tissues, mechanisms of immune evasion, and the development of drug resistance (Zhuang et al. 2021). By providing a more faithful representation of living tumors compared to traditional flat cell cultures, this platform offers valuable insights into the tumor ecosystem, drug resistance, and treatment effectiveness in OSCC (Peng et al. 2025).

7.5 TRADITIONAL IN VIVO MODELS FOR ORAL CANCER

7.5.1 Chemically Induced Models

Chemically induced animal models have become indispensable platforms in oral cancer research, especially in the study of OSCC. These models, which employ carcinogens such as 4-nitroquinoline-1-oxide (4NQO), are particularly valuable for replicating the multistep progression of human oral cancer, from initial dysplasia to invasive carcinoma (Li et al. 2020). OSCC is a complex, aggressive cancer with a high mortality rate, and understanding its pathogenesis is critical for developing effective therapies. These models simulate the stages of oral cancer progression, providing a detailed representation of the disease's development. The use of carcinogens like 4NQO induces OSCC in a dose- and time-dependent manner, which closely mimics the natural progression of the disease in humans. By exposing young animals to carcinogens and monitoring them over time, researchers can study the mechanisms driving oral carcinogenesis and investigate potential therapeutic strategies (Tan et al. 2023).

4NQO, a potent carcinogen, induces oral cancer in animals, starting with dysplasia (abnormal growth of cells) and progressing through hyperplasia (excessive growth of cells) to invasive carcinoma. This progression is crucial in studying OSCC, as it mirrors the stages observed in humans. The carcinogen-induced animal model, therefore, serves as an accurate representation of OSCC's pathogenesis. Carcinogen exposure typically begins in young animals, providing a timeline that mirrors the clinical progression of oral cancer in humans, from the earliest stages of tumor formation to the more advanced stages of invasion. The model allows researchers to monitor tumor development and progression over time, identifying key molecular changes, genetic mutations, and signaling pathways that drive oral cancer. This progression is critical for understanding the biology of the disease and for developing strategies to prevent or treat OSCC effectively. Beyond 4NQO, several other carcinogenic compounds, including 7,12-dimethylbenz(a)anthracene (DMBA) and N-nitrosamines, have been used to induce OSCC in animal models. These carcinogens, when administered in mouse and hamster models, facilitate the study of OSCC across different species (Altuwairgi, Papageorge, and Doku 1995). The hamster cheek pouch model, in particular, is widely used due to its ability to provide clear clinical and pathological evidence of disease progression. DMBA, when applied to the hamster cheek pouch, induces a high frequency of oral lesions that closely resemble human OSCC. This model has proven particularly useful for studying molecular mechanisms and assessing potential treatments. The sequential study of tumor development in these models—from hyperplasia to dysplasia to invasive carcinoma—closely mirrors the natural progression of oral cancer in humans (Wani et al. 2001). To validate the chemically induced models, researchers employ a variety of experimental techniques that enable them to analyze tumor progression at the molecular level. IHC is a key technique used to examine tissue samples, enabling the visualization of protein expression patterns, signaling pathway activation, and TME characterization. Through IHC, researchers can identify critical tumor markers, assess the activation of molecular pathways, and understand the signaling mechanisms that contribute to OSCC progression. Commonly dysregulated pathways, such as the EGFR and Akt signaling pathways, are often examined through IHC and Western blotting, allowing for the quantification of protein expression levels and the identification of signaling molecules that drive cancer progression (Wee and Wang 2017) providing valuable insights into the molecular mechanisms driving the disease. In addition to IHC, molecular assays, including quantitative PCR and gene expression profiling, are essential for studying the genetic changes involved in OSCC progression. These assays allow researchers to detect overexpression of oncogenes or silencing of TSGs, identifying potential biomarkers for early detection and therapeutic monitoring is also crucial (Sasahira et al. 2022). For example, studies have shown that environmental factors like ethanol consumption can accelerate OSCC development by activating key molecular pathways, such as EGFR and Akt, which promote tumor growth and metastasis (Singh and Singh 2022). These findings emphasize the importance of chemically induced models for studying how lifestyle and environmental factors can influence oral cancer development. Transgenic mouse models have further advanced oral cancer research. These genetically engineered models replicate the genetic alterations observed in human HNSCC. For example, transgenic mice with mutations in TSGs such as p53 or oncogenes like EGFR exhibit OSCC development similar to that seen in humans. Transgenic models provide researchers with a powerful tool for studying specific genetic alterations that drive oral cancer, allowing for the identification of molecular pathways involved in tumor formation and metastasis (Tasoulas et al. 2023). Moreover, these models facilitate the development of targeted therapeutic strategies by enabling researchers to manipulate the expression of specific genes and investigate their roles in OSCC. *In vitro* tools complement the findings from animal models, offering additional platforms for studying tumor biology, drug efficacy, and therapeutic target identification. Cell lines derived from transgenic mouse models or chemically induced tumors are frequently used in laboratory settings to test potential drugs or interventions. These *in vitro* systems provide controlled environments for high-throughput screening, allowing researchers to evaluate the effects of various compounds on cancer cell proliferation, migration, and invasion. *In vitro* models also help researchers study the molecular mechanisms of drug resistance, providing a platform for developing more effective cancer treatments (Onaciu et al. 2020). The hamster cheek pouch model, specifically, has played a significant role in studying OSCC. It offers a detailed examination of tumor progression from dysplasia to invasive carcinoma, allowing researchers to explore the role of environmental factors, such as tobacco and alcohol, in promoting oral cancer. Additionally, this model has been used to assess the effectiveness of chemopreventive agents and therapeutic strategies. It provides valuable data for understanding the TME and how it influences tumor growth, invasion, and metastasis (Monti-Hughes et al. 2015).

7.5.2 Xenograft Models

Xenograft models, in which human cancer tissues are implanted into immunodeficient mice, are invaluable for the study of oral cancer. These models have emerged as powerful platforms for studying OSCC, offering critical insights into tumor biology, therapeutic strategies, and molecular mechanisms. Orthotopic xenograft models provide a more accurate representation of the tumor's native microenvironment compared to subcutaneous models. In these models, human cancer cells are implanted directly into the target organ, specifically the oral cavity or tongue of immunodeficient mice. This setup is particularly valuable for studying the invasion patterns, metastatic potential, and interactions between the tumor and its surrounding tissues. As the tumor grows in its natural location, orthotopic xenografts more closely mimic the biology of human cancers, including the complex interactions between tumor cells and host tissues such as the epithelium, stroma, and vasculature. These models are essential for investigating the invasiveness and metastatic behavior of tumors, which are key characteristics of OSCC. Other researchers have utilized orthotopic xenografts to study the invasion and metastasis of OSCC, providing insights into cancer cell invasion into surrounding tissues and the development of lymphatic and distant metastases, often the cause of poor prognosis in OSCC patients. This model is particularly significant for studying therapies targeting tumor invasiveness and metastasis. One major advantage of orthotopic xenografts is their ability to replicate intricate tumor–stroma interactions that influence tumor behavior. However, these models are more technically challenging to establish and require more invasive procedures to implant the tumor cells. Additionally, the proximity to vital structures in the oral cavity raises concerns regarding animal welfare and complications in tumor progression monitoring (Stribbling, Beach, and Ryan 2024). Patient-derived xenograft (PDX) models are regarded as one of the most accurate representations of human cancer biology. In these models, tumor fragments from patients are directly implanted into immunodeficient mice, where they grow and retain the original tumor's genetic and histological characteristics. These models preserve the heterogeneity of the patient's tumor, which is critical for studying cancer biology and therapeutic responses. PDX models are especially valuable for investigating personalized medicine, as they allow researchers to test the efficacy of various treatments on patient-specific tumor models (Liu et al. 2023). They are also instrumental in identifying biomarkers of drug response and resistance, making them a powerful tool for translational cancer research. In oral cancer studies, PDX models provide a more accurate representation of the genetic and molecular characteristics of patient tumors. They have shown strong predictive value for assessing the effectiveness of different drug therapies and have been used to explore mechanisms of drug resistance, a major challenge in OSCC treatment (Janitri et al. 2024). PDX models preserve the original tumor architecture and tumor-host interactions, which are often lost in cell line-based models, making them more physiologically relevant for evaluating therapeutic strategies. Due to their patient-specific nature, PDX models enable the development of more personalized treatment approaches. However, these models present challenges, such as the high cost of generation and maintenance, along with the relatively slow growth rate of certain patient tumors in mice. Whole-exome and RNA sequencing of PDXs have revealed key biomarkers like matrix metalloproteinase-1 (MMP1), which predicts successful engraftment and tumor progression, helping to identify therapeutic targets and understand individual tumor characteristics (Roy, Yang, and Moses 2009). Xenograft models have also been utilized to investigate various therapeutic compounds and their effects on oral cancer. Studies on compounds like Apaziquone have demonstrated significant tumor growth reduction and apoptosis induction in OSCC cells, showcasing the model's potential for evaluating novel treatments (Srivastava et al. 2015). Molecular studies using xenografts have explored diverse therapeutic strategies. Semaphorin 3A (Sema3A) has been found to suppress tumor growth by reducing tumor size, inhibiting cell proliferation, and decreasing angiogenesis. Similarly, honokiol, a natural compound, has shown promise in targeting oral cancer stem cells by suppressing their self-renewal and reducing tumor size (Subramaniam et al. 2021). Subcutaneous xenografts are one of the most commonly used xenograft models in cancer research. In these models, human cancer cells are injected into the subcutaneous tissue of immunodeficient mice, usually in the flank region, where tumors can be easily monitored. The primary advantage of this approach is the ability to observe tumor growth and response to treatments due to the superficial location of the implanted tumor (Obaid and Hasan 2022). Tumor size is measured regularly, providing quantitative data that can be used to assess the efficacy of novel therapeutic interventions. Subcutaneous xenografts are widely employed in preclinical studies to evaluate the effectiveness of new drugs and targeted therapies. In the context of oral cancer, subcutaneous xenografts have been successfully used to study the treatment of OSCC. Researchers have demonstrated the use of subcutaneous xenografts to evaluate the response of OSCC to novel targeted therapies, showing that these models offer a reliable and reproducible means of assessing the therapeutic potential of specific molecular targets in OSCC. Subcutaneous xenografts are particularly beneficial for screening large numbers of potential therapeutic agents in a relatively short period, making them essential in early drug development stages. Despite their advantages, subcutaneous xenografts do not replicate the complex TME of the oral cavity, particularly the tumor's interactions with surrounding tissues such as the tongue, mucosa, and vasculature, which are crucial for understanding the full range of tumor behavior (Obaid and Hasan 2022).

7.5.3 Transgenic Mouse Models

Transgenic mouse models have significantly advanced the understanding of oral cancer by enabling the study of specific genetic alterations and their roles in tumor development. These models have become pivotal in oral cancer research, providing sophisticated platforms for studying mechanisms behind oral cancer initiation, progression, metastasis, and therapy resistance. The ability to introduce specific genetic alterations into mice allows researchers to simulate human cancer more accurately, exploring aspects of cancer biology that would be impossible or unethical to investigate directly in humans (Luo et al. 2018). In oral cancer research, transgenic mouse models are essential for understanding underlying molecular mechanisms and developing targeted therapies. These models involve the manipulation of the mouse genome to induce or mimic genetic alterations found in human cancers. Techniques such as retroviral infection and advanced genome-editing technologies like CRISPR/Cas9 have enabled researchers to replicate specific genetic mutations, facilitating the study of cancer initiation, progression, and therapy response in controlled environments. Lamprecht et al. (2018) emphasized the critical role of transgenic models in unraveling cancer biology, offering insights into mechanisms that drive oral cancer and identifying potential therapeutic targets (Lamprecht et al. 2018). A critical feature of transgenic mouse models is their ability to closely replicate human OSCC, both genetically and histopathologically. In 2019, four mouse oral carcinoma cell lines—MOC-L1 to MOC-L4—were established from tongue tumors in transgenic mice. These cell lines, retaining many genetic and histological features of human OSCC, offer a unique platform for studying tumor biology and the progression of OSCC from early-stage lesions to invasive and metastatic tumors (Chen et al. 2019). These cell lines have been used to investigate various aspects of OSCC, including the molecular signaling pathways driving tumor growth and metastasis. The manipulation of specific genes or pathways known to be involved in OSCC has provided valuable insights into the role of these genetic alterations in cancer development. This approach is essential for identifying potential biomarkers for early detection and prognosis, as well as for developing targeted therapies to improve patient outcomes (Tan et al. 2023). Transgenic mouse models employ a range of techniques to comprehensively evaluate tumor progression. Histological analysis is widely used to examine tissue samples and understand molecular and cellular changes during cancer progression. Through histological staining and IHC, researchers can visualize tumor markers, alterations in tissue architecture, and changes in protein expression. For example, overexpression of EGFR and cyclin D1, commonly seen in OSCC, can be monitored in transgenic models to study their contribution to tumor growth and metastasis (Tratar, Horvat, and Cemazar 2018). Molecular assays, such as PCR and gene expression profiling, are critical for understanding genetic alterations that drive oral cancer. Analyzing the expression of specific genes involved in tumor progression, metastasis, and drug resistance provides insights into the molecular mechanisms underlying oral cancer. Gene expression profiling can identify signature genes that are differentially expressed in OSCC, helping to pinpoint potential therapeutic targets (Deshmukh et al. 2021). Transgenic models can also be subjected to molecular assays to study the effects of specific mutations or gene knockouts on oral cancer progression, informing the development of targeted therapies. Gene expression profiling is particularly valuable for investigating molecular pathways involved in OSCC, such as those regulating cell proliferation, apoptosis, and angiogenesis (Nakagaki et al. 2017). It also provides insights into the molecular mechanisms of metastasis, a major cause of poor prognosis in OSCC patients. Transgenic mouse models have become essential tools in oral cancer research, offering invaluable insights into the genetic and molecular mechanisms driving tumor progression. These models use advanced techniques like histological analysis, IHC, molecular assays, and gene expression profiling to thoroughly evaluate tumor biology and assess therapeutic interventions. Histological and immunohistochemical techniques examine tissue architecture and protein expression, while molecular assays and gene expression profiling help researchers understand genetic alterations contributing to cancer development and metastasis (Tasoulas et al. 2023). Various implantation strategies, including orthotopic, heterotopic, and autochthonous models, replicate different aspects of tumor progression. Orthotopic models, where cancer cells are implanted directly into the oral cavity, closely mimic the TME, allowing for the study of tumor growth, invasion, and metastasis (Gelbard et al. 2008). Heterotopic models, which involve implanting cancer cells into sites outside the oral cavity, such as the subcutaneous space, offer easier monitoring but are less representative of the natural oral cancer environment. Autochthonous models, where tumors develop directly within the mouse, provide unique insights into the genetic and molecular pathways driving oral cancer initiation and progression (Gelbard et al. 2008). Therapeutic interventions in transgenic models are typically evaluated 40 weeks after carcinogen treatment to ensure tumors have reached an appropriate size and stage for assessment. This long-term approach enables the study of therapy efficacy and its impact on tumor growth and metastasis. These models are crucial for identifying new therapeutic targets and assessing the effectiveness of drug candidates in preclinical settings, offering valuable data that can inform clinical development (Liu et al. 2024).

7.5.4 Zebrafish Models

Zebrafish models have emerged as powerful and versatile platforms for studying oral cancer, offering unique advantages in research and drug discovery. These models have gained increasing popularity in recent years due to their cost-effectiveness, high-throughput capabilities, and remarkable ability to replicate various aspects of human cancer biology. By providing insights into cancer progression, metastasis, and treatment responses, zebrafish models have reshaped the landscape of oral cancer research (Hason and Bartůněk 2019). A standout feature of zebrafish models is their transparency, particularly during the embryonic and larval stages. This transparency allows for real-time, noninvasive observation of tumor development, enabling researchers to monitor cellular behaviors such as cancer cell proliferation, invasion, and metastasis. Techniques such as fluorescence microscopy, quantitative PCR (qPCR), and droplet digital PCR (ddPCR) can be employed to analyze these processes with remarkable precision, providing valuable data in a short timeframe (Hason and Bartůněk 2019). Zebrafish models are also highly efficient for high-throughput studies. Their small size, rapid development, and high fecundity make them ideal for large-scale experiments. These features are particularly beneficial in drug discovery, where researchers can test multiple therapeutic agents simultaneously to evaluate their efficacy and toxicity. Moreover, zebrafish embryos and larvae are permeable to small molecules, allowing drugs to be directly administered into their aquatic environment, thereby streamlining the drug testing process (Hason and Bartůněk 2019). Zebrafish models excel in personalized medicine applications, particularly in the study of HNSCC. By injecting patient-derived cancer cells into zebrafish, researchers create patient-specific xenograft models, which enable rapid evaluation of therapeutic efficacy and provide critical insights into how individual tumors respond to specific treatments. This application has proven beneficial, especially when patient-derived material is limited. It bridges the gap between laboratory research and clinical practice, offering a platform for tailoring treatments to individual patients. The evolutionary similarity of zebrafish to humans in cancer-related biological processes further enhances their translational relevance, making them a valuable tool in developing targeted therapies for oral cancer (Al-Samadi et al. 2019). The zebrafish larvae xenograft model has been increasingly applied as a promising *in vivo* tool for cancer research, particularly in drug testing against patient-derived tumors. This model offers several advantages, including a large number of offspring, small size (able to fit in a 96-well plate), short experimental duration, low cost, and the possibility of completing high-throughput testing. These benefits have encouraged researchers to use zebrafish larvae in experiments. Until recently, the zebrafish larvae model has primarily been used to study cancer cell proliferation, metastasis, tumor angiogenesis, and drug testing (Chen et al. 2021). Despite these positive findings, there are still some limitations, particularly regarding the drug response evaluation technique. Currently, imaging is the primary method used to assess tumor size and metastasis. However, imaging techniques are time-consuming and limit the number of zebrafish that can be tested, making it difficult to conduct high-throughput screenings. Published reports of drug testing in zebrafish larvae are currently limited to testing three or four drugs (Chen et al. 2021). Additionally, the zebrafish larvae xenograft model has been used to investigate the effects of sandensolide, an extract from the herb *Sinularia flexibilis*. Sandensolide was shown to induce apoptosis in human OSCC cells, supporting its potential as an adjunctive therapy for oral cancer treatment. These findings contribute to the growing body of evidence regarding the application of zebrafish larvae models in oral cancer research and drug discovery (Yu et al. 2018).

7.5.5 Emerging Technologies in *In vivo* Models

In vivo microbiome studies for oral cancer research have advanced significantly, integrating cutting-edge technologies to explore the intricate interplay between microbial communities and cancer biology. These studies aim to uncover how bacterial populations influence cancer initiation, progression, and response to therapies, ultimately contributing to the development of novel diagnostic and therapeutic approaches. One significant advancement is the use of nanoparticle-based imaging and therapeutics, which enable precise molecular visualization and targeted drug delivery in experimental models. These nanoparticles allow researchers to monitor bacterial interactions with cancer cells and track cancer progression at the cellular level. This technology provides detailed, dynamic insights into the role of microbiota in oral cancer development, offering the potential for more effective and targeted therapeutic strategies (Al-Thani et al. 2024). Another groundbreaking technology is organs-on-chips, which are microfluidic devices that replicate the complex microenvironment of the oral cavity. These miniaturized platforms simulate the physical and biochemical conditions of the human oral mucosa, allowing controlled investigation of bacterial-host interactions and cancer stem cell dynamics. The precision offered by organs-on-chips enables researchers to study how microbial populations affect tumorigenesis and test potential therapeutic interventions under physiologically relevant conditions. This technology bridges the gap between *in vitro* and *in vivo* studies, enhancing the translatability of research findings (Ahmed 2022). The integration of multiomics approaches, including genomics, transcriptomics, and proteomics, has further revolutionized microbiome research in oral cancer. High-throughput sequencing techniques, such as 16S rDNA sequencing, enable comprehensive analysis of microbial community structures and their shifts during cancer progression. These techniques reveal intricate changes in bacterial composition, uncovering potential biomarkers and therapeutic targets that traditional methods might overlook. For example, studies using multiomics approaches have identified specific bacterial taxa associated with increased oral cancer risk, shedding light on the potential mechanisms by which the microbiome contributes to carcinogenesis (Hemmati et al. 2024). Advanced *in vivo* models have also evolved to incorporate complex bacterial communities, allowing for more accurate simulation of the human oral microbiome. These models enable the observation of how specific microbial populations influence various stages of cancer, from initiation to metastasis. Researchers utilize techniques such as fluorescence microscopy and genetic profiling to delve deeper into the molecular mechanisms by which bacteria contribute to oral carcinogenesis. These tools provide valuable insights into how bacterial metabolites, toxins, and signaling molecules interact with host cells to promote or inhibit tumor growth (Radaic and Kapila 2021).

7.6 *IN SILICO* MODELING APPROACHES FOR ORAL CANCER

In silico approaches have revolutionized oral cancer research by providing cost-effective, time-efficient, and high-throughput analysis of complex biological data. These methods, which leverage computational biology, artificial intelligence (AI), and bioinformatics, play a crucial role in decoding the genetic and molecular basis of oral cancer. By utilizing high-throughput sequencing technologies like whole-genome sequencing (WGS) and RNA sequencing (RNA-seq), researchers can identify key mutations in genes such as TP53, PIK3CA, and NOTCH1, which drive OSCC (Radaic and Kapila 2021). Computational modeling and network-based analyses further help map interactions between oncogenes and tumor suppressors, facilitating novel biomarker discoveries for early diagnosis and targeted therapy. *In silico* drug discovery techniques, including molecular docking, virtual screening, and molecular dynamics (MD) simulations, accelerate the identification of potential drugs and predict drug-target interactions, leading to more effective therapies with fewer side effects. Furthermore, AI-driven predictive models integrate genomic, transcriptomic, and clinical data to personalize treatment plans based on individual genetic heterogeneity, while computational modeling of TMEs advances immunotherapy research. AI and machine learning (ML) also enhance diagnostic accuracy by analyzing histopathology slides and radiographs to detect early cancerous changes, alongside bioinformatics-driven biomarker discovery that identifies circulating tumor DNA (ctDNA) and microRNAs for minimally invasive diagnostics. Mathematical and computational models simulate tumor growth and treatment responses, providing insights into angiogenesis, chemotherapy, and radiotherapy, while AI-driven predictive analytics optimize treatment regimens, aiding clinical decision-making (Bi et al. 2019).

7.6.1 Bioinformatics and Computational Tools

Bioinformatics and computational tools were regarded as indispensable in oral cancer research, aiding in gene expression analysis, mutational profiling, and network pathway mapping. Various bioinformatics resources, such as The Cancer Genome Atlas (TCGA), Gene Expression Omnibus (GEO), and Bioconductor, were utilized to facilitate differential gene expression studies, enabling the identification of key biomarkers associated with oral cancer progression. Whole-genome and transcriptome sequencing data were processed using algorithms like DESeq2, edgeR, and Mutect2 to detect mutations playing a role in tumorigenesis (Rosati et al. 2024). Additionally, pathway and network analysis tools such as Cytoscape, STRING, and KEGG were employed to understand the interactions between genes and proteins in cancer development. These approaches were found to have provided significant insights into the molecular underpinnings of oral cancer, paving the way for targeted therapeutic interventions. Computational modeling techniques were also applied to enhance the understanding of tumor growth and treatment responses. Agent-based and multiscale models, implemented using simulators like PhysiCell and BioFVM, were employed to replicate tumor growth dynamics and TME interactions. Radiomics and computational imaging, employing tools such as PyRadiomics and 3D Slicer, were utilized to extract quantitative features from CT, MRI, and PET scans, enabling precise treatment response predictions. These advancements were found to have underscored the potential of *in silico* approaches in improving therapeutic strategies for oral cancer patients (Wang et al. 2015). In personalized medicine, integrative multi-omics analysis was performed to tailor treatment strategies based on individual genetic and clinical profiles. By combining genomics, proteomics, and transcriptomics data, personalized therapeutic targets were identified, ensuring more effective treatment plans. Recent research highlighted the effectiveness of *in silico* approaches in oral cancer studies. For instance, UBE2D1 was identified as an upregulated gene in OSCC using RNA-Seq data analysis (Soni et al. 2024). DESeq2 was employed for differential gene expression analysis, and Cytoscape for protein-protein interaction network analysis. The research concluded that naturally occurring compounds such as Epigallocatechin 3-gallate, Quercetin, Luteolin, Curcumin, and Baicalein could serve as potent inhibitors of UBE2D1, offering promising therapeutic avenues for OSCC treatment (Soni et al. 2024). Similarly, a comprehensive bioinformatics analysis was conducted to identify biomarkers for HNSCC (Cheon et al. 2024). By leveraging the GSE58911 dataset from GEO, 605 differentially expressed genes (DEGs) were identified. Database cross-referencing through GeneCards and the Comparative Toxicogenomics Database (CTD) was performed to validate key biomarkers. Functional pathway analyses and The Cancer Genome Atlas (TCGA) validation were used to confirm the significance of SERPINE1, PLAU, PLAUR, and SERPINB2 in HNSCC. Notably, SERPINE1 was found to promote tumor progression *in vitro*, establishing its potential as a diagnostic biomarker (Wang et al. 2015). The role of deep learning in biomarker discovery was demonstrated by employing an autoencoder-based framework to analyze gene expression profiles from the GSE26549 dataset. Twenty-one key genes, including FUT8, DDR2, ATM, CD247, and ETS1, were identified as associated with oral cancer progression. Hierarchical clustering and Random Forest classifiers were used to stratify patients into high-risk and low-risk groups, emphasizing the predictive power of deep learning models in oral cancer prognosis (Unger et al. 2025). The integration of technology-driven strategies in oral cancer research was emphasized, highlighting the role of omics data and bioinformatics. The review underscored salivaomics as an emerging noninvasive diagnostic approach, leveraging salivary biomarkers for early cancer detection. Bioinformatics tools were acknowledged for their contribution to processing and interpreting large-scale omics data, thereby facilitating the discovery of novel therapeutic targets (Satish et al. 2024).

7.6.2 Molecular Docking and Drug Discovery

Recent research has emphasized the importance of *in silico* molecular docking techniques in identifying potential therapeutic compounds for oral cancer, specifically OSCC. Computational methods such as network pharmacology, molecular docking, and MD simulations have been pivotal in evaluating both natural and synthetic compounds targeting oncogenic pathways in OSCC. These approaches have facilitated the identification of drug candidates with strong binding affinities to cancer-related proteins, offering a foundation for novel therapeutic interventions. Molecular docking and virtual screening are essential in drug discovery, enabling predictions of interactions between drug molecules and cancer-related targets. Software such as AutoDock, Schrödinger, and GOLD has been widely used to perform these docking studies. Additionally, MD simulations using tools like GROMACS and AMBER assess the stability of drug-target complexes, enhancing drug design accuracy. Further, pharmacokinetics and toxicity predictions using ADMET models such as SwissADME, pkCSM, and GastroPlus allow for refining drug candidates before clinical trials (Agu et al. 2023; Sharma et al. 2022). These computational techniques streamline the drug discovery process, reducing time and costs compared to traditional experimental methods. One significant example of computational drug design is the investigation of ursolic acid as a potential treatment for OSCC. Network pharmacology was used to identify potential targets of ursolic acid and related genes in OSCC, using databases like GeneCards, DrugBank, and the Therapeutic Target Database (Huang et al. 2024). This analysis identified 21 common targets between ursolic acid and OSCC, and a protein-protein interaction (PPI) network was constructed to understand these relationships. Molecular docking studies revealed strong binding affinities between ursolic acid and key proteins such as STAT3, RELA, and NFKB1, with binding energies of −8.1 and −7.3 kcal/mol, respectively. These findings were confirmed through *in vitro* experiments, which showed that ursolic acid induced apoptosis and inhibited cell migration in OSCC cells in a dose-dependent manner, demonstrating its anticancer potential (Zhu, Li, and Dong 2025). Another study explored the role of bacterial pathogens in OSCC progression, specifically targeting the FimA protein of *Porphyromonas gingivalis*, a bacterium linked to oral cancer. High-throughput virtual screening (HTVS) was used to identify natural compounds that could inhibit

FimA (Zhu, Li, and Dong 2025). Ligand selection was based on drug-likeness filters and molecular docking scores, followed by MD simulations to assess the stability of the ligand-FimA complexes. Several compounds with high binding affinity for FimA were identified, suggesting a new avenue for disrupting bacterial-mediated carcinogenesis in OSCC (Meng et al. 2011). In addition, a study by Shankar et al. (2023) in *Organic & Biomolecular Chemistry* investigated the synthesis and anticancer potential of novel succinimide derivatives. The researchers synthesized (E)-3-(β-C-glyc osylmethylidene)-N-aryl/alkyl succinimides using a green chemistry approach and characterized them using NMR and mass spectrometry (Hussain et al. 2019; Shankar et al. 2023). Molecular docking studies showed that over fifteen derivatives exhibited stronger binding interactions with the E6 oncoprotein of high-risk HPV16 than the standard anticancer drug 5-FU. These findings provide a basis for further *in vitro* and *in vivo* studies to validate the efficacy of these compounds in OSCC therapy (Dunjic et al. 2024). A separate study by Alqarni et al. (2024) focused on the link between breast cancer and OSCC metastasis, identifying a potential natural inhibitor from *Fucus sargassaceae*. Using network pharmacology and molecular docking, the researchers identified key metastasis-related proteins linking breast cancer to OSCC and selected isonahocol D2 as a potential inhibitor. ADMET analysis confirmed favorable pharmacokinetic properties, making this compound a promising candidate for further preclinical validation. These findings highlight the potential of natural compounds in disrupting metastasis pathways in OSCC (Alqarni et al. 2024). Using molecular docking, the study identified luteolin as a promising compound with strong binding affinity for Cyclin D1, a key protein in OSCC progression. ADMET analysis further supported its potential as a therapeutic agent, demonstrating favorable pharmacokinetic properties. This highlights the role of phytochemicals as natural therapeutic agents in OSCC treatment and the need for further validation through experimental studies (Priyadharshini et al. 2024). Collectively, these studies underscore the significant role of *in silico* molecular docking techniques in drug discovery for OSCC. By integrating computational approaches such as network pharmacology, molecular docking, and MD simulations, researchers have identified several promising drug candidates with strong binding affinities for cancer-related proteins. The combination of computational and experimental validation provides a comprehensive strategy for developing new therapeutic options for OSCC. These findings emphasize the need for additional preclinical and clinical studies to confirm the efficacy of these compounds and explore their potential for personalized medicine in oral cancer treatment.

7.7 MACHINE LEARNING AND AI IN CANCER DETECTION

AI and ML are revolutionizing medical diagnostics, particularly in the detection, classification, and prognosis of oral cancer. By integrating AI-driven tools into clinical workflows, diagnostic accuracy has improved, early detection has been facilitated, and predictive insights into disease progression have been achieved. A study on the application of ML techniques like Support Vector Machine (SVM), XGBoost, and Random Forest (RF) to classify oral cancer stages used patient oral images (Agarwal et al. 2024). The study focused on extracting features such as color, texture, and shape from images to train models. Among the techniques, RF achieved the highest accuracy of 87.1%, surpassing others in performance (Liu, Pu, and Sun 2021).

This highlights the potential of ML algorithms, particularly RF, in supporting clinical decision-making for noninvasive, efficient oral cancer diagnosis. Additionally, a comprehensive review examined AI models used to detect OSCC through oral photographs. The study demonstrated that AI could detect OSCC with high precision, aiding clinicians in identifying cancerous lesions early. However, it pointed out the need for standardized datasets and preprocessing techniques to improve model generalizability (Liu, Pu, and Sun 2021). A novel approach explored AI integration with Raman histology techniques for OSCC detection. In a study using Stimulated Raman Scattering (SRS) microscopy, AI distinguished between cancerous and normal tissues with remarkable accuracy. SRS microscopy offers high-resolution, real-time tissue visualization, making it ideal for cancer diagnostics. By combining SRS with AI, the method enables rapid, noninvasive detection, reducing reliance on traditional biopsy methods, which are invasive and time-consuming (Weber et al. 2024). Beyond detection, ML models have been applied to predict the recurrence of oral tongue squamous cell carcinoma (OTSCC). A study utilizing data from the SEER database developed ML models to forecast recurrence risks in OTSCC patients. These models achieved high accuracy in predicting 5- and 10-year recurrence, allowing for more personalized treatment plans and proactive patient management (Fatapour et al. 2023). This predictive capability helps oncologists tailor interventions and improve patient outcomes. AI also aids in treatment optimization by assisting in radiation therapy planning and predicting patient responses to chemotherapy. AI-driven models analyze large patient datasets to identify treatment regimens most likely to succeed, minimizing adverse effects and improving outcomes. Additionally, AI-based natural language processing (NLP) tools extract insights from electronic health records (EHRs), facilitating personalized cancer care. Despite the progress, challenges remain. One major issue is the interpretability of AI models, as their decision-making processes are often viewed as a "black box." Efforts are underway to develop explainable AI (XAI) models that provide transparent and interpretable predictions, helping clinicians understand AI-driven decisions. Ethical and regulatory concerns also persist. AI models must undergo rigorous validation and regulatory approval to ensure safety and efficacy in patient care (Abdul Rasool Hassan et al. 2025). Bias in training datasets can lead to disparities in diagnosis accuracy across diverse populations, making it essential to ensure dataset diversity and apply bias-mitigation techniques. The future of AI and ML in oral cancer research is promising. Advances in deep learning, particularly convolutional neural networks (CNNs) and transformer models, are expected to further improve cancer detection. The integration of multimodal AI, combining imaging data with genomic and proteomic information, could provide a comprehensive understanding of oral cancer biology, accelerating research and improving diagnostic capabilities (Norori et al. 2021).

7.8 CONCLUSION

OSCC, a major subtype of head and neck cancers that remains a significant global health concern, particularly in low- and middle-income countries. Understanding the multifactorial etiology and molecular mechanisms underlying OSCC is critical for developing more effective diagnostic, therapeutic, and preventive strategies. A range of experimental models have been developed to explore the biology of oral cancer and test new treatments. *In vitro* models, including 2D and advanced 3D cultures (e.g., spheroids, organoids, and tumor-on-a-chip platforms), provide insight into tumor behavior and enable drug screening in a controlled environment. These models are crucial for studying cell–cell and cell–matrix interactions and mimicking the TME more accurately than conventional monolayers. *In vivo* models, including chemically induced tumors, xenografts, and genetically engineered mice, remain fundamental for studying tumor progression and therapeutic responses in complex biological systems. In recent years, zebrafish models have emerged as a valuable alternative due to their transparency, rapid development, cost-effectiveness, and suitability for high-throughput screening. Zebrafish larvae xenografts, in particular, have shown promise in evaluating drug efficacy using patient-derived cells, thereby contributing to advances in personalized medicine. *In silico* approaches using bioinformatics and computational modeling complement these experimental systems by predicting disease behavior and drug interactions, thereby streamlining research efforts. Collectively, these *in vitro, in vivo*, and *in silico* models offer complementary insights and are integral to advancing our understanding of OSCC. Continued innovation and integration of these models will be key to identifying novel biomarkers, optimizing therapies, and ultimately improving patient outcomes in oral cancer.

REFERENCES

Aazmi, Abdellah, Duo Zhang, Corrado Mazzaglia, Mengfei Yu, Zhen Wang, Huayong Yang, Yan Yan Shery Huang, and Liang Ma. 2024. "Biofabrication Methods for Reconstructing Extracellular Matrix Mimetics." *Bioactive Materials* 31: 475–496. doi:10.1016/J.BIOACTMAT.2023.08.018.

Abdul Rasool Hassan, Bassam, Ali Haider Mohammed, Souheil Hallit, Diana Malaeb, and Hassan Hosseini. 2025. "Exploring the Role of Artificial Intelligence in Chemotherapy Development, Cancer Diagnosis, and Treatment: Present Achievements and Future Outlook." *Frontiers in Oncology* 15: 1475893. doi:10.3389/FONC.2025.1475893.

Abuwatfa, Waad H., William G. Pitt, and Ghaleb A. Husseini. 2024. "Scaffold-Based 3D Cell Culture Models in Cancer Research." *Journal of Biomedical Science* 31(1): 1–39. doi:10.1186/S12929-024-00994-Y.

Agarwal, Pinky, Naman Gupta, Yashita Bharadwaj, Anju Yadav, and Pratishtha Mathur. 2024. "Oral Cancer Stage Classification Using Machine Learning." *Procedia Computer Science* 235: 3174–3180. doi:10.1016/J.PROCS.2024.04.300.

Agu, Peter Chinedu, Celestine A. Afiukwa, Obasi Orji, Ernest Ezeh, Ifeanyi Ofoke, Celestine Ogbu, Emmanuel Ugwuja, and Patrick Aja. 2023. "Molecular Docking as a Tool for the Discovery of Molecular Targets of Nutraceuticals in Diseases Management." *Scientific Reports* 13(1): 1–18. doi:10.1038/s41598-023-40160-2.

Ahmed, Tanvir. 2022. "Organ-on-a-Chip Microengineering for Bio-Mimicking Disease Models and Revolutionizing Drug Discovery." *Biosensors and Bioelectronics: X* 11: 100194. doi:10.1016/J.BIOSX.2022.100194.

Almela, Thafar, Lobat Tayebi, and Keyvan Moharamzadeh. 2021. "3D Bioprinting for In vitro Models of Oral Cancer: Toward Development and Validation." *Bioprinting (Amsterdam, Netherlands)* 22: e00132. doi:10.1016/J.BPRINT.2021.E00132.

Alqarni, Abdullah, Jagadish Hosmani, Saeed Alassiri, Ali Mosfer A. Alqahtani, and Hassan Ahmed Assiri. 2024. "A Network Pharmacology Identified Metastasis Target for Oral Squamous Cell Carcinoma Originating from Breast Cancer with a Potential Inhibitor from F. Sargassaceae." *Pharmaceuticals* 17(10): 1309. doi:10.3390/PH17101309.

Al-Samadi, Ahmed, Katja Tuomainen, Anne Kivimäki, Abdelhakim Salem, Sakhr Al-Kubati, Aini Hyytiäinen, Mataleena Parikka, et al. 2019. "PCR-Based Zebrafish Model for Personalised Medicine in Head and Neck Cancer." *Journal of Translational Medicine* 17(1): 1–6. doi:10.1186/S12967-019-1985-1/FIGURES/4.

Al-Thani, Alshayma N., Asma Ghafoor Jan, Mohamed Abbas, Mithra Geetha, and Kishor Kumar Sadasivuni. 2024. "Nanoparticles in Cancer Theragnostic and Drug Delivery: A Comprehensive Review." *Life Sciences* 352: 122899. doi:10.1016/J.LFS.2024.122899.

Altuwairgi, Othman S., Maria B. Papageorge, and H. Chris Doku. 1995. "The Cancer-Promoting Effect of N-Nitrosonornicotine Used in Combination with a Subcarcinogenic Dose of 4-Nitroquinoline-N-Oxide and 7, 12-Dimethylbenz (A) Anthracene." *Journal of Oral and Maxillofacial Surgery* 53(8): 910–913. doi:10.1016/0278-2391(95)90279-1.

American Cancer Society. 2025. *Oral Cavity & Oropharyngeal Cancer Key Statistics 2021 | American Cancer Society*. American Cancer Society, New York. https://www.cancer.org/cancer/types/oral-cavity-and-oropharyngeal-cancer/about/key-statistics.html.

Arutyunyan, Irina, Enar Jumaniyazova, Andrey Makarov, and Timur Fatkhudinov. 2023. "*In vitro* Models of Head and Neck Cancer: From Primitive to Most Advanced." *Journal of Personalized Medicine* 13(11): 1575. doi:10.3390/JPM13111575.

Badwelan, Mohammed, Hasan Muaddi, Abeer Ahmed, Kyungjun T. Lee, and Simon D. Tran. 2023. "Oral Squamous Cell Carcinoma and Concomitant Primary Tumors, What Do We Know? A Review of the Literature." *Current Oncology* 30(4): 3721. doi:10.3390/CURRONCOL30040283.

Bernareggi, Davide, Qi Xie, Briana C. Prager, Jiyoung Yun, Luisjesus S. Cruz, Timothy V. Pham, William Kim, et al. 2022. "CHMP2A Regulates Tumor Sensitivity to Natural Killer Cell-Mediated Cytotoxicity." *Nature Communications* 13(1): 1899. doi:10.1038/S41467-022-29469-0.

Bhattacharya, Anamitra, Kamare Alam, Nakka Sharmila Roy, Kulwinder Kaur, Santanu Kaity, Velayutham Ravichandiran, and Subhadeep Roy. 2023. "Exploring the Interaction between Extracellular Matrix Components in a 3D Organoid Disease Model to Replicate the Pathophysiology of Breast Cancer." *Journal of Experimental & Clinical Cancer Research : CR* 42(1): 343. doi:10.1186/S13046-023-02926-4.

Bi, Wenya Linda, Ahmed Hosny, Matthew B. Schabath, Maryellen L. Giger, Nicolai J. Birkbak, Alireza Mehrtash, Tavis Allison, et al. 2019. "Artificial Intelligence in Cancer Imaging: Clinical Challenges and Applications." *CA: A Cancer Journal for Clinicians* 69(2): 127. doi:10.3322/CAAC.21552.

Brauchle, Eva, Hannah Johannsen, Samantha Nolan, Sibylle Thude, and Katja Schenke-Layland. 2013. "Design and Analysis of a Squamous Cell Carcinoma *in vitro* Model System." *Biomaterials* 34(30): 7401–7407. doi:10.1016/J.BIOMATERIALS.2013.06.016.

Bray, Freddie, Jacques Ferlay, Isabelle Soerjomataram, Rebecca L. Siegel, Lindsey A. Torre, and Ahmedin Jemal. 2018. "Global Cancer Statistics 2018: GLOBOCAN Estimates of Incidence and Mortality Worldwide for 36 Cancers in 185 Countries." *CA: A Cancer Journal for Clinicians* 68(6): 394–424. doi:10.3322/CAAC.21492.

Bray, Freddie, Mathieu Laversanne, Hyuna Sung, Jacques Ferlay, Rebecca L. Siegel, Isabelle Soerjomataram, and Ahmedin Jemal. 2024. "Global Cancer Statistics 2022: GLOBOCAN Estimates of Incidence and Mortality Worldwide for 36 Cancers in 185 Countries." *CA: A Cancer Journal for Clinicians* 74(3): 229–263. doi: 10.3322/caac.21834

Brown, Linda Morris, David P. Check, and Susan S. Devesa. 2011. "Oropharyngeal Cancer Incidence Trends: Diminishing Racial Disparities." *Cancer Causes and Control* 22(5): 753–763. doi:10.1007/S10552-011-9748-1.

Chai, Annie Wai Yeeng, Pei San Yee, Stacey Price, Shi Mun Yee, Hui Mei Lee, Vivian K.H. Tiong, Emanuel Gonçalves, et al. 2020. "Genome-Wide CRISPR Screens of Oral Squamous Cell Carcinoma Reveal Fitness Genes in the Hippo Pathway." *ELife* 9: 1–81. doi:10.7554/ELIFE.57761.

Chaves, Patricia, María Garrido, Javier Oliver, Elisabeth Pérez-Ruiz, Isabel Barragan, and Antonio Rueda-Domínguez. 2023. "Preclinical Models in Head and Neck Squamous Cell Carcinoma." *British Journal of Cancer* 128(10): 1819. doi:10.1038/S41416-023-02186-1.

Chehelgerdi, Mohammad, Matin Chehelgerdi, Milad Khorramian-Ghahfarokhi, Marjan Shafieizadeh, Esmaeil Mahmoudi, Fatemeh Eskandari, Mohsen Rashidi, Asghar Arshi, and Abbas Mokhtari-Farsani. 2024. "Comprehensive Review of CRISPR-Based Gene Editing: Mechanisms, Challenges, and Applications in Cancer Therapy." *Molecular Cancer* 23(1): 9. doi:10.1186/S12943-023-01925-5.

Chen, Xingyu, Yongyun Li, Tengteng Yao, and Renbing Jia. 2021. "Benefits of Zebrafish Xenograft Models in Cancer Research." *Frontiers in Cell and Developmental Biology* 9: 616551. doi:10.3389/FCELL.2021.616551.

Chen, Yi Fen, Chung Ji Liu, Li Han Lin, Chung Hsien Chou, Li Yin Yeh, Shu Chun Lin, and Kuo Wei Chang. 2019. "Establishing of Mouse Oral Carcinoma Cell Lines Derived from Transgenic Mice and Their Use as Syngeneic Tumorigenesis Models." *BMC Cancer* 19(1): 281. doi:10.1186/S12885-019-5486-7.

Cheon, Jaehwan, Byoungjae Kim, Jaehyung Park, Jaemin Shin, and Tae Hoon Kim. 2024. "Unveiling Biomarkers in Head and Neck Squamous Cell Carcinoma through Bioinformatics: The Role of SPP1 and KRT78." *International Journal of Molecular Sciences* 25(22): 12062. doi:10.3390/IJMS252212062.

Close, David A., Allen Xinwei Wang, Stanton J. Kochanek, Tongying Shun, Julie L. Eiseman, and Paul A. Johnston. 2019. "Implementation of the NCI-60 Human Tumor Cell Line Panel to Screen 2260 Cancer Drug Combinations to Generate >3 Million Data Points Used to Populate a Large Matrix of Anti-Neoplastic Agent Combinations (ALMANAC) Database." *SLAS Discovery* 24(3): 242–263. doi:10.1177/2472555218812429.

Colley, H. E., V. Hearnden, A. V. Jones, P. H. Weinreb, S. M. Violette, S. Macneil, M. H. Thornhill, and C. Murdoch. 2011. "Development of Tissue-Engineered Models of Oral Dysplasia and Early Invasive Oral Squamous Cell Carcinoma." *British Journal of Cancer* 105(10): 1582–1592. doi:10.1038/BJC.2011.403.

Conway, David I., Mitana Purkayastha, and Ivor G. Chestnutt. 2018. "The Changing Epidemiology of Oral Cancer: Definitions, Trends, and Risk Factors." *British Dental Journal* 225(9): 867. doi:10.1038/SJ.BDJ.2018.922.

Corrò, Claudia, Laura Novellasdemunt, and Vivian S.W. Li. 2020. "A Brief History of Organoids." *American Journal of Physiology - Cell Physiology* 319(1): C151. doi:10.1152/AJPCELL.00120.2020.

Costea, Daniela Elena, Keerthi Kulasekara, Evelyn Neppelberg, Anne Christine Johannessen, and Olav Karsten Vintermyr. 2006. "Species-Specific Fibroblasts Required for Triggering Invasiveness of Partially Transformed Oral Keratinocytes." *American Journal of Pathology* 168(6): 1889–1897. doi:10.2353/AJPATH.2006.050843.

Dalir Abdolahinia, Elaheh, and Xiaozhe Han. 2023. "The Three-Dimensional *In vitro* Cell Culture Models in the Study of Oral Cancer Immune Microenvironment." *Cancers* 15(17): 4266. doi:10.3390/CANCERS15174266.

De Pieri, Andrea, Yury Rochev, and Dimitrios I. Zeugolis. 2021. "Scaffold-Free Cell-Based Tissue Engineering Therapies: Advances, Shortfalls and Forecast." *NPJ Regenerative Medicine* 6(1): 1–15. doi:10.1038/s41536-021-00133-3.

de Visser, Karin E., and Johanna A. Joyce. 2023. "The Evolving Tumor Microenvironment: From Cancer Initiation to Metastatic Outgrowth." *Cancer Cell* 41(3): 374–403. doi:10.1016/J.CCELL.2023.02.016.

Deshmukh, Anuja, Karthik Nagaraja Rao, Ripu Daman Arora, Nitin M. Nagarkar, Ambesh Singh, and Om Shree Shetty. 2021. "Molecular Insights into Oral Malignancy." *Indian Journal of Surgical Oncology* 13(2): 267. doi:10.1007/S13193-021-01431-4.

Driehuis, Else, Sigrid Kolders, Sacha Spelier, Kadi Lõhmussaar, Stefan M. Willems, Lot A. Devriese, Remco de Bree, et al. 2019. "Oral Mucosal Organoids as a Potential Platform for Personalized Cancer Therapy." *Cancer Discovery* 9(7): 852–871. doi:10.1158/2159-8290.CD-18-1522.

Duan, Chunyan, Mingjia Yu, Jiyuan Xu, Bo Yi Li, Ying Zhao, and Ranjith Kumar Kankala. 2023. "Overcoming Cancer Multi-Drug Resistance (MDR): Reasons, Mechanisms, Nanotherapeutic Solutions, and Challenges." *Biomedicine & Pharmacotherapy* 162: 114643. doi:10.1016/J.BIOPHA.2023.114643.

Dunjic, Momir, Stefano Turini, Lazar Nejkovic, Nenad Sulovic, Sasa Cvetkovic, Marija Dunjic, Katarina Dunjic, and Dina Dolovac. 2024. "Comparative Molecular Docking of Apigenin and Luteolin Versus Conventional Ligands for TP-53, PRb, APOBEC3H, and HPV-16 E6: Potential Clinical Applications in Preventing Gynecological Malignancies." *Current Issues in Molecular Biology* 46(10): 11136. doi:10.3390/CIMB46100661.

Ellington, Taylor D., S. Jane Henley, Virginia Senkomago, Mary Elizabeth O'Neil, Reda J. Wilson, Simple Singh, Cheryll C. Thomas, Manxia Wu, and Lisa C. Richardson. 2020. "Trends in Incidence of Cancers of the Oral Cavity and Pharynx — United States 2007–2016." *MMWR. Morbidity and Mortality Weekly Report* 69(15): 433–438. doi:10.15585/MMWR.MM6915A1.

Fan, Han, Utkan Demirci, and Pu Chen. 2019. "Emerging Organoid Models: Leaping Forward in Cancer Research." *Journal of Hematology & Oncology* 12(1): 142. doi:10.1186/S13045-019-0832-4.

Fang, Ye, and Richard M. Eglen. 2017. "Three-Dimensional Cell Cultures in Drug Discovery and Development." *SLAS Discovery* 22(5): 456–472. doi:10.1177/1087057117696795.

Fatapour, Yasaman, Arash Abiri, Edward C. Kuan, and James P. Brody. 2023. "Development of a Machine Learning Model to Predict Recurrence of Oral Tongue Squamous Cell Carcinoma." *Cancers* 15(10): 7. doi:10.3390/CANCERS15102769.

Gelbard, Alexander, Michael E. Kupferman, Samar A. Jasser, Wantao Chen, Adel K. El-Naggar, Jeffrey N. Myers, and Ehab Y. Hanna. 2008. "An Orthotopic Murine Model of Sinonasal Malignancy." *Clinical Cancer Research : An Official Journal of the American Association for Cancer Research* 14(22): 7348. doi:10.1158/1078-0432.CCR-08-0977.

Geum, Dong-Ho, Young-Chea Roh, Sang-Yong Yoon, Hyo-Geon Kim, Jung-Han Lee, Jae-Min Song, Jae-Yeol Lee, et al. 2013. "The Impact Factors on 5-Year Survival Rate in Patients Operated with Oral Cancer." *Journal of the Korean Association of Oral and Maxillofacial Surgeons* 39(5): 207. doi:10.5125/JKAOMS.2013.39.5.207.

Gorphe, Philippe. 2019. "A Comprehensive Review of Hep-2 Cell Line in Translational Research for Laryngeal Cancer." *American Journal of Cancer Research* 9(4): 644. https://pmc.ncbi.nlm.nih.gov/articles/PMC6511639/.

Hason, Martina, and Petr Bartůněk. 2019. "Zebrafish Models of Cancer—New Insights on Modeling Human Cancer in a Non-Mammalian Vertebrate." *Genes* 10(11): 935. doi:10.3390/GENES10110935.

Hemmati, Mohammad Amin, Marzieh Monemi, Shima Asli, Sina Mohammadi, Behina Foroozanmehr, Dariush Haghmorad, Valentyn Oksenych, and Majid Eslami. 2024. "Using New Technologies to Analyze Gut Microbiota and Predict Cancer Risk." *Cells* 13(23): 1987. doi:10.3390/CELLS13231987.

Hsu, Cheng Ming, Ming Yu Yang, Ming Shao Tsai, Geng He Chang, Yao Hsu Yang, Yao Te Tsai, Ching Yuan Wu, and Shun Fu Chang. 2021. "Dihydroisotanshinone I as a Treatment Option for Head and Neck Squamous Cell Carcinomas." *International Journal of Molecular Sciences* 22(16): 8881. doi:10.3390/IJMS22168881.

Huang, Xiao Long, Yan Sun, Peng Wen, Jun Cheng Pan, and Wei Yang He. 2024. "The Potential Mechanism of Ursolic Acid in the Treatment of Bladder Cancer Based on Network Pharmacology and Molecular Docking." *Journal of International Medical Research* 52(3): 03000605241234006. doi:10.1177/03000605241234006.

Hussain, Fida, Zeeshan Khan, Muhammad Saeed Jan, Sajjad Ahmad, Ashfaq Ahmad, Umer Rashid, Farhat Ullah, Muhammad Ayaz, and Abdul Sadiq. 2019. "Synthesis, in-Vitro α-Glucosidase Inhibition, Antioxidant, in-Vivo Antidiabetic and Molecular Docking Studies of Pyrrolidine-2,5-Dione and Thiazolidine-2,4-Dione Derivatives." *Bioorganic Chemistry* 91: 103128. doi:10.1016/J.BIOORG.2019.103128.

Janitri, Venkatachalababu, Kandasamy Nagarajan ArulJothi, Vijay Murali Ravi Mythili, Sachin Kumar Singh, Parteek Prasher, Gaurav Gupta, Kamal Dua, Rakshith Hanumanthappa, Karthikeyan Karthikeyan, and Krishnan Anand. 2024. "The Roles of Patient-derived Xenograft Models and Artificial Intelligence toward Precision Medicine." *MedComm* 5(10): e745. doi:10.1002/MCO2.745.

Jena, Diptismita, Bijaya K. Padhi, Quazi Syed Zahiruddin, Suhas Ballal, Sanjay Kumar, Mahakshit Bhat, Shilpa Sharma, et al. 2024. "Estimation of Burden of Cancer Incidence and Mortality in India: Based on Global Burden of Disease Study 1990–2021." *BMC Cancer* 24(1): 1278. doi:10.1186/S12885-024-13035-6/FIGURES/10.

Joseph, Lindsay J., Michael Goodman, Kristin Higgins, Rathi Pilai, Suresh S. Ramalingam, Kelly Magliocca, Mihir R. Patel, et al. 2015. "Racial Disparities in Squamous Cell Carcinoma of the Oral Tongue among Women: A SEER Data Analysis." *Oral Oncology* 51(6): 586–592. doi:10.1016/J.ORALONCOLOGY.2015.03.010.

Khalil, Ashraf, and Mark J. Jameson. 2017. "The EGFR Inhibitor Gefitinib Enhanced the Response of Human Oral Squamous Cell Carcinoma to Cisplatin *In vitro*." *Drugs in R&D* 17(4): 545. doi:10.1007/S40268-017-0204-X.

Khorsandi, Danial, Dorsa Rezayat, Serap Sezen, Rafaela Ferrao, Arezoo Khosravi, Atefeh Zarepour, Melika Khorsandi, Mohammad Hashemian, Siavash Iravani, and Ali Zarrabi. 2024. "Application of 3D, 4D, 5D, and 6D Bioprinting in Cancer Research: What Does the Future Look Like?" *Journal of Materials Chemistry B* 12(19): 4584–4612. doi:10.1039/D4TB00310A.

Lamprecht, Peter, Anja Kerstein, Sebastian Klapa, Susanne Schinke, Christian M. Karsten, Xinhua Yu, Marc Ehlers, et al. 2018. "Pathogenetic and Clinical Aspects of Anti-Neutrophil Cytoplasmic Autoantibody-Associated Vasculitides." *Frontiers in Immunology* 9: 680. doi:10.3389/FIMMU.2018.00680.

Laschke, Matthias W., and Michael D. Menger. 2017. "Life Is 3D: Boosting Spheroid Function for Tissue Engineering." *Trends in Biotechnology* 35(2): 133–144. doi:10.1016/J.TIBTECH.2016.08.004.

Li, Hua, John S. Wawrose, William E. Gooding, Levi A. Garraway, Vivian Wai Yan Lui, Noah D. Peyser, and Jennifer R. Grandis. 2014. "Genomic Analysis of Head and Neck Squamous Cell Carcinoma Cell Lines and Human Tumors: A Rational Approach to Preclinical Model Selection." *Molecular Cancer Research* 12(4): 571–582. doi:10.1158/1541-7786.MCR-13-0396.

Li, Jiao, Zhuqiang Jia, Jing Kong, Fuyin Zhang, Shimeng Fang, Xiaojie Li, Wuwei Li, et al. 2016. "Carcinoma-Associated Fibroblasts Lead the Invasion of Salivary Gland Adenoid Cystic Carcinoma Cells by Creating an Invasive Track." *PLoS One* 11(3): e0150247. doi:10.1371/JOURNAL.PONE.0150247.

Li, Qiang, Heng Dong, Guangwen Yang, Yuxian Song, Yongbin Mou, and Yanhong Ni. 2020. "Mouse Tumor-Bearing Models as Preclinical Study Platforms for Oral Squamous Cell Carcinoma." *Frontiers in Oncology* 10: 499929. doi:10.3389/FONC.2020.00212/XML/NLM.

Lin, Charles J., Jennifer R. Grandis, Thomas E. Carey, Susanne M. Gollin, Theresa L. Whiteside, Wayne M. Koch, Robert L. Ferris, and Stephen Y. Lai. 2007. "Head and Neck Squamous Cell Carcinoma Cell Lines: Established Models and Rationale for Selection." *Head and Neck* 29(2): 163–188. doi:10.1002/HED.20478.

Lin, Ruei Zhen, and Hwan You Chang. 2008. "Recent Advances in Three-Dimensional Multicellular Spheroid Culture for Biomedical Research." *Biotechnology Journal* 3(9–10): 1172–1184. doi:10.1002/biot.200700228.

Lingen, Mark W., Elliot Abt, Nishant Agrawal, Anil K. Chaturvedi, Ezra Cohen, Gypsyamber D'Souza, Jo Ann Gurenlian, et al. 2017. "Evidence-Based Clinical Practice Guideline for the Evaluation of Potentially Malignant Disorders in the Oral Cavity: A Report of the American Dental Association." *Journal of the American Dental Association* 148(10): 712–727.e10. doi:10.1016/J.ADAJ.2017.07.032.

Liu, Beilei, Hongyu Zhou, Licheng Tan, Kin To Hugo Siu, and Xin Yuan Guan. 2024. "Exploring Treatment Options in Cancer: Tumor Treatment Strategies." *Signal Transduction and Targeted Therapy* 9(1): 1–44. doi:10.1038/s41392-024-01856-7.

Liu, Yao, Hongbin Pu, and Da Wen Sun. 2021. "Efficient Extraction of Deep Image Features Using Convolutional Neural Network (CNN) for Applications in Detecting and Analysing Complex Food Matrices." *Trends in Food Science & Technology* 113: 193–204. doi:10.1016/J.TIFS.2021.04.042.

Liu, Yihan, Wantao Wu, Changjing Cai, Hao Zhang, Hong Shen, and Ying Han. 2023. "Patient-Derived Xenograft Models in Cancer Therapy: Technologies and Applications." *Signal Transduction and Targeted Therapy* 8(1): 1–24. doi:10.1038/s41392-023-01419-2.

Lobat Tayebi, and Keyvan Moharamzadeh (Eds.). 2025. *Biomaterials for Oral and Dental Tissue Engineering*. Woodhead Publishing, Cambridge. https://www.sciencedirect.com/book/9780081009611/biomaterials-for-oral-and-dental-tissue-engineering.

Loh, Qiu Li, and Cleo Choong. 2013. "Three-Dimensional Scaffolds for Tissue Engineering Applications: Role of Porosity and Pore Size." *Tissue Engineering. Part B, Reviews* 19(6): 485. doi:10.1089/TEN.TEB.2012.0437.

Luo, J. J., C. D. Young, H. M. Zhou, and X. J. Wang. 2018. "Mouse Models for Studying Oral Cancer: Impact in the Era of Cancer Immunotherapy." *Journal of Dental Research* 97(6): 683. doi:10.1177/0022034518767635.

Ma, Xin, Qin Wang, Guozheng Li, Hui Li, Shouping Xu, and Da Pang. 2023. "Cancer Organoids: A Platform in Basic and Translational Research." *Genes & Diseases* 11(2): 614. doi:10.1016/J.GENDIS.2023.02.052.

Markopoulos, Anastasios K. 2012. "Current Aspects on Oral Squamous Cell Carcinoma." *The Open Dentistry Journal* 6(1): 126–130. doi:10.2174/1874210601206010126.

Mauk, Michael G., Barry L. Ziober, Zongyuan Chen, Jason A. Thompson, and Haim H. Bau. 2007. "Lab-on-a-Chip Technologies for Oral-Based Cancer Screening and Diagnostics: Capabilities, Issues, and Prospects." *Annals of the New York Academy of Sciences* 1098: 467–475. doi:10.1196/ANNALS.1384.025.

Mehta, Geeta, Amy Y. Hsiao, Marylou Ingram, Gary D. Luker, and Shuichi Takayama. 2012. "Opportunities and Challenges for Use of Tumor Spheroids as Models to Test Drug Delivery and Efficacy." *Journal of Controlled Release : Official Journal of the Controlled Release Society* 164(2): 192. doi:10.1016/J.JCONREL.2012.04.045.

Meng, Xuan-Yu, Hong-Xing Zhang, Mihaly Mezei, and Meng Cui. 2011. "Molecular Docking: A Powerful Approach for Structure-Based Drug Discovery." *Current Computer-Aided Drug Design* 7(2): 146. doi:10.2174/157340911795677602.

Moharamzadeh, K., H. Colley, C. Murdoch, V. Hearnden, W. L. Chai, I. M. Brook, M. H. Thornhill, and S. MacNeil. 2012. "Tissue-Engineered Oral Mucosa." *Journal of Dental Research* 91(7): 642–650. doi:10.1177/0022034511435702.

Monti-Hughes, Andrea, Romina F. Aromando, Miguel A. Pérez, Amanda E. Schwint, and Maria E. Itoiz. 2015. "The Hamster Cheek Pouch Model for Field Cancerization Studies." *Periodontology* 67(1): 292–311. doi:10.1111/PRD.12066.

Mun, Siwon, Hyun Jin Lee, and Pilnam Kim. 2024. "Rebuilding the Microenvironment of Primary Tumors in Humans: A Focus on Stroma." *Experimental & Molecular Medicine* 56(3): 527. doi:10.1038/S12276-024-01191-5.

Nagesh, M., S. Gowtham, B. Bharadwaj, Mohsin Ali, Arjun Kumar Goud, and Sara Siddiqua. 2023. "Evolution of TNM Classification for Clinical Staging of Oral Cancer: The Past, Present and the Future." *Journal of Maxillofacial & Oral Surgery* 22(3): 710. doi:10.1007/S12663-023-01915-6.

Nakagaki, Takafumi, Miyuki Tamura, Kenta Kobashi, Ryota Koyama, Hisayo Fukushima, Tomoko Ohashi, Masashi Idogawa, et al. 2017. "Profiling Cancer-Related Gene Mutations in Oral Squamous Cell Carcinoma from Japanese Patients by Targeted Amplicon Sequencing." Oncotarget 8(35): 59113. doi:10.18632/ONCOTARGET.19262.

Nayak, Pallavi, Valeria Bentivoglio, Michela Varani, and Alberto Signore. 2023. "Three-Dimensional In vitro Tumor Spheroid Models for Evaluation of Anticancer Therapy: Recent Updates." *Cancers* 15(19): 4846. doi:10.3390/CANCERS15194846.

Nikonorova, Varvara G., Vladimir V. Chrishtop, Vladimir A. Mironov, and Artur Y. Prilepskii. 2023. "Advantages and Potential Benefits of Using Organoids in Nanotoxicology." Cells 12(4): 610. doi:10.3390/CELLS12040610.

Norori, Natalia, Qiyang Hu, Florence Marcelle Aellen, Francesca Dalia Faraci, and Athina Tzovara. 2021. "Addressing Bias in Big Data and AI for Health Care: A Call for Open Science." *Patterns* 2(10): 100347. doi:10.1016/J.PATTER.2021.100347.

Obaid, Girgis, and Tayyaba Hasan. 2022. "Subcutaneous Xenograft Models for Studying PDT *in vivo*." *Methods in Molecular Biology (Clifton, N.J.)* 2451: 127. doi:10.1007/978-1-0716-2099-1-10.

Onaciu, Anca, Raluca Munteanu, Vlad Cristian Munteanu, Diana Gulei, Lajos Raduly, Richard Ionut Feder, Radu Pirlog, et al. 2020. "Spontaneous and Induced Animal Models for Cancer Research." *Diagnostics* 10(9): 660. doi:10.3390/DIAGNOSTICS10090660.

Park, Su-Bin, Woo Jung, Hyung Kim, Hwa-Young Yu, Yong Kim, and Junghyun Kim. 2022. "Esculetin Has Therapeutic Potential via the Proapoptotic Signaling Pathway in A253 Human Submandibular Salivary Gland Tumor Cells." *Experimental and Therapeutic Medicine* 24(2): 11460. doi:10.3892/ETM.2022.11460.

Patriarca, Carlo, Roberto Maria Macchi, Anja K. Marschner, and Hakan Mellstedt. 2012. "Epithelial Cell Adhesion Molecule Expression (CD326) in Cancer: A Short Review." *Cancer Treatment Reviews* 38(1): 68–75. doi:10.1016/J.CTRV.2011.04.002.

Peng, Zheng, Xiaolan Lv, Hao Sun, Lina Zhao, and Shigao Huang. 2025. "3D Tumor Cultures for Drug Resistance and Screening Development in Clinical Applications." *Molecular Cancer* 24(1): 93. doi:10.1186/S12943-025-02281-2.

Priyadharshini, G., G. Sukumaran, E. Dilipan, and Pratibha Ramani. 2024. "Targeting Oral Cancer: *In silico* Docking Studies of Phytochemicals on Oncogenic Molecular Markers." *Asian Pacific Journal of Cancer Prevention* 25(6): 2069–2075. doi:10.31557/APJCP.2024.25.6.2069.

Radaic, Allan, and Yvonne L. Kapila. 2021. "The Oralome and Its Dysbiosis: New Insights into Oral Microbiome-Host Interactions." *Computational and Structural Biotechnology Journal* 19: 1335. doi:10.1016/J.CSBJ.2021.02.010.

Rivera, César, and Bernardo Venegas. 2014. "Histological and Molecular Aspects of Oral Squamous Cell Carcinoma (Review)." *Oncology Letters* 8(1): 7. doi:10.3892/OL.2014.2103.

Rosati, Diletta, Maria Palmieri, Giulia Brunelli, Andrea Morrione, Francesco Iannelli, Elisa Frullanti, and Antonio Giordano. 2024. "Differential Gene Expression Analysis Pipelines and Bioinformatic Tools for the Identification of Specific Biomarkers: A Review." *Computational and Structural Biotechnology Journal* 23: 1154. doi:10.1016/J.CSBJ.2024.02.018.

Roy, Roopali, Jiang Yang, and Marsha A. Moses. 2009. "Matrix Metalloproteinases as Novel Biomarkers and Potential Therapeutic Targets in Human Cancer." *Journal of Clinical Oncology* 27(31): 5287. doi:10.1200/JCO.2009.23.5556.

Rusling, James F. 2013. "Multiplexed Electrochemical Protein Detection and Translation to Personalized Cancer Diagnostics." *Analytical Chemistry* 85(11): 5304. doi:10.1021/AC401058V.

Saglam-Metiner, Pelin, Sultan Gulce-Iz, and Cigir Biray-Avci. 2019. "Bioengineering-Inspired Three-Dimensional Culture Systems: Organoids to Create Tumor Microenvironment." *Gene* 686: 203–212. doi:10.1016/J.GENE.2018.11.058.

Sankaranarayanan, Rengaswamy, Kunnambath Ramadas, Hemantha Amarasinghe, Sujha Subramanian, and Newell Johnson. 2015. "Oral Cancer: Prevention, Early Detection, and Treatment." *Disease Control Priorities, Third Edition (Volume 3): Cancer* 3: 85–99. doi:10.1596/978-1-4648-0349-9_CH5.

Sant, Shilpa, and Paul A. Johnston. 2017. "The Production of 3D Tumor Spheroids for Cancer Drug Discovery." *Drug Discovery Today: Technologies* 23: 27–36. doi:10.1016/j.ddtec.2017.03.002.

Sasahira, Tomonori, Miyako Kurihara-Shimomura, Yudai Shimojjukoku, Kaori Shima, and Tadaaki Kirita. 2022. "Searching for New Molecular Targets for Oral Squamous Cell Carcinoma with a View to Clinical Implementation of Precision Medicine." *Journal of Personalized Medicine* 12(3): 413. doi:10.3390/JPM12030413.

Satish, Kshreeraja S., Kamatchi Sundara Saravanan, Dominic Augustine, Ganesan Rajalekshmi Saraswathy, Sowmya Sarah, Samar Saeed Khan, Vanishri C. Haragannavar, et al. 2024. "Leveraging Technology-Driven Strategies to Untangle Omics Big Data: Circumventing Roadblocks in Clinical Facets of Oral Cancer." *Frontiers in Oncology* 13: 1183766. doi:10.3389/FONC.2023.1183766.

Sawant, Sharada, Harsh Dongre, Archana Kumari Singh, Shriya Joshi, Daniela Elena Costea, Snehal Mahadik, Chetan Ahire, et al. 2016. "Establishment of 3D Co-Culture Models from Different Stages of Human Tongue Tumorigenesis: Utility in Understanding Neoplastic Progression." *PLoS One* 11(8): 0160615. doi:10.1371/JOURNAL.PONE.0160615.

Shankar, Bhawani, Tejveer Singh, Banty Kumar, Aditi Arora, Sumit Kumar, and Brajendra K. Singh. 2023. "Solvent-Free Synthesis and in-Silico Molecular Docking Study of (E)-3-(β-C-Glycosylmethylidene)-N-Aryl/Alkyl Succinimides." *Organic & Biomolecular Chemistry* 21(47): 9398–9409. doi:10.1039/D3OB01252B.

Sharma, Yash, Amritpal Kaur, Rupa Mishra, Sharvari Kulkarni, Mausumi Bharadwaj, and Kumud Bala. 2022. "Antiproliferative Efficacy of the Antioxidant Bioactive Compounds of Defatted Seeds of Azadirachta Indica and Momordica Charantia against the Regulatory Function of Tumor Suppressor Gene Inducing Oral Carcinoma." *Journal of Biomolecular Structure and Dynamics* 41: 1–15. doi:10.1080/07391102.2022.2083015.

Singh, Suchitra, and Ajay Kumar Singh. 2022. "Porphyromonas Gingivalis in Oral Squamous Cell Carcinoma: A Review." *Microbes and Infection* 24(3): 104925. doi:10.1016/J.MICINF.2021.104925.

Sitarski, Anna M., Heather Fairfield, Carolyne Falank, and Michaela R. Reagan. 2018. "3D Tissue Engineered *in Vitro* Models of Cancer in Bone." *ACS Biomaterials Science and Engineering* 4(2): 324–336. doi:10.1021/ACSBIOMATERIALS.7B00097.

Soares, Andrey Coatrini, Juliana Coatrini Soares, Valquiria Cruz Rodrigues, Heveline Dal Magro Follmann, Lidia Maria Rebolho Batista Arantes, Ana Carolina Carvalho, Matias Eliseo Melendez, et al. 2018. "Microfluidic-Based Genosensor to Detect Human Papillomavirus (HPV16) for Head and Neck Cancer." *ACS Applied Materials and Interfaces* 10(43): 36757–36763. doi:10.1021/ACSAMI.8B14632.

Soni, Unnati, Anirudh Singh, Ramendra Soni, Sintu Kumar Samanta, Pritish Kumar Varadwaj, and Krishna Misra. 2024. "Identification of Candidate Target Genes of Oral Squamous Cell Carcinoma Using High-Throughput RNA-Seq Data and *in silico* Studies of Their Interaction with Naturally Occurring Bioactive Compounds." *Journal of Biomolecular Structure and Dynamics* 42(15): 8024–8044. doi:10.1080/07391102.2023.2242515.

Sowmya, S. V., Dominic Augustine, Shazia Mushtaq, Hosam Ali Baeshen, Heba Ashi, Reem Nabil Hassan, Mohammed Alshahrani, and Shankargouda Patil. 2024. "Revitalizing Oral Cancer Research: Crispr-Cas9 Technology the Promise of Genetic Editing." *Frontiers in Oncology* 14: 1383062. doi:10.3389/FONC.2024.1383062.

Srivastava, Gunjan, Raj Thani Somasundaram, Paul G. Walfish, and Ranju Ralhan. 2015. "Anticancer Activity of Apaziquone in Oral Cancer Cells and Xenograft Model: Implications for Oral Cancer Therapy." *PLoS One* 10(7): e0133735. doi:10.1371/JOURNAL.PONE.0133735.

Stribbling, Stephen M., Callum Beach, and Anderson J. Ryan. 2024. "Orthotopic and Metastatic Tumour Models in Preclinical Cancer Research." *Pharmacology & Therapeutics* 257: 108631. doi:10.1016/J.PHARMTHERA.2024.108631.

Subramaniam, Dharmalingam, Sivapriya Ponnurangam, Satish Ramalingam, Deep Kwatra, Prasad Dandawate, Scott J. Weir, Shahid Umar, Roy A. Jensen, and Shrikant Anant. 2021. "Honokiol Affects Stem Cell Viability by Suppressing Oncogenic YAP1 Function to Inhibit Colon Tumorigenesis." *Cells* 10(7): 1607. doi:10.3390/CELLS10071607.

Sun, Aria, Dileep Sharma, Siu Wai Choi, Poornima Ramamurthy, and Peter Thomson. 2023. "Oral Cancer in Australia: Rising Incidence and Worsening Mortality." *Journal of Oral Pathology and Medicine* 52(4): 328–334. doi:10.1111/JOP.13421.

Surappa, Sushruta, Priyanka Multani, Ugur Parlatan, Prima Dewi Sinawang, Jussuf Kaifi, Demir Akin, and Utkan Demirci. 2023. "Integrated 'Lab-on-a-Chip' Microfluidic Systems for Isolation, Enrichment, and Analysis of Cancer Biomarkers." *Lab on a Chip* 23(13): 2942. doi:10.1039/D2LC01076C.

Suryaprakash, Ravi Teja Chitturi, Omar Kujan, Kate Shearston, and Camile S. Farah. 2020. "Three-Dimensional Cell Culture Models to Investigate Oral Carcinogenesis: A Scoping Review." *International Journal of Molecular Sciences* 21(24): 1–17. doi:10.3390/IJMS21249520.

Sutera, S., F. Astolfi, A. Trapani, F. D'Amore, and G. Lodi. 2013. "Oral Squamous Cell Carcinoma Presenting as a Cervical Lymphadenopathy." *Annali Di Stomatologia* 4(Suppl 2): 42. https://pmc.ncbi.nlm.nih.gov/articles/PMC3860238/.

Tan, Yunhan, Zhihan Wang, Mengtong Xu, Bowen Li, Zhao Huang, Siyuan Qin, Edouard C. Nice, Jing Tang, and Canhua Huang. 2023. "Oral Squamous Cell Carcinomas: State of the Field and Emerging Directions." *International Journal of Oral Science* 15(1): 1–23. doi:10.1038/s41368-023-00249-w.

Tasoulas, Jason, Sonal Srivastava, Xiaonan Xu, Valentina Tarasova, Anastasios Maniakas, Florian A. Karreth, and Antonio L. Amelio. 2023. "Genetically Engineered Mouse Models of Head and Neck Cancers." *Oncogene* 42(35): 2593. doi:10.1038/S41388-023-02783-7.

Tranby, Eric P., Lisa J. Heaton, Scott L. Tomar, Abigail L. Kelly, Gulielma Leonard Fager, Mary Backley, and Julie Frantsve-Hawley. 2022. "Oral Cancer Prevalence, Mortality, and Costs in Medicaid and Commercial Insurance Claims Data." *Cancer Epidemiology, Biomarkers & Prevention* 31(9): 1849. doi:10.1158/1055-9965.EPI-22-0114.

Tratar, Ursa Lampreht, Simon Horvat, and Maja Cemazar. 2018. "Transgenic Mouse Models in Cancer Research." *Frontiers in Oncology* 8: 268. doi:10.3389/FONC.2018.00268.

Unger, Michaela, Chiara M. L. Loeffler, Laura Žigutytė, Srividhya Sainath, Tim Lenz, Julien Vibert, Andreas Mock, et al. 2025. "Deep Learning for Biomarker Discovery in Cancer Genomes." *bioRxiv*, 2025: 631471. doi:10.1101/2025.01.06.631471.

Valavanidis, Athanasios, Thomais Vlachogianni, and Konstantinos Fiotakis. 2009. "Tobacco Smoke: Involvement of Reactive Oxygen Species and Stable Free Radicals in Mechanisms of Oxidative Damage, Carcinogenesis and Synergistic Effects with Other Respirable Particles." *International Journal of Environmental Research and Public Health* 6(2): 445. doi:10.3390/IJERPH6020445.

von Bültzingslöwen, Inger, Thomas P. Sollecito, Philip C. Fox, Troy Daniels, Roland Jonsson, Peter B. Lockhart, David Wray, et al. 2007. "Salivary Dysfunction Associated with Systemic Diseases: Systematic Review and Clinical Management Recommendations." *Oral Surgery, Oral Medicine, Oral Pathology, Oral Radiology and Endodontology* 103(Suppl.): S57.e1–S57.e15. doi:10.1016/j.tripleo.2006.11.010.

Wang, Yin, and Hui Sun. 2021. "Polymeric Nanomaterials for Efficient Delivery of Antimicrobial Agents." *Pharmaceutics* 13(12): 2108. doi:10.3390/PHARMACEUTICS13122108.

Wang, Zhihui, Joseph D. Butner, Romica Kerketta, Vittorio Cristini, and Thomas S. Deisboeck. 2015. "Simulating Cancer Growth with Multiscale Agent-Based Modeling." *Seminars in Cancer Biology* 30: 70–78. doi:10.1016/j.semcancer.2014.04.001.

Wani, Manish K., Robert H. Yarber, Asif Ahmed, Arne Hengesteg, and Kevin Thomas Robbins. 2001. "Cancer Induction in the DMBA Hamster Cheek Pouch: A Modified Technique Using a Promoter." *Laryngoscope* 111(2): 204–206. doi:10.1097/00005537-200102000-00003.

Watters, Carolina, Sabrina Brar, and Tom Pepper. 2024. *Cancer of the Oral Mucosa.* StatPearls Publishing, Treasure Island, FL. https://www.ncbi.nlm.nih.gov/books/NBK565867/.

Weber, Andreas, Kathrin Enderle-Ammour, Konrad Kurowski, Marc C. Metzger, Philipp Poxleitner, Martin Werner, René Rothweiler, et al. 2024. "AI-Based Detection of Oral Squamous Cell Carcinoma with Raman Histology." *Cancers* 16(4): 689. doi:10.3390/CANCERS16040689.

Wee, Ping, and Zhixiang Wang. 2017. "Epidermal Growth Factor Receptor Cell Proliferation Signaling Pathways." *Cancers* 9(5): 52. doi:10.3390/CANCERS9050052.

Yakavets, Ilya, Samuel Jenard, Aurelie Francois, Yulia Maklygina, Victor Loschenov, Henri Pierre Lassalle, Gilles Dolivet, and Lina Bezdetnaya. 2019. "Stroma-Rich Co-Culture Multicellular Tumor Spheroids as a Tool for Photoactive Drugs Screening." *Journal of Clinical Medicine* 8(10): 1686. doi:10.3390/JCM8101686.

Yu, Chung I., Chung Yi Chen, Wangta Liu, Po Chih Chang, Chiung Wei Huang, Kuang Fen Han, In Pin Lin, Mei Ying Lin, and Chien Hsing Lee. 2018. "Sandensolide Induces Oxidative Stress-Mediated Apoptosis in Oral Cancer Cells and in Zebrafish Xenograft Model." *Marine Drugs* 16(10): 387. doi:10.3390/MD16100387.

Zhang, Xiaojun, Mazharul Karim, Md Mahedi Hasan, Jacob Hooper, Riajul Wahab, Sourav Roy, and Taslim A. Al-Hilal. 2022. "Cancer-on-a-Chip: Models for Studying Metastasis." *Cancers* 14(3): 648. doi:10.3390/CANCERS14030648.

Zhu, Junmin. 2010. "Bioactive Modification of Poly(Ethylene Glycol) Hydrogels for Tissue Engineering." *Biomaterials* 31(17): 4639. doi:10.1016/J.BIOMATERIALS.2010.02.044.

Zhu, Ling, Zhiyao Li, and Jing Dong. 2025. "Application of Network Pharmacology, Computational Molecular Docking and Experimental Techniques to Study the Anticancer Effects of Ursolic Acid on Oral Squamous Carcinoma Cells." *Indian Journal of Pharmaceutical Education and Research* 59(2): 826–840. doi:10.5530/IJPER.20250379/FEED/.

Zhuang, Pei, Yi Hua Chiang, Maria Serafim Fernanda, and Mei He. 2021. "Using Spheroids as Building Blocks Towards 3D Bioprinting of Tumor Microenvironment." *International Journal of Bioprinting* 7(4): 444. doi:10.18063/IJB.V7I4.444.

Zidarič, Tanja, Marko Milojević, Lidija Gradišnik, Karin Stana Kleinschek, Uroš Maver, and Tina Maver. 2020. "Polysaccharide-Based Bioink Formulation for 3D Bioprinting of an *In vitro* Model of the Human Dermis." *Nanomaterials* 10(4): 733. doi:10.3390/NANO10040733.

Ziober, Barry L., Michael G. Mauk, Erica M. Falls, Zongyuan Chen, Amy F. Ziober, and Haim H. Bau. 2008. "Lab-on-a-Chip for Oral Cancer Screening and Diagnosis." *Head and Neck* 30(1): 111–121. doi: 10.1002/hed.20680.

8 Liver Cancer
Epidemiology, Prevalence and Advanced Modeling Techniques

Rohit Bhardwaj, Nidhi Sharma, and Satyendra Kumar Rajput

ABBREVIATIONS

CAFs	Cancer-associated fibroblasts
CCl4	Carbon tetrachloride
CSCs	Cancer stem cells
DECM	Decellularized and solubilized extracellular matrix
DEN	Diethylnitrosamine
ECM	Extracellular matrix
GEM	Genetically engineered mouse
HCC	Hepatocellular carcinoma
HSC	Hepatic stellate cells
HUVEC	Human umbilical vein endothelial cells
ICC	Intrahepatic cholangiocarcinoma
iPSC	Induced pluripotent stem cell
iPSCs	Induced pluripotent stem cells
LSECs	Liver sinusoidal endothelial cells
LX-2	Hepatic stellate cells
MWA	Microwave ablation
OoC	Organ-on-chip
PDX	Patient-derived xenografts
PDXs	Patient-Derived Xenografts
PHH	Primary human hepatocytes
RFA	Radiofrequency ablation
TACE	Transarterial chemoembolization
TARE	Transarterial radioembolization
TME	Tumor microenvironment

DOI: 10.1201/9781003583349-9

8.1 INTRODUCTION

Cancer represents a critical global health burden, impacting millions of individuals and contributing substantially to worldwide mortality rates (Bray et al., 2024). Among the many forms of this disease, liver cancer stands out due to its rapidly increasing incidence and impact on public health (Liu et al., 2024). Estimates suggest that in India, the current incidence (2.15 per 100,000), occurrence (2.27 per 100,000), and mortality (2.21 per 100,000) rates of hepatocellular carcinoma (HCC) are lesser than universal averages. However, these rates are increasing more rapidly in India than worldwide. Among Indians, HCC incidence, prevalence, and mortality are currently higher in males than in females. The Northeastern states report the highest rates of incidence, prevalence, and mortality for HCC, while Western states—specifically Gujarat, Maharashtra, Goa, and Kerala—are emerging as new hotspots, showing higher annual increases in incidence, prevalence, and mortality rates (Giri & Singh, 2024). Together, these two types account for approximately 85% of all liver cancer cases, with HCC representing the majority (70%) and intrahepatic cholangiocarcinoma (ICC) comprising around 15% (Chidambaranathan et al., 2021). In recent decades, especially in the United States, the cases of HCC and ICC are increasing rapidly, largely attributed to changing patterns of chronic liver disease and improvements in diagnostic practices. HCC, in particular, has emerged as one of the most frequently diagnosed cancers, with its rising incidence mirrored by increasing cancer-related deaths (Massarweh & El-Serag, 2017). A common feature among many HCC risk factors is liver injury, which often progresses to cirrhosis, establishing a favorable environment for cancer development (Sharma et al., 2021). However, the epidemiology of liver disease is evolving. Severe liver diseases, including nonalcoholic fatty liver disease (NAFLD), have highlighted the role of metabolic factors in liver cancer development, even in cases without cirrhosis. Despite this shift, patients diagnosed with HCC still face poor long-term survival outcomes, emphasizing the pressing need for more effective therapies (Figure 8.1) (Sharma et al., 2024). Similarly, ICC, the second most common primary hepatic malignancy, has demonstrated an upward trend in the United States, though its exact epidemiology remains less understood. Risk factors for ICC, such as bile stasis and inflammation of the bile ducts, overlap significantly with those of HCC. ICC, however, often carries an even poorer prognosis, likely due to its high metastatic potential and lack of effective systemic therapies (Boortalary & Loren, 2022). With many ICC cases incidentally identified during HCC screening, understanding the complex interplay of shared and unique risk factors between HCC and ICC remains a research priority. Treatment options for liver cancer have evolved, emphasizing the need for a multidisciplinary approach that includes surgical, medical, and radiation oncology, as well as hepatology and interventional radiology (Oh & Sinn, 2024). The prominent therapeutic approaches for HCC in early-stage and well-selected patients are surgery and liver transplantation. For patients who cannot undergo surgery, they are treated with common ablation methods, including radiofrequency ablation (RFA) or microwave ablation (MWA), while transarterial radioembolization (TARE) or stereotactic body radiotherapy (SBRT) provide other localized treatments. Adjuvant therapies such as atezolizumab plus bevacizumab have shown promise in reducing recurrence risk after curative treatments, while locoregional therapies like transarterial chemoembolization (TACE) and TARE are widely used for intermediate-stage HCC (Sharma et al., 2021). Despite these advances, significant challenges remain in treating liver cancer. The frequent coexistence of underlying liver disease complicates treatment selection, as many therapies must balance efficacy with the preservation of liver function. The heterogeneous nature of liver cancer, with various etiologies and disease burdens, further complicates standardized treatment protocols (Ito & Nguyen, 2023). Moreover, although locoregional and systemic therapies have expanded, survival benefits are often limited, especially for advanced-stage disease. In light of these challenges, there is a growing need for innovative therapeutic strategies that can effectively target liver cancer while managing the unique complexities posed by underlying liver conditions. This chapter explores both traditional and advanced models in liver cancer research, covering computational approaches, two-dimensional (2D) cell cultures, animal models, and patient-derived xenografts (PDXs) to highlight their uses and limitations (Table 8.1, Figure 8.2). It then examines cutting-edge three-dimensional (3D) liver cancer models. The chapter concludes by discussing the advantages and challenges of in vitro models, future directions such as AI integration, and the essential connection between research models and clinical applications.

TABLE 8.1
Summary of Preclinical Models Utilized in HCC Research

S. No.	Modelling Techniques	Advantages	Disadvantages	References
1.I	Two-dimensional cell lines	Easy maintenance, cost-effective, high reproducibility, commercially available (all varieties).	Cultures lack a three-dimensional structure, exhibit poor retention of the original tumor's histological and mutational characteristics, and consist of a single cell type with minimal cell-cell interactions.	Fukuyama et al. (2021)
2.	Organoids	Forms a three-dimensional structure, can be derived from patients, predicts clinical outcomes, supports mechanisms of action studies, is cost-effective, and is well-suited for high-throughput screening.	Takes a long time to develop, lacks specific cell types, such as immune cells, and presents technical challenges.	Huch et al. (2015), Sato et al. (2009)

(Continued)

TABLE 8.1 (*Continued*)
Summary of Preclinical Models Utilized in HCC Research

S. No.	Modelling Techniques	Advantages	Disadvantages	References
3.	Organs on Chips	Resembles the biochemical and physical properties of the in vivo tumor microenvironment (TME), allows precise control of the TME, and enables coculture of multiple cell types.	Does not replicate the extracellular matrix (ECM) found in vivo, involves high-cost materials, and lacks standardization in design, manufacturing, and application.	Deng et al. (2019), Ehrlich et al. (2019)
4.	Tissue Slice Cultures	Preserves the complex tissue structure found in vivo, allows for multiple readouts to be analyzed, and serves as a suitable alternative to animal models.	Requires a lengthy and complex preparation process for culturing, limited duration for studying chronic diseases.	Wu et al. (2018), Doornebal et al. (2022a), Palma et al. (2019)
5.	Bio-Scaffold	Provides physical support for cell proliferation and migration, mimics the in vivo extracellular matrix (ECM), and is used for drug screening.	Heterogeneous material composition, challenges in verifying material properties, and reduced reproducibility.	Langhans (2018), Hoarau-Véchot et al. (2018)
6.	Chemically and Dietarily induced	Can be administered orally, offers a reproducible model of HCC progression, exhibits greater heterogeneity, easily integrates with other methods that promote HCC, and serves as an effective tool for modeling NASH/NAFLD.	Species differences from humans, with HCC development time varying based on sex, age, and strain, and not suitable for high-throughput screening.	Vesselinovitch et al. (1984), Newell et al. (2008), Vucur et al. (2010),
7.	Xenograft	Enables real-time monitoring of tumor growth, demonstrates cell-cell interactions, and offers a relatively cost-effective and straightforward procedure for heterotopic engraftment. Orthotopic engraftment occurs in the natural liver microenvironment, closely mimicking HCC development.	Heterotopic implantation lacks an accurate physiological representation due to the absence of the microenvironment, requires an immunodeficient host, and is not an ideal model. Orthotopic engraftment is challenging to perform, demands high maintenance, and is not suitable for high-throughput screening.	Olson et al. (2018), Jung (2014), Qi et al. (2019)

(*Continued*)

TABLE 8.1 (*Continued*)
Summary of Preclinical Models Utilized in HCC Research

S. No.	Modelling Techniques	Advantages	Disadvantages	References
8.Q	Patient-Derived Xenograft	Used to test personalized therapies, preserving the characteristics of the donor tumor. It recapitulates the original tumor, reflecting its morphology, genetic mutations, and correlating with patient outcomes.	Species differences from humans, use of immunodeficient mice, high maintenance costs, and limited suitability for high-throughput screening, along with constraints due to the availability of patient samples.	Jung et al. (2018), Lai et al. (2017), Blumer et al. (2019), Xu et al. (2020)
9.	Genetically Modified	Facilitates the study of hepatocarcinogenesis, includes the natural tumor microenvironment (TME), and allows for the testing of targeted gene therapies.	Expensive, with species differences from humans; manipulated genes may be crucial for embryo development; and single-gene HCC models typically have a longer latency period for tumor formation.	Heindryckx et al. (2009), Li et al. (2012)

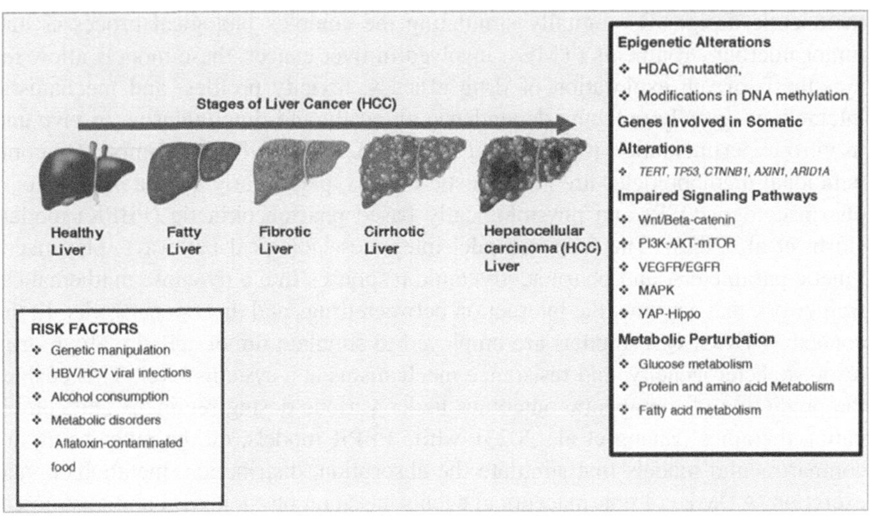

FIGURE 8.1 The development of HCC is a complex, multistep process in which normal liver cells gradually transform into malignant, invasive cancer cells. Key molecular factors involved in this progression include viral infections, genetic mutations, and dysregulated signaling pathways. Abbreviations used in the figure are as follows: HBV (Hepatitis B Virus), HCV (Hepatitis C Virus), TERT (Telomerase Reverse Transcriptase), TP53 (Tumor Protein 53), CTNNB1 (Catenin Beta 1), ARID1A (AT-rich Interaction Domain 1A), HDAC (Histone Deacetylase), PI3K (Phosphoinositide 3-Kinase), MAPK (Mitogen-Activated Protein Kinase), VEGFR (Vascular Endothelial Growth Factor Receptor), and EGFR (Epidermal Growth Factor Receptor). These factors collectively contribute to the progression from normal liver tissue to malignant tumor formation in HCC.

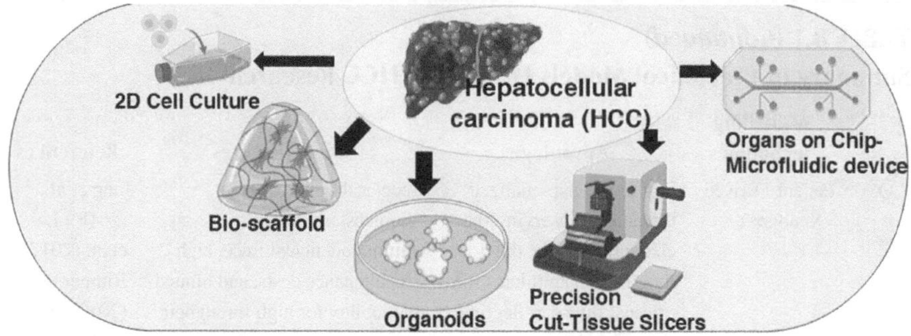

FIGURE 8.2 A schematic representation of various in vitro models used to study HCC. These models are essential for understanding the molecular mechanisms underlying HCC and testing potential therapeutic strategies. By utilizing these models, researchers can simulate the TME and better replicate the complexity of HCC for experimental studies.

8.2 *IN SILICO* MODELS

In silico modeling has become an indispensable tool in liver cancer research, providing advanced computational frameworks that facilitate drug development, pharmacokinetic prediction, systems-level biological analysis, and personalized therapeutic design. By digitally simulating the complex biological processes and tumor microenvironments (TMEs) involved in liver cancer, these models allow for hypothesis-driven exploration of drug efficacy, toxicity profiles, and mechanistic interactions, greatly reducing dependence on costly and time-intensive in vivo and in vitro experimentation (Carvalho et al., 2024). Among the most impactful computational methodologies are mechanistic models, particularly quantitative systems pharmacology (QSP) and physiologically based pharmacokinetic (PBPK) models (Joshi et al., 2023). The former model integrates biological pathways, pharmacokinetic parameters, and pharmacodynamic responses into a dynamic mathematical framework that captures the interaction between drugs and disease networks. In the context of HCC, QSP models are employed to simulate tumor cell signaling, drug response heterogeneity, and resistance mechanisms at a systems level. This enables the prediction of therapeutic outcomes under various dosing regimens and combination therapies (Zhang et al., 2023), while PBPK models, on the other hand, are compartmental models that simulate the absorption, distribution, metabolism, and excretion (ADME) of pharmaceutical agents based on physiological and anatomical parameters (Zhuang & Lu, 2016). Given the liver's central role in drug metabolism and its altered architecture in HCC, PBPK modeling allows researchers to accurately forecast hepatic drug clearance, bioavailability, and potential toxicological effects in both healthy and diseased states. These models are particularly useful for extrapolating preclinical findings to human scenarios and optimizing dosage in liver-impaired populations. Complementing these mechanistic approaches are network-based models, which construct molecular interaction maps using curated biological data or omics-derived insights. These models enable researchers to decipher the topological and functional organization of signaling networks disrupted in HCC (Lucena-Padros et al., 2024). By integrating data from transcriptomic, proteomic, and metabolomic sources, network models can highlight central hub proteins, critical feedback loops, and redundant pathways, thereby identifying robust drug targets and diagnostic biomarkers (Reitz et al., 2023). They are especially valuable for elucidating the molecular interplay underlying oncogenic pathways like Wnt/β-catenin, PI3K/Akt/mTOR, and ER stress-mediated responses, all of which are frequently dysregulated in liver carcinogenesis. Machine learning (ML) algorithms further enhance the capabilities of in silico studies by enabling pattern recognition and predictive modeling from large-scale datasets. Supervised learning techniques can be trained on clinical and molecular data to classify tumor subtypes, predict patient survival, and stratify responders to specific therapies. Unsupervised approaches, such as clustering and dimensionality reduction, uncover latent structures within omics data, aiding in the discovery of novel HCC subpopulations or treatment vulnerabilities. Importantly, ML models are now being incorporated into clinical decision support systems to guide real-time therapeutic recommendations based on individualized patient data (Quazi, 2022).

Additionally, in silico virtual screening platforms facilitate the high-throughput evaluation of thousands of chemical entities—both synthetic compounds and phytoconstituents—against molecular targets implicated in HCC (Shukla & Kumar, 2023). Structure-based methods, such as molecular docking and molecular dynamics simulations, allow for rapid identification of lead molecules with high binding affinity toward oncogenic proteins. These approaches have significantly accelerated the discovery of inhibitors for key regulatory elements in cancer pathways, including kinases, transcription factors, and chaperone proteins involved in endoplasmic reticulum (ER) stress responses (Agu et al., 2023).

8.3 TRADITIONAL RESEARCH MODELS

8.3.1 Primary Human Hepatocytes

Primary human hepatocytes (PHH) are widely regarded as the gold standard for in vitro research on biotransformation, toxicity, and drug-induced liver injury, owing to their strong resemblance to the in vivo phenotype (Wilkening et al., 2003; Collins et al., 2019; Guo et al., 2020; Vilas-Boas et al., 2019). The liver is a key target for chemical-induced damage, particularly from carcinogens (Arellanes-Robledo et al., 2017). In HCC research, PHHs (from human or rodent origins) are mainly used to explore the initiation and progression of cancer. While PHH has been employed in genotoxicity assays, their use is less common compared to rodent hepatocytes. More recently, PHH has been instrumental in uncovering the molecular pathways and genomic alterations induced by carcinogens (Seo et al., 2021). Additionally, PHH have been utilized to investigate the molecular and cellular mechanisms underlying HCC initiation via in vitro transformation, which involves lentiviral transduction of oncogenic Harvey-RAS, simian virus 40 (SV40) small T antigen, and SV40 large T antigen (Pez et al., 2019). Despite their advantages, PHH have limitations, including significant donor-to-donor variability (Collins et al., 2019) and progressive dedifferentiation, which reduces their viability for long-term culture (Kiamehr et al., 2019). Dedifferentiation begins during the isolation process. To mitigate this, researchers use sandwich cultures or spheroid models of PHH. While spheroid models have not been applied to genotoxicity studies, sandwich cultures have been used to explore the epigenetic effects of aflatoxin B1 (AFB1) in the initiation of HCC and to assess how chemopreventive agents may influence AFB1-induced genotoxicity (Romualdo et al., 2021).

8.3.2 Human Hepatoma Cell Lines

These cell lines provide a valuable tool for understanding liver physiology, drug metabolism, toxicity screening, and disease pathogenesis, specifically HCC. Typically grown as a monolayer on culture plates, hepatic cell lines are derived from human liver cancer tissues and represent important resources for liver research. Thirty HCC cell lines are documented in the literature to date, which are documented in Table 8.2 and Figure 8.3, though many of their genetic characteristics remain unexplored. Common human HCC cell lines include Hep3B, Huh7, SMMC-7721, SNU449, SNU182, MHCC97H, and PLC/PRF/5 (Qiu et al., 2016). Notably, each cell line exhibits variable responses to identical treatments, thereby reflecting the heterogeneity observed in HCC patients' responses to available drugs (Yildiz, 2018). This variation underscores the need for selecting appropriate cell lines based on specific research objectives (Fukuyama et al., 2021). While 2D cell lines are invaluable for liver cancer research, a significant limitation lies in their inability to replicate the spatial arrangement of a structure in 3D and the intricacy of a tumor. Since these cells are generated from poorly differentiated tumors, they do not allow for the study of early, preneoplastic stages of hepatocarcinogenesis. To address these gaps, complementary models—such as organoids, organ-on-chip (OoC) systems, tissue slices, and bio-scaffolds—are increasingly integrated with traditional 2D models. These advanced models provide a more comprehensive approach to studying tumor molecular characteristics and microenvironmental interactions, thus improving our understanding of HCC's complex biology.

8.3.3 Cocultures

The human liver comprises approximately 70%–80% hepatocytes and 5%–6% non-parenchymal cells (Kupffer cells, hepatic stellate cells (HSC), and sinusoidal endothelial cells) (Cheng et al., 2021). Cocultures can be established by combining hepatic cell lines with non-parenchymal hepatocytes to replicate this in vivo cellular heterogeneity. These coculture systems may employ transwell chambers to facilitate communication between the two cell populations (Ma et al., 2023). Co-culturing fibroblasts or tumor-specific neutrophils can generate tumor-specific microenvironments that closely resemble those in vivo. This approach results in a more accurate in vitro model, which reflects the evolving cancer characteristics of HCC cells influenced by interactions with other cell types (Ying et al., 2023). Additionally, coculture systems are valuable in immune cell therapy studies and in understanding the interactions between HCC cells and the immune system (Jeong & Kang, 2023). For example, co-culturing HepG2 cells with natural killer cells has been shown to produce anti-proliferative effects, while M1 macrophages reduce HCC cell viability,

TABLE 8.2
HCC Cell Lines that are Publicly Available and Documented in the Literature as Established in Vitro Experimental Models

Cell Line	Source	Appearnce	Cause and Key Features	Uses
HepG2	European male, 15 years old	Epithelial	Initially believed to be an HCC cell line, it has been identified as originating from a hepatoblastoma.	3D cell culture, cancer studies High-throughput screening, toxicity testing
Hep3B	African male, 8 years old	Epithelial	HCC containing an integrated HBV genome fragment (2–3 kb in size).	3D cell culture, cancer research, high-throughput screening, infectious disease research, STD research, toxicology
HUH7	Asian male, 57 years old	Epithelial	Originating from a well-differentiated HCC; characterized by the presence of multiple plasma proteins; lack TP53 expression.	3D cell culture, cancer research, in vitro drug metabolism and toxicology evaluations
SMMC-7721	Asian female, 30 years old	Epithelial-like	Contaminated; initially believed to originate from HCC, but later identified as a HeLa derivative.	3D cell culture, Bioproduction
SNU182	Asian male, 24 years old	Epithelial	Pleomorphic; tumor grade: III–IV; HBV DNA identified through Southern blot hybridization; HBV genomic RNA not expressed.	3D cell culture, cancer research, infectious disease research, STD research
SNU387	Asian female, 41 years old	Epithelial	Pleomorphic; tumor grade: IV–V; HBV DNA detected via Southern blot hybridization.	3D cell culture, Cancer research, Infectious disease research, STD research
SNU449	Asian male, 52 years old	Epithelial; diffusely spreading cells	HCC; Grade II–III/IV; HBV DNA detected by Southern blot hybridization; HBV genomic RNA was not expressed	3D cell culture, cancer research
SNU475	Asian male, 43 years old	Epithelial	Derived from primary HCC; tumor grade: II–IV/V; HBV DNA identified through Southern blot hybridization; HBV genomic RNA not expressed.	3D cell culture, cancer research, infectious disease research, STD research
MHCC97H	Asian male, 39 years old	Epithelial	Primary HCC with high metastatic potential.	3D cell culture, Cancer research
PLC/PRF75	African male, 24 years old	Epithelial	HCC-derived, with the presence of hepatitis B surface antigen (HBsAg) and reduced TP53 expression.	3D cell culture, cancer research, infectious disease research, STD research
HLF	Asian male, 68 years old	Epithelial	Originating from a well-differentiated HCC.	3D cell culture, cancer research, in vitro drug metabolism and toxicology evaluations

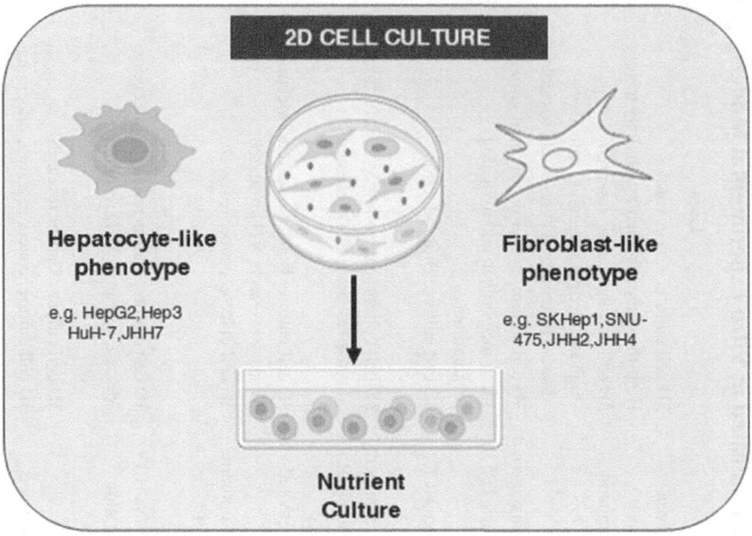

FIGURE 8.3 A schematic representation of the 2D in vitro model used to study HCC. This model typically involves cultured cancer cells grown in a monolayer on a flat surface, allowing researchers to examine cellular behavior, gene expression, and responses to treatments in a controlled environment.

invasion, migration, and therapy resistance. In contrast, M2 macrophages induce cancer invasiveness in liver cancer (HCC) cocultures. In order to generate an effective model (in vitro) of angiogenesis, it is required to coculture HepG2 and endothelial cells together.

8.3.4 Models Derived from Stem Cells

Emerging research suggests that liver cancer, particularly HCC, originates from cancer stem cells (CSCs), which have regenerative abilities and contribute to the diverse cell populations within HCC tumors. With their capacity to initiate tumor formation, CSCs play a critical role in driving chemoresistance, metastasis, and tumor recurrence following treatment (Wang et al., 2018). Considering the oncogenic scenario occurring in CSCs is essential for developing targeted therapies against HCC. Induced pluripotent stem cell (iPSC) technology serves as a powerful tool to model this reprogramming process, shedding light on pathways involved in tumor initiation and progression (Chehelgerdi et al., 2023). Research indicates that iPSCs offer potential therapeutic avenues by targeting genes related to CSCs. Specifically, liver cancer cells can be reprogrammed into induced pluripotent CSCs to model liver-specific CSCs and study their behavior. Techniques include retroviral reprogramming with four key stem cell transcription factors—KLF4, Sox2, Myc, and Oct4 (Aguirre et al., 2023). Alongside NANOG and LIN28, these factors are associated with HCC and poor clinical outcomes (Balzeau et al., 2017). Additionally, CSC-like cells, which are extracted directly from primary tumors or HCC cell lines, are valuable for exploring the molecular and epigenetic mechanisms underlying HCC (Wang et al., 2018). Beyond modeling tumor formation, stem cells are employed to evaluate patient-specific therapies that target reprogramming processes and to identify molecular targets that could interrupt tumor initiation.

8.4 ANIMAL MODELS

Animal models serve as an essential tool to bridge the gap between in vitro and clinical studies, providing scientists with the opportunity to explore various facets of cancer biology within the living organisms (Figure 8.4). Given the genetic similarities between rodents and humans, along with advances in gene-editing technologies, *in vivo* mouse models have become an ideal system for investigating the intricate biology of hepatocarcinogenesis (Brown et al., 2018; Zheng et al., 2023). They contribute to the investigation of potential treatment options through medication testing

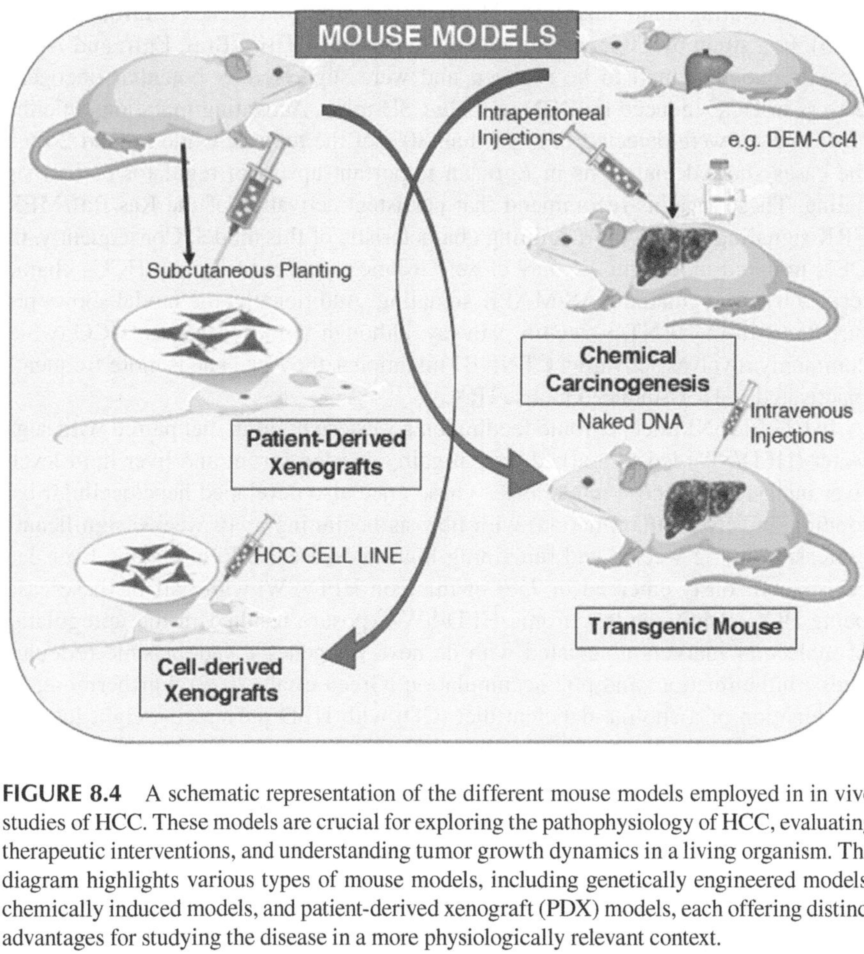

FIGURE 8.4 A schematic representation of the different mouse models employed in in vivo studies of HCC. These models are crucial for exploring the pathophysiology of HCC, evaluating therapeutic interventions, and understanding tumor growth dynamics in a living organism. The diagram highlights various types of mouse models, including genetically engineered models, chemically induced models, and patient-derived xenograft (PDX) models, each offering distinct advantages for studying the disease in a more physiologically relevant context.

and facilitate an understanding of the interactions between hepatic tumor cells and their surrounding microenvironment. A thorough understanding of the phenotypic characteristics of various mouse strains is crucial in guiding experimental design decisions. For example, strains including C3H, CBA, and DBA/2 are prone to developing liver tumors either spontaneously or in response to chemical induction, while others, including C57BL/6, BALB/c, and A/J, exhibit resistance to similar oncogenic stimuli (Maronpot, 2009; Rogers, 2018; Blidisel et al., 2021).

8.4.1 Chemically and Diet-Induced HCC Models

The use of liver carcinogen diethylnitrosamine (DEN) is a well-established and extensively utilized model for inducing hepatocarcinogenesis in mice (Vesselinovitch et al., 1984). In this model, a single intraperitoneal injection of DEN is administered to 14-day-old mice, which leads to DNA modifications via ethylation, initiating the formation of neoplastic foci (Newell et al., 2008). Typically, HCCs develop within 40 weeks, with a marked male predominance. DEN-induced tumors frequently exhibit activating mutations in the Hras or Braf proto-oncogenes (Maronpot et al., 1995). In a study by Connor et al. (2018), four genes—Hras, Braf, Egfr, and Apc—were frequently found to be mutated and were suggested as potential oncogenic drivers in HCC induced by DEN in male C3H mice. Activating mutations in either *Hras* or *Braf* were detected in more than 80% of the tumors, while around 20% of the cases showed mutations in *Egfr*, an important upstream regulator of Ras signaling. These insights recommend that persistent activation of the Ras/Raf/MEK/ERK signaling pathway is a defining characteristic of this model. Consequently, the DEN-initiated mouse model may closely mimic a subset of human HCCs characterized by dysregulated RAS/MAPK signaling. Additionally, the model shows perturbations in the WNT/β-catenin pathway, although unlike in human HCC (which commonly involves activating CTNNB1 mutations), the Apc gene is more frequently inactivated in DEN-induced mouse HCC.

In C57BL/6NJ mice, chronic feeding on a western high-fat diet paired with sugar water (HFD/SW) led to marked weight gain, elevated serum and liver lipid levels, liver injury, and glucose intolerance. These mice also developed hepatocellular ballooning and liver inflammation, with fibrosis beginning at 16 weeks, significantly increasing by 32 weeks, and remaining high at 54 weeks. Remarkably, liver cancer spontaneously emerged in 75% of mice on HFD/SW, with half of these cases being HCC. Additionally, chronic HFD/SW exposure resulted in the upregulation of molecular markers associated with de novo lipogenesis, endoplasmic reticulum stress, inflammation, and p62 accumulation (Green et al., 2022). Furthermore, the combination of a choline-deficient diet (CD) with HFD prevented weight loss and promoted HCC development in roughly 25% of mice, primarily through an NF-κB signaling-dependent mechanism (Wolf et al., 2014). However, a major limitation of this model is the relatively low incidence of HCC, with limited insights into why only some mice progress to HCC while most remain tumor-free. An ideal preclinical model for nonalcoholic steatohepatitis (NASH) would be straightforward and reflect human risk factors (obesity, insulin resistance, and dyslipidemia), while closely matching human disease in terms of histology, metabolic characteristics, clinical outcomes, gene expression patterns, lipid accumulation, and activation of human-relevant pathways. The western diet model, characterized by high cholesterol, saturated fat, and fructose, has been shown to induce NASH with distinct hepatocyte ballooning and progressive fibrosis after six months. Asgharpour et al. (2016) developed a diet-induced model for NAFLD-related HCC in an isogenic strain from the cross of 129S1/SvImJ and C57BL/6J mice. In these mice, consumption of a high-fat diet along with unrestricted intake of fructose- and glucose-enriched water results in a sequential development of liver conditions—starting with steatosis, followed by steatohepatitis, advancing fibrosis, and ultimately HCC. Although the majority of individuals with NAFLD experience only fat accumulation in the liver, a notable fraction progresses to NASH, which can further evolve into cirrhosis, liver cancer, and elevated risk of liver-related death. In a related model using DEN as a tumor initiator and phenobarbital as a promoter, chromosomal instability and activating mutations in the β-catenin gene were linked to tumor progression (Aleksic et al., 2011). Carbon tetrachloride (CCl_4) has also been widely used for decades to induce liver injury and fibrosis in mice (Scholten et al., 2015). Notably, susceptibility to CCl_4 varies significantly across strains: BALB/c mice are particularly prone to developing fibrosis, while FVB/N mice show resistance to the fibrogenic effects of CCl_4. While C57BL/6 mice typically exhibit moderate levels of liver fibrosis, they are still commonly employed in fibrosis research involving CCl_4 treatment. Chronic administration of CCl_4 combined with a choline-deficient, L-amino acid-defined (CD) diet for up to nine months induces NASH with fibrosis and HCC (De Minicis et al., 2014). Recently, Tsuchida et al. (2018) showed that combining this dietary regimen with low-dose weekly administration of CCl_4 significantly hastens the development of stage 3 liver fibrosis by 12 weeks and HCC by 24 weeks. This experimental setup results in hepatic abnormalities that closely mirror the histopathological, immune-related, and gene expression profiles observed in human NASH-associated HCC, thereby offering a highly relevant model for investigating liver cancer in a clinically meaningful framework.

8.4.2 Tumor Transplant Models of HCC: Orthotopic and Heterotopic Techniques

HCC xenograft models are typically established by introducing cancer cell lines or transplanting tumor fragments. Depending on the implantation site, these models are classified as either orthotopic or heterotopic. In heterotopic models, HCC cells are usually injected beneath the skin of mice, allowing researchers to investigate tumor development, invasive behavior, and responses to treatment. This approach is simple, reproducible, and allows for easy monitoring and measurement of tumor volume due to its accessible location (Zheng et al., 2023). In contrast, orthotopic models involve implanting human malignant cells into their original anatomical site, offering a closer mimicry of the TME and accurately reflecting the influence of vascularization and stromal interactions. However, these models are technically more complex (Heinrich et al., 2021). Implantation of human HCC cell lines is limited to immunodeficient mice to avoid immune-mediated rejection of the tumor. As these mice lack a functional immune system, it limits studies on immune responses within the human HCC xenograft model. Additionally, the transplanted cells are clonally homogeneous and fully malignant, which constrains investigations into tumor heterogeneity and preneoplastic stages (Cigliano et al., 2024). The heterotopic subcutaneous tumor location further fails to replicate the essential HCC–stroma interactions typical of liver tissue and does not accurately model intrahepatic or distant metastasis development. For tumor immunology and immunotherapy studies, syngeneic models are preferred, as they use recipient mice with a fully functional immune system. Notably, Qi and his fellow researchers recently developed an orthotopic HCC model in immunocompetent C57BL/6J mice by inoculating histologically normal oncogenic hepatocytes and treating them with CCl_4 to simulate human HCC initiation and progression in the context of liver fibrosis/cirrhosis (Qi et al., 2019). Similarly, Reiberger and experimentalists generated a syngeneic orthotopic HCC model by administering CCl_4 to induce cirrhosis, replicating the features of human HCC (Reiberger et al., 2015).

8.4.3 Patient-Derived Xenografts (PDXs)

Unlike traditional cell line-derived xenografts, PDX models serve as a more realistic reflection of human cancer biology as they are derived directly from patient tumors without prior selection in tissue culture. Following the National Cancer Institute's discontinuation of the (NCI)-60 panel of tumor cell lines in 2016, PDX models emerged as a more reliable option for cancer research, created by transplanting cancerous tissue from patient tumors into immunodeficient mice, such as NOD-SCID and NSG mice, significantly improving xenograft success rates (Jung et al., 2018). Important models maintain the genetic, molecular, and histopathologic characteristics of the original tumor, as well as the interactions between the tumor and its stromal components, making them valuable for predicting a patient's therapeutic response (Lai et al., 2017). Additionally, the transplanted tumors can be surgically removed and propagated in culture for further analysis in vitro. However, several limitations must be considered while producing PDX models. First, developing PDX models demands significant financial investment and a considerable amount of time. Which can be a barrier to personalized therapies, especially for aggressive cancers. Second, the engraftment success rate of PDXs can be challenging to achieve. For example, the successful establishment of PDX mouse models from undifferentiated HCCs occurs only about 20% of the time (Blumer et al., 2019). In contrast, higher success was seen with samples from patients with O blood type, TNM stage III–IV, and elevated levels of CEA or CA199 (Xu et al., 2020). Furthermore, only surgically resected tissue be used for implantation, meaning that the full spectrum of the disease may not be represented (Calderaro et al., 2019). As with HCC cell line-derived xenografts, PDX models are transplanted into immunodeficient hosts, limiting the ability to study the effects of immune cells on tumor progression (Jin et al., 2010). Additionally, the role of stromal influences on treatment may be represented in PDX models. Finally, while aggressive tumors are more likely to successfully revive in vitro, this is rarely achievable with benign or less aggressive tumors (Cao et al., 2019).

8.4.4 Mouse Model with Targeted Genetic Alterations

The creation of genetically engineered mouse (GEM) models has been a significant advancement in liver cancer research. Numerous GEM models have been developed, offering a valuable tool for studying tumor progression in an immunocompetent living organism (Newell et al., 2008). Various techniques, such as Cre-Loxp recombination and CRISPR-Cas9, have generated these models. These approaches enable the creation of models that replicate the genetic and epigenetic changes seen in HCC, facilitating the identification of the sequence of events that occur during the preneoplastic, early, and late stages of tumor development (Heindryckx et al., 2009). Typically, GEM models are produced by triggering oncogene expression or disabling tumor suppressor mechanisms to trigger tumor formation. Notably, HCC GEM models have been developed with single or multiple gene mutations, as well as models that express partial segments of the HBV (hepatitis B virus) and HCV (hepatitis C virus) genomes (Li et al., 2012; Frese & Tuveson, 2007).

8.5 INTRODUCTION TO 3D LIVER CANCER MODELS

Traditional 3D cell culture is an advanced technique that utilizes specific cultivation methods or creates complex cell culture systems featuring spatial 3D configurations using matrices, scaffolds, and specialized materials. This approach is categorized into scaffold-supported and scaffold-independent 3D culture techniques.

8.5.1 Scaffolds Dependent 3D Models

This method employs artificially constructed substrates that imitate the extracellular matrix (ECM), creating a stable surface for liver cancer cells to adhere and grow. Within these scaffolds, cells can proliferate and migrate, adopting characteristics typical of in vivo tumors. To form a cell-free matrix or hydrogel, these scaffolds are mainly constructed from natural materials like collagen, hyaluronic acid, and gelatin, or synthetic polymers such as polycaprolactone, polyethylene glycol, and polylactic acid (Drury & Mooney, 2003). Natural materials are known for their superior biocompatibility and minimal toxicity, whereas synthetic polymers provide enhanced manipulability and multifunctional properties. Furthermore, 3D porous scaffolds created from biomaterials offer essential physical structural support, facilitating both in vitro cell culture and in vivo tissue regeneration.

Hydrogels are commonly employed in the 3D culture of liver cancer cells due to their biocompatibility and close resemblance to the ECM. A plant-derived nanofibrillated cellulose hydrogel was used to culture the HepG2 liver cancer cell line in 3D (Bhattacharya et al., 2012). The hydrogel enhanced cell proliferation and differentiation, demonstrating significant promise for applications in drug testing, tissue engineering, and cell therapy (Bhattacharya et al., 2012). Greene & Lin encapsulated Huh7 liver cancer cells within a hydrogel made from gelatin, which increased cellular metabolic activity (Greene & Lin, 2015). Decellularized matrices can retain the microstructure and biochemical composition of the natural ECM, offering a tissue-specific environment that supports the growth of both normal and cancerous liver cells (Peng et al., 2018). Additionally, Calitz and colleagues embedded Huh7 cells, hepatic stellate cells (LX-2), and human umbilical vein endothelial cells in a mixture of collagen and fibrinogen to create a 3D liver cancer model. This biomimetic model demonstrated substantially enhanced tumor proliferation, migration, and chemotherapy resistance compared to a 2D model (Calitz et al., 2020).

Porous micro scaffolds made from hydrogels and other biomaterials have also been applied to create 3D liver cancer models, including hydrogel microspheres, interlayers, porous microspheres, and 3D bioprinted hydrogel structures (Liaw et al., 2018). These porous scaffolds offer a large area for cell adherence and promote intracellular communication. In a study by Moscato and team (2015), the combination of gelatin and polyvinyl alcohol was repeatedly freeze-dried until it was converted to porous scaffolds with 30–150 μm pore size. HepG2 liver cancer cells were grown within these 3D scaffolds, forming tumor spheroids that sustained cell viability for up to 21 days, with necrotic regions developing in the center. In a study conducted by Wang et al. (2020), microfluidic techniques were employed to fabricate porous microspheres composed of poly(lactic-co-glycolic acid) (PLGA). These microspheres had an approximate diameter of 395 μm, with pore dimensions varying between 10 and 60 μm. HepG2 cells were dynamically cultured within these porous microspheres, and HUVEC cells were introduced to encourage tumor vascularization and interconnectivity between the microspheres. This model enabled the formation of tumor microtissues, displaying significantly higher IC_{50} values than those in conventional 2D cultures, which suggests improved drug resistance and a closer approximation to actual tumor tissue behavior. A study conducted by Leung et al. grew liver cancer cell lines PLC/PRF/5 and HepG2 within chitosan–alginate scaffolds to create three-dimensional (3D) liver cancer models (Leung et al., 2010). The results demonstrated an upregulation of tumor-associated markers such as IL-8, bFGF, VEGF, and GPC-3 within the 3D culture systems. Furthermore, HCC cells grown in 3D scaffolds exhibited notably higher resistance to doxorubicin treatment in comparison to those maintained in 2D cultures or matrix-based gels.

The construction of liver cancer 3D models using novel material scaffolds provides a tumor matrix environment that more closely resembles *in vivo* conditions, facilitating cell growth and proliferation (Nikolova & Chavali, 2019). The scaffold's porous architecture facilitates the diffusion of oxygen and nutrients, supports drug transport, and assists in the removal of metabolic waste. 3D scaffold-based systems hold substantial value and offer promising applications for exploring the molecular mechanisms of liver cancer, analyzing cellular dynamics within the TME, and advancing the discovery and evaluation of anti-cancer therapeutics. The development of liver cancer 3D models using innovative material scaffolds creates a tumor matrix environment that better replicates in vivo conditions, supporting cell growth and proliferation (Nikolova & Chavali, 2019). The porous structure of these scaffolds allows for effective oxygen and nutrient exchange, drug administration, and waste removal. These 3D scaffold models hold substantial importance and offer promising applications in researching liver cancer molecular biology, understanding cell interactions within the TME, and advancing anti-cancer drug development and screening.

8.5.1.1 Scaffolds Free 3D Models

A scaffold-free 3D liver cancer model utilizes advanced culturing strategies to promote the natural clustering of cells, resulting in the formation of tumor spheroids. In these spheroids, cancer cells are capable of producing their own tissue matrix. Several established techniques are employed for developing such models, including rotating wall bioreactors, hanging drop systems, suspension culture methods, and the more recent integration of microfluidic platforms. These approaches primarily focus on enhancing tumor cell self-organization and facilitating their aggregation into spheroidal shapes, which improves intercellular communication. Additionally, the use of a rotating wall vessel bioreactor has enabled the coculture of liver cancer cells with fragments of healthy liver tissue under dynamic 3D rotational conditions. Observations from this method revealed that cancer cells clustered around the normal tissue, forming spheroids that closely resemble the invasive and metastatic behavior observed in liver cancer progression. This technique offers an innovative in vitro platform for investigating the biological mechanisms underlying liver cancer (Zhu et al., 2023). Similarly, Salinas and the team applied the rotating culture technique to create a 3D spherical breast cancer model that included fibroblasts and endothelial cells. This model enabled long-term culture, allowing for the study of cell-cell interactions and examining how endothelial cells may influence drug responses within the TME. Tang and the team utilized a rotational bioreactor to cultivate liver cancer cells on a biomolecular framework, thereby establishing a 3D model of hepatic cancer (in vitro) characterized by maximum invasiveness. Morphological assessment and biochemical evaluations indicated that this 3D system effectively replicated multiple pathological traits observed in clinical liver cancer cases, such as cellular structure, tissue ultrastructure, distinct gene expression profiles, and programmed cell death (Tang et al., 2011). In a similar study by Shah and his colleagues, they utilized hepatoma cells (HepG2) to develop a similar model using hanging drop culture to investigate tumor spheroid developmental features and evaluate the suitability of the model for gene toxicity testing. Other researchers cocultured liver cancer cells with fibroblasts to form spheroids using the hanging drop technique, then embedded the spheroids in a collagen hydrogel to create a 3D liver cancer model (Yip & Cho, 2013). Drug testing revealed enhanced drug resistance in the 3D model, making it a valuable preclinical tool for drug development. Compared to conventional 2D liver cancer cell models, this 3D model showed a significant increase in the expression of unfavorable predictive markers, including those associated with tumor generation and progression. Additionally, research has shown that using magnetic nanoparticles to create a magnetic environment for cells enables magnetic suspension culture, which helps preserve cell characteristics during 3D tissue formation (Caleffi et al., 2021). Other studies have enhanced cell adhesion and tumor spheroid formation by introducing nanofibers into the cell suspension (Lee et al., 2021).

8.5.2 Organ-on-a-Chip Model

OoC technology has emerged to be a promising substitute for traditional 2D and 3D in vitro models, addressing many of their limitations. These platforms leverage advancements in microfluidics, microfabrication, and tissue engineering to create patient-specific models. OoCs can be described as biomimetic systems that replicate key aspects of human physiology, including in vivo-like microenvironments and cell-to-cell and cell-ECM interactions (Srivastava et al., 2024). By integrating patient-derived cells, OoCs enable the development of personalized models for studying biological processes and are increasingly used in drug screening applications (Jodat et al., 2018). Researchers generated a chip system based on liver designed to precisely mimic the oncogenic environment. This was achieved by incorporating components from decellularized liver matrices into gelatin methacryloyl, significantly enhancing the functionality of this tumor-on-a-chip model (Lu et al., 2018). The platform was subsequently employed to evaluate dose-dependent drug responses involving sorafenib and acetaminophen. A similar research study was conducted in which OoC and microfluidics technology were applied to create an oxygen-deprived tumor model for testing cytotoxicity induced by agents like paclitaxel, also exploring the associated mechanisms of drug resistance. OoC technology has also been applied to produce a systemic organ network focusing on the gut–liver axis. This model evaluated the effect of gut microbiota on irinotecan metabolism, specifically its contribution to transforming the inactive metabolite SN-38G into the toxic form, i.e., SN-38. This setup provides insight into the complex interactions between gut microbiota and pharmaceuticals (Lucchetti et al., 2024). Polidoro and colleagues developed an advanced cholangiocarcinoma chip model derived from patient samples, which incorporates essential elements of the TME, such as malignant cells, cancer-associated fibroblasts (CAFs), vascular endothelial cells, and infiltrating immune cells. This configuration recreates a physiological-like Chitosan–Alginate–Alginate (CCA) niche, offering a precise, high-throughput tool for screening anticancer drugs. Such a model could also serve as a valuable resource for research in HCC (Polidoro et al., 2024).

8.5.3 3D Bioprinting

3D bioprinting is an innovative approach in tissue engineering that adapts the concepts of software-based rapid simulation used in conventional 3D printing. It employs bioinks made of hydrogels, cells, growth factors, and other substances to create complex biological microstructures, layer by layer, on a substrate or culture medium (Mirshafiei et al., 2024). By precisely controlling factors, such as cell types and quantities, ECM composition, and the accumulation of cell-laden scaffolds, 3D bioprinting can replicate certain aspects of complex tissue structures with biologically functional properties. This method has become a pivotal tool for developing intricate in vitro tissue models. Different 3D bioprinting approaches include extrusion-based, laser-assisted, inkjet-based, magnetic, coaxial, and acoustic bioprinting, each with distinct principles (Tripathi et al., 2023).

Extrusion-based bioprinting is one of the most widely used techniques, involving the controlled deposition of cells via a fluid dispensing system, which enables the combination of bioinks and cells for the printing process (Tetsuka & Shin, 2020). This method can even print complex tissues by utilizing multiple channels and different bioinks, offering advantages such as cost-effectiveness and high cell viability (Zhuang et al., 2023). Derived from conventional 2D inkjet printing technology, inkjet-based bioprinting involves the deposition of bioink in droplet form, enabling accurate regulation of tissue pattern size and volume, thus offering a simple and efficient method. This technique can be categorized into two types: on-demand and continuous inkjet bioprinting, depending on whether the droplet ejection occurs intermittently or in a continuous stream. However, there are several limitations associated with inkjet bioprinting. First, the precision of droplet deposition can be compromised by issues such as nozzle blockages and cell accumulation within the printing compartment. Second, inkjet printing needs bioinks with high charge conduction, which complicates the selection of appropriate inks. Additionally, the cell viability in inkjet printing tends to be suboptimal. Unlike other techniques, laser-assisted bioprinting utilizes the mechanism of laser-associated transfer, enabling highly accurate placement of cells and biomaterials with superior resolution. The absence of nozzles in this approach contributes to enhanced cell survival rates. However, the high cost of the necessary equipment remains a limiting factor. The choice of hydrogels in bioinks is critical, as they create the microenvironment needed for cell proliferation and differentiation (Huang et al., 2017). Thus, factors such as biocompatibility, flowability, polymer characteristics, and biodegradability must be carefully considered when selecting hydrogels for bioinks (Decante et al., 2021).

Bioprinting technology enables the precise placement of cells and biomaterials, effectively mimicking the composition and spatial complexity of the TME (Liu et al., 2020). This makes it a potential approach for creating reliable and robust 3D in vitro liver cancer models. For example, Mao employed a hybrid hydrogel formulation consisting of gelatin, alginate, and Matrigel to fabricate 3D constructs of intrahepatic cholangiocarcinoma using cells obtained from patients. These models demonstrated significantly higher cell proliferation and tumor marker expression compared to traditional 2D models, as well as enhanced resistance to chemotherapy and targeted drugs (Mao et al., 2020). Similarly, Changcan applied bioprinting to develop 3D models of cholangiocarcinoma, incorporating tumor cells along with tumor-associated endothelial cells, fibroblasts, and macrophages for coculture (Li et al., 2022). The 3D bioprinted models exhibited improved metastasis, increased expression of tumor-related genes, and enhanced drug resistance, with stromal cells supporting the activity of tumor cells. Xie and his collaborators, combined patient-derived HCC cells with gelatin and alginate, using 3D bioprinting to create personalized in vitro models of liver cancer (Xie et al., 2021). Their investigation revealed that the models preserved the features of the original liver cancer cells, involving reliable biomarker expression and stable mRNA expression patterns. Additionally, the models effectively replicated the outcomes of anticancer drug screenings, offering valuable insights for the personalized therapies of HCC patients.

Processed ECM stripped of cellular components (dECM) is regarded as a valuable bioink because of its tissue-specific composition. Nonetheless, the limited gelation capability of ECM poses challenges to the accuracy of 3D bioprinting. De Santis and team developed a tissue-specific composite biological ink made from the natural polymer alginate, which is augmented with dECM to enhance its biological activity. The inclusion of ECM supports the survival of primary human progenitor cells during 3D bioprinting, aids in tissue-specific cell differentiation, and promotes full-layer vascularization of implants in vivo, while reducing allogeneic reactions (De Santis et al., 2021). An innovative bioprinting approach was developed for dECM-incorporated hepatocyte spheroids, which simultaneously strengthens both cell-cell and cell-ECM interactions. The dECM was uniformly distributed throughout the bioprinted spheres, and its incorporation significantly boosted the liver function of hepatocyte spheroids (Kim et al., 2023). Overall, dECM-based biological inks can be used to recreate tissue and organ-specific microenvironments that influence cell behavior and tissue development in vitro. To increase the biomimicry and intricacy of models, the aggregation of 3D printing and microfluidic technology has gained significant attention in biomanufacturing and drug screening research. With advancements in printing technologies and the development of novel bioinks, extrusion-based printing has emerged as a viable method for constructing OoC models (Yi et al., 2017). Microfluidic channels can be created by printing bioinks, such as hydrogels, followed by the addition of cell-laden bioink mixtures into the pre-formed channels (Bertassoni et al., 2014; Bhise et al., 2016). This approach enriches the effectiveness and consistency of drug screening chip systems, marking a crucial step toward unifying 3D printing with microfluidic technologies. 3D printing is used to create liver cancer cell spheroids with precise control over cell number and spheroid size, while microfluidic chips offer a more biomimetic in vitro environment for the cells (Ouyang et al., 2015). Li and team combined the benefits of 3D bioprinting, microfluidics, and hydrogel technology to create a liver cancer 3D bioprinting-microfluidic model with controllable size (Li et al., 2019). The system was applied in drug testing studies involving the new anti-CD147 monoclonal antibody, Metuzumab, providing important understanding for advancing sophisticated in vitro liver cancer platforms and evaluating antibody-based therapies. Additionally, 3D bioprinting has accelerated advancements in tissue and organ regeneration engineering. Previous research has demonstrated that bioprinting technology can be used to create liver tissue models composed of primary hepatocytes, endothelial cells, and fibroblasts, which could have future applications in liver tissue regeneration, artificial livers, and liver transplantation (Lee et al., 2016). Indeed, 3D bioprinting of tumor models can provide a more accurate replication of the complex TME and treatment response in vivo, offering significant potential for personalized treatment of liver cancer and the development of anticancer drugs (Meng et al., 2019). Despite advancements, most existing 3D bioprinted liver cancer models are still unable to entirely mimic the full spectrum of tumor physiology, particularly aspects such as the vascular and immune microenvironments (Kronemberger et al., 2021).

8.5.4 Liver Carcinoma-Derived Organoids

Systems developed using primary cell lines obtained directly from individuals with HCC patients are limited due to challenges in establishing human cell lines from fresh patient tumors, as freshly isolated HCC cells often exhibit low viability. Additionally, the inherent variability in tumors across patients can significantly influence experimental outcomes, making patient-derived models both essential and challenging to establish. The development of 3D culture systems has enabled researchers to better simulate the structural complexity and functionality of mammalian tissues in vitro by creating organ-like structures, termed "organoids" (Huch & Koo, 2015). They are defined as 3D cultures of primary tissues that preserve the spatial architecture, physiological characteristics, and properties of adult organs, including inherent organization, diverse lineage potential, and tissue architecture. Organoids have been developed for various organs, including the intestine, liver, pancreas, stomach, and kidney, providing valuable insights into tissue development, organogenesis, and stem cell behavior (Sato et al., 2009).

Organoids emerged as promising disease models, with cancer organoids becoming indispensable tools for studying cancer development, metastasis, drug screening, and therapy resistance, with the ultimate goal of developing personalized cancer treatments (Fan et al., 2019). Notably, organoids can be maintained long-term in culture, cryopreserved, and genetically manipulated, offering significant advantages similar to those of 2D cell cultures. However, organoids still lack critical in vivo components such as organ crosstalk, stromal tissue, and immune cells, which limits their ability to fully replicate the TME (Neal et al., 2018). To address these limitations, coculture systems have been developed to facilitate the study of tumor interactions with the TME components. In 2017, Laura and her colleagues successfully generated organoids resembling primary liver cancer (PLC) using tumor tissues obtained from PLC patients. This pioneering work demonstrated that PLC-like organoids could be maintained long-term and could replicate various aspects of the parent tumor, including tissue structure, gene expression, genomic landscape, and in vivo tumorigenicity (Broutier et al., 2017). Furthermore, $Lgr5^+$ liver stem cells and human induced pluripotent stem cells (iPSCs) have been used to generate liver organoids. Single Lgr^{5+} liver stem cells from mice can be expanded as epithelial organoids in vitro, where they differentiate into functional hepatocytes both in vitro and in vivo (Huch et al., 2015).

Human liver tumoroids can also be developed from surgical samples, offering a valuable model for early-stage HCC cases where surgical resection is possible (Broutier et al., 2017). However, late-stage HCC cases, which are often inoperable, pose a challenge for organoid development. In 2018, Sandro Nuciforo and colleagues successfully established HCC tumoroids from needle biopsy samples from advanced liver cancer patients. These tumoroids maintained the primary tumor's genetic diversity, morphology, and gene expression, making them effective models for studying HCC heterogeneity. In addition to human-derived samples, HCC organoids can also be generated from chemically induced tumors in animal models (Nuciforo et al., 2018). Additionally, DEN-induced liver tumors in mice were used to culture liver tumor organoids. These organoids showed a success rate of around 70% but could fail due to early-stage necrosis (Cao et al., 2019). Successfully cultured organoids can undergo long-term expansion and initiate tumors when transplanted into immunodeficient mice. Organoids also provide a unique opportunity to study tumor-initiating cells that represent the cancer cell population driving tumor initiation, relapse, metastasis, and chemotherapy resistance in liver cancer (Xu et al., 2022). Recent advancements have combined organoid technology with CRISPR/Cas9 gene editing, allowing researchers to investigate the effects of specific genetic mutations on cancer biology within an organoid model. Studies have demonstrated that this combined approach can offer mechanistic insights into cancer gene function in a context that more closely mimics human biology (Saborowski et al., 2019; Artegiani et al., 2019). This approach addresses some limitations identified in the work of Broutier and

Nuciforo cancer organoids were typically derived only from poorly differentiated tumors (Broutier et al., 2017; Nuciforo et al., 2018). While PLC-derived organoids preserve histological features, gene expression, and the tumor, enabling differentiation between tumor subtypes, organoid models still face limitations. Challenges include high costs, restricted availability of patient-derived tissues, and the need for more advanced organoid systems that replicate liver cell diversity and heterogeneity. Moreover, organoids lack essential components found in in vivo systems, such as vasculature and immune cells, which are critical for accurately modeling the liver's complex biology.

8.5.5 Precision-Cut Tissue Slices (PCTSs)

Precision-Cut Tissue Slices (PCTSs) are thin, live sections of tissue prepared with specialized cutting techniques. These slices offer an ex vivo model that closely replicates the characteristics of the original tissue, preserving cellular diversity and histoarchitecture, including liver-infiltrating immune cells, to reflect the complexity of the tumor and its microenvironment (Dimou et al., 2022). PCTSs are cost-effective, reproducible, and maintain cellular viability for up to 5 days under standard culture conditions, extendable to 15 days with optimal conditions (Doornebal et al., 2022b). Liver tumor-derived PCTSs provide a useful platform for preclinical drug efficacy testing (Palma et al., 2019). For instance, sorafenib has been tested for liver fibrosis along with other compounds (Westra et al., 2016), while regorafenib has been used to assess the reproducibility of cryopreservation protocols with HCC tissue slices (Zhang et al., 2022). PCTSs have been employed in comparative studies of HCC and surrounding liver tissue to examine apoptosis and proliferation levels after selective COX_2 inhibition (Kern et al., 2006). Additionally, Zimmermann and colleagues utilized the PCTS HCC model as a predictive tool for assessing the safety of oncolytic vaccine viruses. Liver slices have also been applied to study the phenotypic traits of two rare, inherited disorders affecting urea metabolism (Zimmermann et al., 2009). Research has also explored the progression of metabolic-associated steatotic liver disease (MASLD), which may lead to HCC. Notably, a study led by Li and collaborators aimed to extend the incubation period of PCTSs to analyze different MASLD stages, establishing a robust and rapid ex vivo model to aid in identifying new therapeutic targets (Li et al., 2024). PCTSs could further advance immunotherapy for HCC. Multiple studies have explored PCTSs as a preclinical model to capture patient variability and develop effective immunotherapies for PLC (Doornebal et al., 2022b). These models show promise for incorporating immune checkpoint inhibitors due to the availability of immune-competent HCC specimens, supporting testing of combination therapies and personalized treatment approaches.

8.6 LIMITATIONS OF 3D MODELS OF LIVER CANCER IN VITRO

Regardless of the utilization of the scaffold technology, constructing a 3D model of PLC through cell culturing increases the dimensionality of cells, providing a more accurate representation of the in vivo tumor environment (Abuwatfa et al., 2024). Furthermore, the 3D model promotes cell-cell interactions, and coculturing with stromal cells, such as tumor-associated fibroblasts and endothelial cells, allowing for a deeper understanding of the TME and the interactions between liver cancer cells and stromal cells (Fontana et al., 2021). Despite their advantages, traditional 3D cell culture methods have some limitations too. In scaffold-based 3D models, while natural ECM-derived materials offer binding sites for cells and enable tumor cells to exhibit behavior similar to in vivo conditions, these natural materials are often affected by variability in species source, individual differences, and the methods used for separation and purification. As a result, there can be significant batch-to-batch differences, impacting the consistency and repeatability of studies (Brancato et al., 2020). On the other hand, synthetic material scaffolds that offer greater structural control and stability, often lack the necessary binding sites for cells or growth factors, requiring chemical modification to address this limitation. Both natural and synthetic scaffolds have their unique advantages, and combining the two may enhance the ECM simulation of tumors in vivo. For scaffold-free 3D models, the complexity of the process limits the uniformity and reproducibility of spheroid formation. Additionally, traditional 3D cell culture methods need to be streamlined to facilitate high-throughput screening, which is crucial for personalized therapy and drug testing (Yun et al., 2022).

8.7 CONCLUSION AND FUTURE PERSPECTIVE

The rising global incidence of liver cancer, particularly HCC, underscores the urgent need for advanced modeling techniques to enhance our understanding of its pathogenesis and to develop effective therapeutic strategies. Traditional in vitro models, such as 2D cell cultures, along with animal models, have been instrumental in advancing liver cancer research. However, these models come with significant limitations, including the inability of 2D cultures to replicate the complex TME and the lack of full translational relevance in animal models due to interspecies differences. In this context, in silico models offer a promising computational approach for simulating tumor behavior, drug responses, and predicting therapeutic outcomes, contributing significantly to systems-level understanding and drug repurposing efforts. However, their success in clinical translation remains limited due to the complexity of accurately modeling the dynamic liver TME and the variability in patient-specific data. The development of 3D cell culture models has emerged as a transformative approach, as these models closely mimic the in vivo architecture and cellular interactions of liver tissue, offering more accurate insights into tumor behavior and drug response. Future directions in liver cancer research should focus on optimizing these 3D models, incorporating patient-specific cells, and advancing toward bioengineered liver tissue platforms, such as organ-on-a-chip technologies. These advancements promise to enhance preclinical studies and provide a more comprehensive understanding of HCC progression, offering new avenues for targeted therapy development and personalized medicine in liver cancer.

CONFLICT OF INTEREST

Authors declare no conflict of interest.

REFERENCES

Abuwatfa, W.H., Pitt, W.G., & Husseini, G.A. (2024). Scaffold-based 3D cell culture models in cancer research. *Journal of Biomedical Science*, 31(1), 7–46. https://doi.org/10.1186/s12929-024-00994-y

Agu, P.C., Afiukwa, C.A., Orji, O.U., Ezeh, E.M., Ofoke, I.H., Ogbu, C.O., Ugwuja, E.I., & Aja, P.M. (2023). Molecular docking as a tool for the discovery of molecular targets of nutraceuticals in diseases management. *Scientific Reports*, 13(1), 13398–13416. https://doi.org/10.1038/s41598-023-40160-2

Aguirre, M., Escobar, M., Forero Amézquita, S., Cubillos, D., Rincón, C., Vanegas, P., Tarazona, M.P., Atuesta Escobar, S., Blanco, J.C., & Celis, L.G. (2023). Application of the Yamanaka transcription factors Oct4, Sox2, Klf4, and c-Myc from the laboratory to the clinic. *Genes*, 14(9), 1697–1709. https://doi.org/10.3390/genes14091697

Aleksic, K., Lackner, C., Geigl, J.B., Schwarz, M., Auer, M., Ulz, P., Fischer, M., Trajanoski, Z., Otte, M., & Speicher, M.R. (2011). Evolution of genomic instability in diethylnitrosamine-induced hepatocarcinogenesis in mice. *Hepatology*, 53(3), 895–904.

Arellanes-Robledo, J., Hernández, C., Camacho, J., & Pérez-Carreón, J.I. (2017). In vitro models of HCC. *Liver Pathophysiology*, 13, 563–579. https://doi.org/10.1016/B978-0-12-804274-8.00042-4

Artegiani, B., Van Voorthuijsen, L., Lindeboom, R.G., Seinstra, D., Heo, I., Tapia, P., López-Iglesias, C., Postrach, D., Dayton, T., Oka, R., & Hu, H. (2019). Probing the tumor suppressor function of BAP1 in CRISPR-engineered human liver organoids. *Cell Stem Cell*, 24(6), 927–943. https://doi.org/10.1016/j.stem.2019.04.017

Asgharpour, A., Cazanave, S.C., Pacana, T., Seneshaw, M., Vincent, R., Banini, B.A., Kumar, D.P., Daita, K., Min, H.K., Mirshahi, F., & Bedossa, P. (2016). A diet-induced animal model of non-alcoholic fatty liver disease and hepatocellular cancer. *Journal of Hepatology*, 65(3), 579–588. https://doi.org/10.1016/j.jhep.2016.05.005

Balzeau, J., Menezes, M.R., Cao, S., & Hagan, J.P. (2017). The LIN28/let-7 pathway in cancer. *Frontiers in Genetics*, 8, 31–43. https://doi.org/10.3389/fgene.2017.00031

Bertassoni, L.E., Cecconi, M., Manoharan, V., Nikkhah, M., Hjortnaes, J., Cristino, A.L., Barabaschi, G., Demarchi, D., Dokmeci, M.R., Yang, Y., & Khademhosseini, A. (2014). Hydrogel bioprinted microchannel networks for vascularization of tissue engineering constructs. *Lab on a Chip*, 14(13), 2202–2211. https://doi.org/10.1039/c4lc00030g

Bhattacharya, M., Malinen, M.M., Lauren, P., Lou, Y.R., Kuisma, S.W., Kanninen, L., Lille, M., Corlu, A., GuGuen-Guillouzo, C., Ikkala, O., & Laukkanen, A. (2012). Nanofibrillar cellulose hydrogel promotes three-dimensional liver cell culture. *Journal of Controlled Release*, 164(3), 291–298. https://doi.org/10.1016/j.jconrel.2012.06.039

Bhise, N.S., Manoharan, V., Massa, S., Tamayol, A., Ghaderi, M., Miscuglio, M., Lang, Q., Shrike Zhang, Y., Shin, S.R., Calzone, G., & Annabi, N. (2016). A liver-on-a-chip platform with bioprinted hepatic spheroids. *Biofabrication,* 8(1), 014101. https://doi.org/10.1088/1758-5090/8/1/014101

Blidisel, A., Marcovici, I., Coricovac, D., Hut, F., Dehelean, C.A., & Cretu, O.M. (2021). Experimental models of hepatocellular carcinoma—a preclinical perspective. *Cancers*, 13(15), 3651–3678. https://doi.org/10.3390/cancers13153651

Blumer, T., Fofana, I., Matter, M.S., Wang, X., Montazeri, H., Calabrese, D., Coto-Llerena, M., Boldanova, T., Nuciforo, S., Kancherla, V., & Tornillo, L. (2019). Hepatocellular carcinoma xenografts established from needle biopsies preserve the characteristics of the originating tumors. *Hepatology Communications*, 3(7), 971–986. https://doi.org/10.1002/hep4.1365

Boortalary, T., & Loren, D. (2022). Diagnosis and evaluation of cholangiocarcinoma. In Doria, C., & Rogart, J.N. (Eds.), *Hepato-Pancreato-Biliary Malignancies: Diagnosis and Treatment in the 21st Century* (pp. 1–28). Springer, Cham. http://doi.org/10.1007/978-3-030-37482-2_15-1

Brancato, V., Oliveira, J.M., Correlo, V.M., Reis, R.L., & Kundu, S.C. (2020). Could 3D models of cancer enhance drug screening? *Biomaterials*, 232, 119744. https://doi.org/10.1016/j.biomaterials.2019.119744

Bray, F., Laversanne, M., Sung, H., Ferlay, J., Siegel, R.L., Soerjomataram, I., & Jemal, A. (2024). Global cancer statistics 2022: GLOBOCAN estimates of incidence and mortality worldwide for 36 cancers in 185 countries. *CA: A Cancer Journal for Clinicians*, 74(3), 229–263. https://doi.org/10.3322/caac.21834

Broutier, L., Mastrogiovanni, G., Verstegen, M.M., Francies, H.E., Gavarró, L.M., Bradshaw, C.R., Allen, G.E., Arnes-Benito, R., Sidorova, O., Gaspersz, M.P., & Georgakopoulos, N. (2017). Human primary liver cancer–derived organoid cultures for disease modeling and drug screening. *Nature Medicine*, 23(12), 1424–1435. https://doi.org/10.1038/nm.4438

Brown, Z.J., Heinrich, B., & Greten, T.F. (2018). Mouse models of hepatocellular carcinoma: An overview and highlights for immunotherapy research. *Nature Reviews Gastroenterology & Hepatology*, 15(9), 536–554. https://doi.org/10.1038/s41575-018-0033-6

Calderaro, J., Ziol, M., Paradis, V., & Zucman-Rossi, J. (2019). Molecular and histological correlations in liver cancer. *Journal of Hepatology*, 71(3), 616–630. https://doi.org/10.1016/j.jhep.2019.06.001

Caleffi, J.T., Aal, M.C.E., Gallindo, H.D.O.M., Caxali, G.H., Crulhas, B.P., Ribeiro, A.O., Souza, G.R., & Delella, F.K. (2021). Magnetic 3D cell culture: State of the art and current advances. *Life Sciences*, 286, 120028. https://doi.org/10.1016/j.lfs.2021.120028

Calitz, C., Pavlović, N., Rosenquist, J., Zagami, C., Samanta, A., & Heindryckx, F. (2020). A biomimetic model for liver cancer to study tumor-stroma interactions in a 3D environment with tunable bio-physical properties. *Journal of Visualised Experiments*, 162(3791), 61606. https://doi.org/10.3791/61606

Cao, W., Liu, J., Wang, L., Li, M., Verstegen, M.M., Yin, Y., Ma, B., Chen, K., Bolkestein, M., Sprengers, D., & Van Der Laan, L.J. (2019). Modeling liver cancer and therapy responsiveness using organoids derived from primary mouse liver tumors. *Carcinogenesis*, 40(1), 145–154. https://doi.org/10.1093/carcin/bgy129

Carvalho, V., Ferreira, M., Rodrigues, R.O., Teixeira, S.F., & Lima, R.A. (2024). Computational and experimental advances in liver-on-a-chip technology for cancer research: A systematic review. *Biophysical Reviews*, 1–17(1), 151–167. https://doi.org/10.1007/s12551-024-01260-z

Chehelgerdi, M., Behdarvand Dehkordi, F., Chehelgerdi, M., Kabiri, H., Salehian-Dehkordi, H., Abdolvand, M., Salmanizadeh, S., Rashidi, M., Niazmand, A., Ahmadi, S., & Feizbakhshan, S. (2023). Exploring the promising potential of induced pluripotent stem cells in cancer research and therapy. *Molecular Cancer*, 22(1), 1–111. https://doi.org/10.1186/s12943-023-01873-0

Cheng, Q.N., Yang, X., Wu, J.F., Ai, W.B., & Ni, Y.R., 2021. Interaction of non-parenchymal hepatocytes in the process of hepatic fibrosis. *Molecular Medicine Reports*, 23(5), 1–11. https://doi.org/10.3892/mmr.2021.12003

Chidambaranathan, R.S., Fisher, P.B., & Sarkar, D. (2021). Hepatocellular carcinoma (HCC): Epidemiology, etiology and molecular classification. *Advances in Cancer Research*, 149, 1–61.https://doi.org/10.1016/bs.acr.2020.10.001

Cigliano, A., Liao, W., Deiana, G.A., Rizzo, D., Chen, X., & Calvisi, D.F. (2024). Preclinical models of hepatocellular Carcinoma: Current utility, limitations, and challenges. *Biomedicines*, 12(7), 1624–1655. https://doi.org/10.3390/biomedicines12071624

Collins, S.D., Yuen, G., Tu, T., Budzinska, M.A., Spring, K.J., Bryant, K., & Shackel, N.A. (2019). In vitro models of the liver: Disease modeling, drug discovery and clinical applications. *Hepatocellular Carcinoma*, 2019, 47–67. https://doi.org/10.15586/hepatocellularcarcinoma.2019.ch3

Connor, F., Rayner, T.F., Aitken, S.J., Feig, C., Lukk, M., Santoyo-Lopez, J., & Odom, D.T. (2018). Mutational landscape of a chemically-induced mouse model of liver cancer. *Journal of Hepatology*, 69(4), 840–850. https://doi.org/10.1016/j.jhep.2018.06.009

De Minicis, S., Agostinelli, L., Rychlicki, C., Sorice, G.P., Saccomanno, S., Candelaresi, C., Giaccari, A., Trozzi, L., Pierantonelli, I., Mingarelli, E., & Marzioni, M. (2014). HCC development is associated to peripheral insulin resistance in a mouse model of NASH. *PLoS One*, 9(5), 97136–97147. https://doi.org/10.1371/journal.pone.0097136

De Santis, M.M., Alsafadi, H.N., Tas, S., Bölükbas, D.A., Prithiviraj, S., Da Silva, I.A., Mittendorfer, M., Ota, C., Stegmayr, J., Daoud, F., & Königshoff, M. (2021). Extracellular-matrix-reinforced bioinks for 3D bioprinting human tissue. *Advanced Materials*, 33(3), 2005476–2005485. https://doi.org/10.1002/adma.202005476

Decante, G., Costa, J.B., Silva-Correia, J., Collins, M.N., Reis, R.L., & Oliveira, J.M. (2021). Engineering bioinks for 3D bioprinting. *Biofabrication*, 13(3), 032001. https://doi.org/10.1088/1758-5090/abec2c

Deng, J., Wei, W., Chen, Z., Lin, B., Zhao, W., Luo, Y., & Zhang, X. (2019). Engineered liver-on-a-chip platform to mimic liver functions and its biomedical applications: A review. *Micromachines*, 10(10), 676. https://doi.org/10.3390/mi10100676

Dimou, P., Trivedi, S., Liousia, M., D'souza, R.R., & Klampatsa, A. (2022). Precision-cut tumor slices (PCTS) as an ex vivo model in immunotherapy research. *Antibodies,* 11(2), 26. https://doi.org/10.3390/antib11020026

Doornebal, E.J., Harris, N., Riva, A., Jagatia, R., Pizanias, M., Prachalias, A., Menon, K., Preziosi, M., Zamalloa, A., Miquel, R., & Zen, Y. (2022a). Human immunocompetent model of neuroendocrine liver metastases recapitulates patient-specific tumour microenvironment. *Frontiers in Endocrinology*, 13, 909180–909211. https://doi.org/10.3389/fendo.2022.909180

Doornebal, E.J., Jagatia, R., Rastovic, U., Harris, N., Martinez Lyons, A., Miquel, R., Zen, Y., Zamalloa, A., Malik, F., & Prachalias, A. (2022b). Patient-derived precision cut tissue slices from primary liver cancer as a suitable platform for preclinical drug testing. *SSRN Electron Journal,* 2022, 1–35. https://doi.org/10.2139/ssrn.4290636

Drury, J.L., & Mooney, D.J. (2003). Hydrogels for tissue engineering: Scaffold design variables and applications. *Biomaterials*, 24(24), 4337–4351. https://doi.org/10.1016/S0142-9612(03)00340-5

Ehrlich, A., Duche, D., Ouedraogo, G., & Nahmias, Y. (2019). Challenges and opportunities in the design of liver-on-chip microdevices. *Annual Review of Biomedical Engineering*, 21(1), 219–239. https://doi.org/10.1146/annurev-bioeng-060418-052305

Fan, H., Demirci, U., & Chen, P. (2019). Emerging organoid models: Leaping forward in cancer research. *Journal of Hematology & Oncology*, 12(1), 142–152. https://doi.org/10.1186/s13045-019-0832-4

Fontana, F., Marzagalli, M., Sommariva, M., Gagliano, N., & Limonta, P. (2021). In vitro 3D cultures to model the tumor microenvironment. *Cancers*, 13(12), 2970–2986. https://doi.org/10.3390/cancers13122970

Frese, K.K., & Tuveson, D.A. (2007). Maximizing mouse cancer models. *Nature Reviews Cancer*, 7(9), 654–658. https://doi.org/10.1038/nrc2192

Fukuyama, K., Asagiri, M., Sugimoto, M., Tsushima, H., Seo, S., Taura, K., Uemoto, S., & Iwaisako, K. (2021). Gene expression profiles of liver cancer cell lines reveal two hepatocyte-like and fibroblast-like clusters. *PLoS One*, 16(2), 1–17. https://doi.org/10.1371/journal.pone.0245939

Giri, S., & Singh, A. (2024). Epidemiology of hepatocellular carcinoma in India–An updated review for 2024. *Journal of Clinical and Experimental Hepatology*, 14, 101447. https://doi.org/10.1016/j.jceh.2024.101447

Green, C.D., Weigel, C., Brown, R.D., Bedossa, P., Dozmorov, M., Sanyal, A.J., & Spiegel, S. (2022). A new preclinical model of western diet-induced progression of non-alcoholic steatohepatitis to hepatocellular carcinoma. *The FASEB Journal*, 36(7), 1–15. https://doi.org/10.1096/fj.202200346R

Greene, T., & Lin, C.C. (2015). Modular cross-linking of gelatin-based thiol–norbornene hydrogels for in vitro 3D culture of hepatocellular carcinoma cells. *ACS Biomaterials Science & Engineering*, 1(12), 1314–1323. https://doi.org/10.1021/acsbiomaterials.5b00436

Guo, X., Seo, J.E., Li, X., & Mei, N. (2020). Genetic toxicity assessment using liver cell models: Past, present, and future. *Journal of Toxicology and Environmental Health, Part B*, 23(1), 27–50. https://doi.org/10.1080/10937404.2019.1692744

Heindryckx, F., Colle, I., & Van Vlierberghe, H. (2009). Experimental mouse models for hepatocellular carcinoma research. *International Journal of Experimental Pathology*, 90(4), 367–386. https://doi.org/10.1111/j.1365-2613.2009.00656.x

Heinrich, M.A., Mostafa, A.M., Morton, J.P., Hawinkels, L.J., & Prakash, J. (2021). Translating complexity and heterogeneity of pancreatic tumor: 3D in vitro to in vivo models. *Advanced Drug Delivery Reviews*, 174, 265–293. https://doi.org/10.1016/j.addr.2021.04.018

Hoarau-Véchot, J., Rafii, A., Touboul, C., & Pasquier, J. (2018). Halfway between 2D and animal models: Are 3D cultures the ideal tool to study cancer-microenvironment interactions?. *International Journal of Molecular Sciences*, 19(1), 181–205. https://doi.org/10.3390/ijms19010181

Huang, Y., Zhang, X.F., Gao, G., Yonezawa, T., & Cui, X. (2017). 3D bioprinting and the current applications in tissue engineering. *Biotechnology Journal*, 12(8), 1600734. https://doi.org/10.1002/biot.201600734

Huch, M., & Koo, B.K. (2015). Modeling mouse and human development using organoid cultures. *Development*, 142(18), 3113–3125. https://doi.org/10.1242/dev.118570

Huch, M., Gehart, H., Van Boxtel, R., Hamer, K., Blokzijl, F., Verstegen, M.M., Ellis, E., Van Wenum, M., Fuchs, S.A., de Ligt, J., & Van De Wetering, M. (2015). Long-term culture of genome-stable bipotent stem cells from adult human liver. *Cell*, 160(1), 299–312. https://doi.org/10.1016/j.cell.2014.11.050

Ito, T., & Nguyen, M.H. (2023). Perspectives on the underlying etiology of HCC and its effects on treatment outcomes. *Journal of Hepatocellular Carcinoma*, 10, 413–428. https://doi.org/10.2147/JHC.S347959

Jeong, S.R., & Kang, M. (2023). Exploring tumor–immune interactions in co-culture models of T cells and tumor organoids derived from patients. *International Journal of Molecular Sciences*, 24(19), 14609–14703. https://doi.org/10.3390/ijms241914609

Jin, K., Teng, L., Shen, Y., He, K., Xu, Z., & Li, G. (2010). Patient-derived human tumour tissue xenografts in immunodeficient mice: A systematic review. *Clinical and Translational Oncology*, 12, 473–480. https://doi.org/10.1007/s12094-010-0540-6

Jodat, Y.A., Kang, M.G., Kiaee, K., Kim, G.J., Martinez, A.F., Rosenkranz, A., Bae, H., & Shin, S.R. (2018). Human-derived organ-on-a-chip for personalized drug development. *Current Pharmaceutical Design*, 24(45), 5471–5486. https://doi.org/10.2174/1381612825666190308150055

Joshi, A., Ramanujan, S., & Jin, J.Y. (2023). The convergence of pharmacometrics and quantitative systems pharmacology in pharmaceutical research and development. *European Journal of Pharmaceutical Sciences*, 182, 106380–106385. https://doi.org/10.1016/j.ejps.2023.106380

Jung, J. (2014). Human tumor xenograft models for preclinical assessment of anticancer drug development. *Toxicological Research*, 30(1), 1–5. https://doi.org/10.5487/TR.2014.30.1.001

Jung, J., Seol, H.S., & Chang, S. (2018). The generation and application of patient-derived xenograft model for cancer research. *Cancer Research and Treatment: Official Journal of Korean Cancer Association*, 50(1), 1–10. https://doi.org/10.4143/crt.2017.307

Kern, M.A., Haugg, A.M., Eiteneuer, E., Konze, E., Drebber, U., Dienes, H.P., Breuhahn, K., Schirmacher, P., & Kasper, H.U. (2006). Ex vivo analysis of antineoplastic agents in precision-cut tissue slices of human origin: Effects of cyclooxygenase-2 inhibition in hepatocellular carcinoma. *Liver International*, 26(5), 604–612. https://doi.org/10.1111/j.1478-3231.2006.01268.x

Kiamehr, M., Heiskanen, L., Laufer, T., Düsterloh, A., Kahraman, M., Kakela, R., Laaksonen, R., & Aalto-Setala, K. (2019). Dedifferentiation of primary hepatocytes is accompanied with reorganization of lipid metabolism indicated by altered molecular lipid and miRNA profiles. *International Journal of Molecular Sciences*, 20(12), 2910–2924. https://doi.org/10.3390/ijms20122910

Kim, M.K., Jeong, W., Jeon, S., & Kang, H.W. (2023). 3D bioprinting of dECM-incorporated hepatocyte spheroid for simultaneous promotion of cell-cell and-ECM interactions. *Frontiers in Bioengineering and Biotechnology*, 11, 1305023–1305037. https://doi.org/10.3389/fbioe.2023.1305023

Kronemberger, G.S., Miranda, G.A., Tavares, R.S., Montenegro, B., Kopke, Ú.D.A., & Baptista, L.S. (2021). Recapitulating tumorigenesis in vitro: Opportunities and challenges of 3D Bioprinting. *Frontiers in Bioengineering and Biotechnology*, 9, 682498–682516. https://doi.org/10.3389/fbioe.2021.682498

Lai, Y., Wei, X., Lin, S., Qin, L., Cheng, L., & Li, P. (2017). Current status and perspectives of patient-derived xenograft models in cancer research. *Journal of Hematology & Oncology*, 10, 1–14. https://doi.org/10.1186/s13045-017-0470-7

Langhans, S.A. (2018). Three-dimensional in vitro cell culture models in drug discovery and drug repositioning. *Frontiers in Pharmacology*, 9, 6–20. https://doi.org/10.3389/fphar.2018.00006

Lee, J., Lee, S., Kim, S.M., & Shin, H. (2021). Size-controlled human adipose-derived stem cell spheroids hybridized with single-segmented nanofibers and their effect on viability and stem cell differentiation. *Biomaterials Research*, 25(1), 14–28. https://doi.org/10.1186/s40824-021-00215-9

Lee, J.W., Choi, Y.J., Yong, W.J., Pati, F., Shim, J.H., Kang, K.S., Kang, I.H., Park, J., & Cho, D.W. (2016). Development of a 3D cell printed construct considering angiogenesis for liver tissue engineering. *Biofabrication*, 8(1), 015007. https://doi.org/10.1088/1758-5090/8/1/015007

Leung, M., Kievit, F.M., Florczyk, S.J., Veiseh, O., Wu, J., Park, J.O., & Zhang, M. (2010). Chitosan-alginate scaffold culture system for hepatocellular carcinoma increases malignancy and drug resistance. *Pharmaceutical Research*, 27, 1939–1948. https://doi.org/10.1007/s11095-010-0198-3

Li, C., Jin, B., Sun, H., Wang, Y., Zhao, H., Sang, X., Yang, H., & Mao, Y. (2022). Exploring the function of stromal cells in cholangiocarcinoma by three-dimensional bioprinting immune microenvironment model. *Frontiers in Immunology*, 13, 941289. https://doi.org/10.3389/fimmu.2022.941289

Li, M., Larsen, F.T., Van Den Heuvel, M.C., Gier, K., Gorter, A.R., Oosterhuis, D., Bijzet, J., De Meijer, V.E., Ravnskjaer, K., Nagelkerke, A., & Olinga, P. (2024). Metabolic dysfunction-associated steatotic liver disease in a dish: Human precision-cut liver slices as a platform for drug screening and interventions. *Nutrients*, 16(5), 626–644. https://doi.org/10.3390/nu16050626

Li, Y., Tang, Z.Y., & Hou, J.X. (2012). Hepatocellular carcinoma: Insight from animal models. *Nature reviews Gastroenterology & hepatology*, 9(1), 32–43. https://doi.org/10.1038/nrgastro.2011.196

Li, Y., Zhang, T., Pang, Y., Li, L., Chen, Z.N., & Sun, W. (2019). 3D bioprinting of hepatoma cells and application with microfluidics for pharmacodynamic test of Metuzumab. *Biofabrication*, 11(3), 034102–034115. https://doi.org/10.1088/1758-5090/ab256c

Liaw, C.Y., Ji, S., & Guvendiren, M. (2018). Engineering 3D hydrogels for personalized in vitro human tissue models. *Advanced Healthcare Materials*, 7(4), 1701165. https://doi.org/10.1002/adhm.201701165

Liu, C., Qin, T., Huang, Y., Li, Y., Chen, G., & Sun, C. (2020). Drug screening model meets cancer organoid technology. *Translational Oncology*, 13(11), 100840–100847. https://doi.org/10.1016/j.tranon.2020.100840

Liu, H., Li, J., Zhu, S., Zhang, X., Zhang, F., Zhang, X., Zhao, G., Zhu, W., & Zhou, F. (2024). Long-term trends in incidence, mortality and burden of liver cancer due to specific etiologies in Hubei Province. *Scientific Reports*, 14(1), 4924–4934. https://doi.org/10.1038/s41598-024-53812-8

Lu, S., Cuzzucoli, F., Jiang, J., Liang, L.G., Wang, Y., Kong, M., Zhao, X., Cui, W., Li, J., & Wang, S. (2018). Development of a biomimetic liver tumor-on-a-chip model based on decellularized liver matrix for toxicity testing. *Lab on a Chip*, 18(22), 3379–3392. https://doi.org/10.1039/C8LC00852C

Lucchetti, M., Aina, K.O., Grandmougin, L., Jäger, C., Pérez Escriva, P., Letellier, E., Mosig, A.S., & Wilmes, P. (2024). An organ-on-chip platform for simulating drug metabolism along the gut–liver axis. *Advanced Healthcare Materials*, 13, 2303943. https://doi.org/10.1002/adhm.202303943

Lucena-Padros, H., Bravo-Gil, N., Tous, C., Rojano, E., Seoane-Zonjic, P., Fernández, R.M., Ranea, J.A., Antiñolo, G., & Borrego, S. (2024). Bioinformatics prediction for network-based integrative multi-omics expression data analysis in Hirschsprung disease. *Biomolecules*, 14(2), 164–192. https://doi.org/10.3390/biom14020164

Ma, Y., Hu, L., Tang, J., Guo, W., Feng, Y., Liu, Y., & Ang, F. (2023). Three-dimensional cell co-culture liver models and their applications in pharmaceutical research. *International Journal of Molecular Sciences*, 24(7), 6248–6266. https://doi.org/10.3390/ijms24076248

Mao, S., He, J., Zhao, Y., Liu, T., Xie, F., Yang, H., Mao, Y., Pang, Y., & Sun, W. (2020). Bioprinting of patient-derived in vitro intrahepatic cholangiocarcinoma tumor model: Establishment, evaluation and anti-cancer drug testing. *Biofabrication*, 12(4), 045014. https://doi.org/10.1088/1758-5090/aba0c3

Maronpot, R.R. (2009). Biological basis of differential susceptibility to hepatocarcinogenesis among mouse strains. *Journal of Toxicologic Pathology*, 22(1), 11–33. https://doi.org/10.1293/tox.22.11

Maronpot, R.R., Fox, T., Malarkey, D.E., & Goldsworthy, T.L. (1995). Mutations in the ras proto-oncogene: Clues to etiology and molecular pathogenesis of mouse liver tumors. *Toxicology*, 101(3), 125–156. https://doi.org/10.1016/0300-483X(95)03112-S

Massarweh, N.N., & El-Serag, H.B. (2017). Epidemiology of hepatocellular carcinoma and intrahepatic cholangiocarcinoma. *Cancer Control*, 24(3), 1073274817729245–1073274817729256. https://doi.org/10.1177/107327481772924

Meng, F., Meyer, C.M., Joung, D., Vallera, D.A., McAlpine, M.C., & Panoskaltsis-Mortari, A. (2019). 3D bioprinted in vitro metastatic models via reconstruction of tumor microenvironments. *Advanced Materials*, 31(10), 1806899–1806918. https://doi.org/10.1002/adma.201806899

Mirshafiei, M., Rashedi, H., Yazdian, F., Rahdar, A., & Baino, F. (2024). Advancements in tissue and organ 3D bioprinting: Current techniques, applications, and future perspectives. *Materials & Design*, 240, 112853–112882. https://doi.org/10.1016/j.matdes.2024.112853

Moscato, S., Ronca, F., Campani, D., & Danti, S. (2015). Poly (vinyl alcohol)/gelatin hydrogels cultured with HepG2 cells as a 3D model of hepatocellular carcinoma: A morphological study. *Journal of Functional Biomaterials*, 6(1), 16–32. https://doi.org/10.3390/jfb6010016

Neal, J.T., Li, X., Zhu, J., Giangarra, V., Grzeskowiak, C.L., Ju, J., Liu, I.H., Chiou, S.H., Salahudeen, A.A., Smith, A.R., & Deutsch, B.C. (2018). Organoid modeling of the tumor immune microenvironment. *Cell*, 175(7), 1972–1988. https://doi.org/10.1016/j.cell.2018.11.021

Newell, P., Villanueva, A., Friedman, S.L., Koike, K., & Llovet, J.M. (2008). Experimental models of hepatocellular carcinoma. *Journal of Hepatology*, 48(5), 858–879. https://doi.org/10.1016/j.jhep.2008.01.008

Nikolova, M.P., & Chavali, M.S. (2019). Recent advances in biomaterials for 3D scaffolds: A review. *Bioactive Materials*, 4, 271–292. https://doi.org/10.1016/j.bioactmat.2019.10.005

Nuciforo, S., Fofana, I., Matter, M.S., Blumer, T., Calabrese, D., Boldanova, T., Piscuoglio, S., Wieland, S., Ringnalda, F., Schwank, G., & Terracciano, L.M. (2018). Organoid models of human liver cancers derived from tumor needle biopsies. *Cell Reports*, 24(5), 1363–1376. https://doi.org/10.1016/j.celrep.2018.07.001

Oh, J.H., & Sinn, D.H. (2024). Multidisciplinary approach for hepatocellular carcinoma patients: Current evidence and future perspectives. *Journal of Liver Cancer*, 24(1), 47–56. https://doi.org/10.17998/jlc.2024.02.27

Olson, B., Li, Y., Lin, Y., Liu, E.T., & Patnaik, A. (2018). Mouse models for cancer immunotherapy research. *Cancer Discovery*, 8(11), 1358–1365. https://doi.org/10.1158/2159-8290.CD-18-0044

Ouyang, L., Yao, R., Mao, S., Chen, X., Na, J., & Sun, W. (2015). Three-dimensional bioprinting of embryonic stem cells directs highly uniform embryoid body formation. *Biofabrication*, 7(4), 044101. https://doi.org/10.1088/1758-5090/7/4/044101

Palma, E., Doornebal, E.J., & Chokshi, S. (2019). Precision-cut liver slices: A versatile tool to advance liver research. *Hepatology International*, 13, 51–57. https://doi.org/10.1007/s12072-018-9913-7

Peng, W.C., Logan, C.Y., Fish, M., Anbarchian, T., Aguisanda, F., Álvarez-Varela, A., Wu, P., Jin, Y., Zhu, J., Li, B., & Grompe, M. (2018). Inflammatory cytokine TNFα promotes the long-term expansion of primary hepatocytes in 3D culture. *Cell*, 175(6), 1607–1619. https://doi.org/10.1016/j.cell.2018.11.012

Pez, F., Gifu, P., Degli-Esposti, D., Fares, N., Lopez, A., Lefrançois, L., Michelet, M., Rivoire, M., Bancel, B., Sylla, B.S., & Herceg, Z. (2019). In vitro transformation of primary human hepatocytes: Epigenetic changes and stemness properties. *Experimental Cell Research*, 384(2), 111643–111683. https://doi.org/10.1016/j.yexcr.2019.111643

Polidoro, M.A., Ferrari, E., Soldani, C., Franceschini, B., Saladino, G., Rosina, A., Mainardi, A., D'Autilia, F., Pugliese, N., Costa, G., & Donadon, M. (2024). Cholangiocarcinoma-on-a-chip: A human 3D platform for personalised medicine. *JHEP Reports*, 6(1), 100910–100923. https://doi.org/10.1016/j.jhepr.2023.100910

Qi, X., Schepers, E., Avella, D., Kimchi, E.T., Kaifi, J.T., Staveley-O'Carroll, K.F., & Li, G. (2019). An oncogenic hepatocyte-induced orthotopic mouse model of hepatocellular cancer arising in the setting of hepatic inflammation and fibrosis. *Journal of Visualized Experiments*, 151, 59368. https://doi.org/10.3791/59368

Qiu, Z., Zou, K., Zhuang, L., Qin, J., Li, H., Li, C., Zhang, Z., Chen, X., Cen, J., Meng, Z., & Zhang, H. (2016). Hepatocellular carcinoma cell lines retain the genomic and transcriptomic landscapes of primary human cancers. *Scientific Reports*, 6(1), 27411–27424. https://doi.org/10.1038/srep27411

Quazi, S. (2022). Artificial intelligence and machine learning in precision and genomic medicine. *Medical Oncology*, 39(8), 120. https://doi.org/10.1007/s12032-022-01711-1

Reiberger, T., Chen, Y., Ramjiawan, R.R., Hato, T., Fan, C., Samuel, R., Roberge, S., Huang, P., Lauwers, G.Y., Zhu, A.X., & Bardeesy, N. (2015). An orthotopic mouse model of hepatocellular carcinoma with underlying liver cirrhosis. *Nature Protocols*, 10(8), 1264–1274. https://doi.org/10.1038/nprot.2015.080

Reitz, C.J., Kuzmanov, U., & Gramolini, A.O. (2023). Multi-omic analyses and network biology in cardiovascular disease. *Proteomics*, 23(21–22), 2200289. https://doi.org/10.1002/pmic.202200289

Rogers, A.B. (2018). Stress of strains: Inbred mice in liver research. *Gene Expression*, 19(1), 61–68. https://doi.org/10.3727/105221618X15337408678723

Romualdo, G.R., Leroy, K., Costa, C.J.S., Prata, G.B., Vanderborght, B., Da Silva, T.C., Barbisan, L.F., Andraus, W., Devisscher, L., Câmara, N.O.S., & Vinken, M. (2021). In vivo and in vitro models of hepatocellular carcinoma: Current strategies for translational modeling. *Cancers*, 13(21), 5583–5640. https://doi.org/10.3390/cancers13215583

Saborowski, A., Wolff, K., Spielberg, S., Beer, B., Hartleben, B., Erlangga, Z., Becker, D., Dow, L.E., Marhenke, S., Woller, N., & Unger, K. (2019). Murine liver organoids as a genetically flexible system to study liver cancer in vivo and in vitro. *Hepatology Communications*, 3(3), 423–436. https://doi.org/10.1002/hep4.1312

Sato, T., Vries, R.G., Snippert, H.J., Van De Wetering, M., Barker, N., Stange, D.E., Van Es, J.H., Abo, A., Kujala, P., Peters, P.J., & Clevers, H. (2009). Single Lgr5 stem cells build crypt-villus structures in vitro without a mesenchymal niche. *Nature*, 459(7244), 262–265. https://doi.org/10.1038/nature07935

Scholten, D., Trebicka, J., Liedtke, C., & Weiskirchen, R. (2015). The carbon tetrachloride model in mice. *Laboratory Animals*, 49, 4–11. https://doi.org/10.1177/0023677215571192

Seo, J.E., Davis, K., Malhi, P., He, X., Bryant, M., Talpos, J., Burks, S., Mei, N., & Guo, X. (2021). Genotoxicity evaluation using primary hepatocytes isolated from rhesus macaque (*Macaca mulatta*). *Toxicology*, 462, 152936–152970. https://doi.org/10.1016/j.tox.2021.152936

Sharma, N., Biswas, S., Al-Dayan, N., Alhegaili, A.S., & Sarwat, M. (2021). Antioxidant role of kaempferol in prevention of hepatocellular carcinoma. *Antioxidants*, 10(9), 1419–1436. https://doi.org/10.3390/antiox10091419

Sharma, N., Gupta, M., Anand, P., Akhter, Y., Al-Dayan, N., Majed, H.A., Biswas, S., Ali, S., & Sarwat, M. (2024). Mechanistic insight into the autophagic and apoptotic activity of kaempferol on liver cancer cells. *OncoTargets and Therapy*, 17, 579–601. https://doi.org/10.2147/OTT.S460359

Shukla, A.K., & Kumar, A. (2023). Virtual screening of orally active lead compounds of pearl millet and their structural activity against target protein of COVID-19. *Russian Journal of Bioorganic Chemistry*, 49, S53–S70. https://doi.org/10.1134/S1068162023080058

Srivastava, S.K., Foo, G.W., Aggarwal, N., & Chang, M.W. (2024). Organ-on-chip technology: Opportunities and challenges. *Biotechnology Notes,* 18(2), 195–244. https://doi.org/10.1016/j.biotno.2024.01.001

Tang, J., Cui, J., Chen, R., Guo, K., Kang, X., Li, Y., Gao, D., Sun, L., Xu, C., Chen, J., & Tang, Z. (2011). A three-dimensional cell biology model of human hepatocellular carcinoma in vitro. *Tumor Biology*, 32, 469–479. https://doi.org/10.1007/s13277-010-0140-7

Tetsuka, H., & Shin, S.R. (2020). Materials and technical innovations in 3D printing in biomedical applications. *Journal of Materials Chemistry B*, 8(15), 2930–2950. https://doi.org/10.1039/D0TB00034E

Tripathi, S., Mandal, S.S., Bauri, S., & Maiti, P. (2023). 3D bioprinting and its innovative approach for biomedical applications. *MedComm*, 4(1), 194–232. https://doi.org/10.1002/mco2.194

Tsuchida, T., Lee, Y.A., Fujiwara, N., Ybanez, M., Allen, B., Martins, S., Fiel, M.I., Goossens, N., Chou, H.I., Hoshida, Y., & Friedman, S.L. (2018). A simple diet-and chemical-induced murine NASH model with rapid progression of steatohepatitis, fibrosis and liver cancer. *Journal of Hepatology*, 69(2), 385–395. https://doi.org/10.1016/j.jhep.2018.03.011

Vesselinovitch, S.D., Koka, M., Mihailovich, N., & Rao, K.V.N. (1984). Carcinogenicity of diethylnitrosamine in newborn, infant, and adult mice. *Journal of Cancer Research and Clinical Oncology*, 108(1), 60–65. https://doi.org/10.1007/BF00390974

Vilas-Boas, V., Cooreman, A., Gijbels, E., Van Campenhout, R., Gustafson, E., Ballet, S., Annaert, P., Cogliati, B., & Vinken, M. (2019). Primary hepatocytes and their cultures for the testing of drug-induced liver injury. *Advances in Pharmacology*, 85, 1–30. https://doi.org/10.1016/bs.apha.2018.08.001

Vucur, M., Roderburg, C., Bettermann, K., Tacke, F., Heikenwalder, M., Trautwein, C., & Luedde, T. (2010). Mouse models of hepatocarcinogenesis: What can we learn for the prevention of human hepatocellular carcinoma?. *Oncotarget*, 1(5), 373–378. https://doi.org/10.18632/oncotarget.170

Wang, N., Wang, S., Li, M.Y., Hu, B.G., Liu, L.P., Yang, S.L., Yang, S., Gong, Z., Lai, P.B., & Chen, G.G. (2018). Cancer stem cells in hepatocellular carcinoma: An overview and promising therapeutic strategies. *Therapeutic Advances in Medical Oncology*, 10, 1–25. https://doi.org/10.1177/1758835918816287

Wang, Y., Kankala, R.K., Zhang, J., Hao, L., Zhu, K., Wang, S., Zhang, Y.S., & Chen, A. (2020). Modeling endothelialized hepatic tumor microtissues for drug screening. *Advanced Science*, 7(21), 1–13. https://doi.org/10.1002/advs.202002002

Westra, I.M., Mutsaers, H.A., Luangmonkong, T., Hadi, M., Oosterhuis, D., De Jong, K.P., Groothuis, G.M., & Olinga, P. (2016). Human precision-cut liver slices as a model to test antifibrotic drugs in the early onset of liver fibrosis. *Toxicology in Vitro*, 35, 77–85. https://doi.org/10.1016/j.tiv.2016.05.012

Wilkening, S., Stahl, F., & Bader, A. (2003). Comparison of primary human hepatocytes and hepatoma cell line Hepg2 with regard to their biotransformation properties. *Drug Metabolism and Disposition*, 31(8), 1035–1042. https://doi.org/10.1124/dmd.31.8.1035

Wolf, M.J., Adili, A., Piotrowitz, K., Abdullah, Z., Boege, Y., Stemmer, K., Ringelhan, M., Simonavicius, N., Egger, M., Wohlleber, D., & Lorentzen, A. (2014). Metabolic activation of intrahepatic CD8+ T cells and NKT cells causes nonalcoholic steatohepatitis and liver cancer via cross-talk with hepatocytes. *Cancer Cell*, 26(4), 549–564. https://doi.org/10.1016/j.ccell.2014.09.003

Wu, X., Roberto, J.B., Knupp, A., Kenerson, H.L., Truong, C.D., Yuen, S.Y., Brempelis, K.J., Tuefferd, M., Chen, A., Horton, H., & Yeung, R.S. (2018). Precision-cut human liver slice cultures as an immunological platform. *Journal of Immunological Methods*, 455, 71–79. https://doi.org/10.1016/j.jim.2018.01.012

Xie, F., Sun, L., Pang, Y., Xu, G., Jin, B., Xu, H., Lu, X., Xu, Y., Du, S., Wang, Y., & Feng, S. (2021). Three-dimensional bio-printing of primary human hepatocellular carcinoma for personalized medicine. *Biomaterials*, 265, 120416. https://doi.org/10.1016/j.biomaterials.2020.120416

Xu, H., Jiao, D., Liu, A., & Wu, K. (2022). Tumor organoids: Applications in cancer modeling and potentials in precision medicine. *Journal of Hematology & Oncology*, 15(1), 58–78. https://doi.org/10.1186/s13045-022-01278-4

Xu, W., Zhao, Z.Y., An, Q.M., Dong, B., Lv, A., Li, C.P., Guan, X.Y., Tian, X.Y., Wu, J.H., & Hao, C.Y. (2020). Comprehensive comparison of patient-derived xenograft models in hepatocellular carcinoma and metastatic liver cancer. *International Journal of Medical Sciences*, 17(18), 3073–3081. https://doi.org/10.7150/ijms.46686

Yi, H.G., Lee, H., & Cho, D.W. (2017). 3D printing of organs-on-chips. *Bioengineering*, 4(1), 10–32. https://doi.org/10.3390/bioengineering4010010

Yildiz, G. (2018). Integrated multi-omics data analysis identifying novel drug sensitivity-associated molecular targets of hepatocellular carcinoma cells. *Oncology Letters*, 16(1), 113–122. https://doi.org/10.3892/ol.2018.8634

Ying, F., Chan, M.S.M., & Lee, T.K.W. (2023). Cancer-associated fibroblasts in hepatocellular carcinoma and cholangiocarcinoma. *Cellular and Molecular Gastroenterology and Hepatology*, 15(4), 985–999. https://doi.org/10.1016/j.jcmgh.2023.01.006

Yip, D., & Cho, C.H. (2013). A multicellular 3D heterospheroid model of liver tumor and stromal cells in collagen gel for anti-cancer drug testing. *Biochemical and Biophysical Research Communications*, 433(3), 327–332. https://doi.org/10.1016/j.bbrc.2013.03.008

Yun, C., Kim, S.H., & Jung, Y.S. (2022). Current research trends in the application of in vitro three-dimensional models of liver cells. *Pharmaceutics*, 15(1), 54–76. https://doi.org/10.3390/pharmaceutics15010054

Zhang, S., Deshpande, A., Verma, B.K., Wang, H., Mi, H., Yuan, L., Ho, W.J., Jaffee, E.M., Zhu, Q., Anders, R.A., & Yarchoan, M. (2023). Informing virtual clinical trials of hepatocellular carcinoma with spatial multi-omics analysis of a human neoadjuvant immunotherapy clinical trial. *bioRxiv*, 8, 553000. https://doi.org/10.1101/2023.08.11.553000

Zhang, Y., Wang, Z.Y., Jing, H.S., Zhang, H.D., Yan, H.X., Fan, J.X., & Zhai, B. (2022). A pre-clinical model combining cryopreservation technique with precision-cut slice culture method to assess the in vitro drug response of hepatocellular carcinoma. *International Journal of Molecular Medicine*, 49(4), 1–9. https://doi.org/10.3892/ijmm.2022.5107

Zheng, H.C., Xue, H., & Yun, W.J. (2023). An overview of mouse models of hepatocellular carcinoma. *Infectious Agents and Cancer*, 18(1), 49. https://doi.org/10.1186/s13027-023-00524-9

Zhu, L., Cheng, C., Liu, S., Yang, L., Han, P., Cui, T., & Zhang, Y. (2023). Advancements and application prospects of three-dimensional models for primary liver cancer: A comprehensive review. *Frontiers in Bioengineering and Biotechnology*, 11, 1343177–1343188. https://doi.org/10.3389/fbioe.2023.1343177

Zhuang, X., & Lu, C. (2016). PBPK modeling and simulation in drug research and development. *Acta Pharmaceutica Sinica B*, 6(5), 430–440. https://doi.org/10.1016/j.apsb.2016.04.004

Zhuang, X., Deng, G., Wu, X., Xie, J., Li, D., Peng, S., Tang, D., & Zhou, G. (2023). Recent advances of three-dimensional bioprinting technology in hepato-pancreato-biliary cancer models. *Frontiers in Oncology*, 13, 1143600–1143613. https://doi.org/10.3389/fonc.2023.1143600

Zimmermann, M., Armeanu, S., Smirnow, I., Kupka, S., Wagner, S., Wehrmann, M., Rots, M.G., Groothuis, G.M., Weiss, T.S., Königsrainer, A., & Gregor, M. (2009). Human precision-cut liver tumor slices as a tumor patient-individual predictive test system for oncolytic measles vaccine viruses. *International Journal of Oncology*, 34(5), 1247–1256. https://doi.org/10.3892/ijo_00000253

Index

Note: **Bold** page numbers refer to tables and *italic* page numbers refer to figures.

C

D

E

P

Q

R

For Product Safety Concerns and Information please contact our
EU representative GPSR@taylorandfrancis.com Taylor & Francis
Verlag GmbH, Kaufingerstraße 24, 80331 München, Germany